The Biology of Chameleons

The Biology of Chameleons

Edited by KRYSTAL A. TOLLEY and ANTHONY HERREL

UNIVERSITY OF CALIFORNIA PRESS
Berkeley Los Angeles London

University of California Press, one of the most distinguished university presses in the United States, enriches lives around the world by advancing scholarship in the humanities, social sciences, and natural sciences. Its activities are supported by the UC Press Foundation and by philanthropic contributions from individuals and institutions. For more information, visit www.ucpress.edu.

University of California Press
Berkeley and Los Angeles, California

University of California Press, Ltd.
London, England

Library of Congress Cataloging-in-Publication Data

The biology of chameleons / edited by Krystal Tolley and Anthony Herrel.
pages cm.
Includes bibliographical references and index.
ISBN 978-0-520-27605-5 (cloth : alk. paper)
1. Chameleons. I. Tolley, Krystal. II. Herrel, Anthony.
QL666.L23B56 2013
597.95'6—dc23 2013026609

Manufactured in the United States of America
22 21 20 19 18 17 16 15
10 9 8 7 6 5 4 3 2

The paper used in this publication meets the minimum requirements of ANSI/NISO Z39.48-1992 (R 2002) (*Permanence of Paper*). ♾

Cover illustration: *Trioceros johnstoni* from the Rwenzori Mountains, Uganda. Photo by Michele Menegon.

CONTENTS

Contributors viii

Foreword xi

1 Biology of the Chameleons: An Introduction 1
Krystal A. Tolley and Anthony Herrel

2 Chameleon Anatomy 7
Christopher V. Anderson and Timothy E. Higham

2.1 Musculoskeletal Morphology 7

2.2 External Morphology and Integument 37

2.3 Sensory Structures 43

2.4 Visceral Systems 50

3 Chameleon Physiology 57
Anthony Herrel

3.1 Neurophysiology 57

3.2 Muscle Physiology 59

3.3 Metabolism, Salt, and Water Balance 60

3.4 Temperature 61

3.5 Skin Pigmentation, Color Change, and the Role of Ultraviolet Light 61

3.6 Developmental Physiology 62

4 Function and Adaptation of Chameleons 63
Timothy E. Higham and Christopher V. Anderson

4.1 Locomotion 64

4.2 Feeding 72

5 Ecology and Life History of Chameleons 85
G. John Measey, Achille Raselimanana, and Anthony Herrel

5.1 Habitat 86

5.2 Life-History Traits 97

5.3 Foraging and Diet 104

5.4 Predators 109

6 Chameleon Behavior and Color Change 115
Devi Stuart-Fox

6.1 Sensory Systems and Modes of Communication 116

6.2 Color Change 117

6.3 Social and Reproductive Behavior 120

6.4 Sexual Dimorphism: Body Size and Ornamentation 126

6.5 Antipredator Behavior 126

7 Evolution and Biogeography of Chameleons 131
Krystal A. Tolley and Michele Menegon

7.1 Evolutionary Relationships 131

7.2 Diversity and Distribution 134

7.3 Regional Diversity 138

7.4 Patterns of Alpha Diversity 146

7.5 Patterns of Beta Diversity 147

8 Overview of the Systematics of the Chamaeleonidae 151
Colin R. Tilbury

8.1 Evolution of Methodology in Chameleon Taxonomy 153

8.2 Current Status of Taxonomy of the Chamaeleonidae 155

8.3 Subfamilial Groupings within Chamaeleonidae 155

8.4 Overview of Extant Genera 158

9 Fossil History of Chameleons 175
Arnau Bolet and Susan E. Evans

9.1 Phylogenetic Relationships of Iguania and Acrodonta 175

9.2 Fossil Record of Acrodonta 179

9.3 Origins of Acrodonta 187

9.4 Origins of Chamaeleonidae 190

10 Chameleon Conservation 193
Richard Jenkins, G. John Measey, Christopher V. Anderson, and Krystal A. Tolley

10.1 Conservation Status of Chameleons 193

10.2 Trade in Chameleons 201

10.3 Chameleons and Global Change 211

10.4 The Way Forward 214

Appendix 217

Abbreviations 223

References 225

Photo Credits 267

Index 269

CONTRIBUTORS

CHRISTOPHER V. ANDERSON
Department of Integrative Biology
University of South Florida, USA and
Department of Ecology and Evolutionary Biology, Brown University, Providence, Rhode Island, USA

ARNAU BOLET
Institut Català de Paleontologia Miquel Crusafont and Universitat Autònoma de Barcelona
Sabadell, Spain

SUSAN E. EVANS
Research Department of Cell and Developmental Biology
College London
London, United Kingdom

ANTHONY HERREL
Centre National de la Recherche Scientifique and Muséum National d'Histoire Naturelle
Paris, France

TIMOTHY E. HIGHAM
Department of Biology
University of California
Riverside, California

RICHARD JENKINS
Durrell Institute of Conservation and Ecology
School of Anthropology and Conservation
The University of Kent and IUCN Global Species Programme
Cambridge, United Kingdom

G. JOHN MEASEY
Department of Zoology
Nelson Mandela Metropolitan University
Port Elizabeth, South Africa

MICHELE MENEGON
Tropical Biodiversity Section
Museo Tridentino di Scienze Naturali
Trento, Italy

ACHILLE RASELIMANANA
Department of Animal Biology
University of Antananarivo and Association Vahatra
Antananarivo, Madagascar

DEVI M. STUART-FOX
Zoology Department
The University of Melbourne
Australia

COLIN R. TILBURY
Evolutionary Genomics Group
University of Stellenbosch
South Africa

KRYSTAL A. TOLLEY
South African National Biodiversity Institute
Cape Town, South Africa

FOREWORD

In putting together this book, we stand on the shoulders of others. The extensive bibliography presented here spans centuries, and the resulting body of literature is based on the work of researchers who dedicated their minds to a deeper understanding of chameleons. We have taken pieces of this great puzzle and have made a start at constructing the whole picture, but there are many glaring gaps. In some respects, it seems there are too many pieces missing and the emerging picture is only a hazy nebula of unclear, scattered, and fragmented bits. But the excitement that comes with the challenge of scientific thought, of asking the questions "why" and "how," is what compels us to keep looking for the missing pieces. For chameleons, the many missing pieces are the why and how of their remarkable evolutionary radiation, and we must keep questioning, even if we never complete the puzzle.

Although this book is built on the works of others, putting together this volume has been a group effort of the authors, all of whom enthusiastically came to the party. Each author brought their own expertise, and together we have made something more than any one of us could have done alone. It has been an extraordinary experience working with this team. As editors, we expected to be herding cats, but on the contrary, the process was surprisingly smooth. Of course, each of the chapters was reviewed by our peers, all of whom invariably provided positive and constructive criticism on the content. It is surprising how many things we missed initially, and we owe much to our colleagues for taking time to review and comment on these chapters: Salvidor Bailon, Bill Branch, Angus Carpenter, Jack Conrad, Frank Glaw, Rob James, Charles Klaver, Lance McBrayer, John Poynton, Phil Stark, Andrew Turner, James Vonesh, Bieke Vanhooydonck, and Martin Whiting. We are grateful to several friends and colleagues who permitted complimentary use of their photos, including Bill Branch, Marius Burger, Tania Fouche, Adnan Moussalli, Devi Stuart-Fox, and Michele Menegon. We also owe much to Chuck Crumly for eagerly taking on the initial responsibility of producing this book, as well as the National Research Foundation of South Africa and Centre National de la Recherche Scientifique and Groupement de Recherche

International for providing the funds that allowed the editors of this volume to collaborate and to aspire. The follow-up production team at UC Press (Lynn Meinhardt, Ruth Weinberg, Kate Hoffman, Blake Edgar, and Deepti Agarwal) were excellent in providing advice and assistance throughout the process. In all, this has been a brilliant experience, despite initial reservations in taking on such a big project. It's clear that the ease of putting this together was due to an outstanding team of authors, all of whom are passionate about their subject and have not forgotten how to ask "why."

ONE

Biology of the Chameleons

An Introduction

KRYSTAL A. TOLLEY and ANTHONY HERREL

Chameleons so easily capture the imagination, and have done so for centuries. In scientific writings, they first appear with Aristotle (350 BC), and although they were recognized as being similar to lizards in some ways, he also likened them to fish, baboons, and crocodiles. Since then, they have since been the subject of strange myths, amusing tales, and nature documentaries and have even taken the form of popular cartoon characters. But what makes them so fascinating that they have infiltrated the common psyche more than other reptiles? Indeed, they are set apart to such a degree that many people are unaware they are even lizards at all. They are simply, "chameleons." In this book, we draw together and review the body of literature that covers chameleons over several centuries by exploring and summarizing our knowledge on this intriguing group of lizards.

Chameleons are highly specialized animals characterized by a suite of morphological, physiological and functional adaptations. Much of what makes chameleons immediately identifiable as such are these morphological specializations. Among these, the laterally compressed body, the prehensile feet and tail, the independently moveable eyes (Figs. 1.1 and 1.2 in color insert), and a long tongue capable of being projected from the mouth are the most striking. As detailed in Chapters 2 (Chameleon Anatomy) and 3 (Chameleon Physiology), many of these features are intimately related to the unique behavior and lifestyle of chameleons as highly specialized arboreal predators. The sensory system of chameleons, for example, is highly tuned toward visual stimuli, as chameleons rely on visual signals in both a social and a feeding-related context. As part of the specialization of the visual system, chameleons have a negatively powered lens and use accommodation cues to judge distance, features unique among lizards (Fig. 1.3 in color insert). Because of their visual specialization, the auditory system is less developed and shows lower sensitivity as compared with that of other lizards. Little is known about the olfactory (smell), vomerolfactory, and gustatory (taste) systems in chameleons, but all appear reduced as compared with other lizards.

Associated with the predominantly arboreal habitat of chameleons comes a number of functional specializations such as their prehensile feet and tails. Chapter 4, Function and Adaptation of Chameleons, examines how these specializations allow chameleons to effectively explore the three-dimensional (mostly) arboreal habitat. However, as a consequence of their specializations, chameleons are the slowest of all lizards, with sprint speeds about 10 times slower than those of other lizards. The slow locomotion of chameleons is a result of the contractile capacities of the locomotor muscles, changes in limb posture that allows them to move effectively on narrow substrates, and a lower overall muscle mass. In contrast to the limb muscles, the tongue muscles of chameleons are anything but slow; they produce high forces for their cross-sectional area. These unusually fast, ballistic tongues permit them to capture a wide variety of prey. The supercontractile tongue-retractor muscles are unique among vertebrates and allow them to reel in even large vertebrate prey.

Chameleons are found among widely varying thermal regimes and climatic conditions, including hot and dry desert habitats, tropical rainforests, Mediterranean climates, and high-altitude environments. Similar to other lizards, however, chameleons carefully regulate body temperatures using behavioral thermoregulation and color change to maintain temperatures close to their preferred temperature of around 30 to 32°C. The preferred temperatures of chameleons are low as compared with those of most other diurnal lizards. The temperature invariant function of the ballistic tongue protraction and their low preferred temperatures may consequently have allowed chameleons to invade high mountain habitats rarely accessible to other lizards.

Ecologically, chameleons have taken control of the arboreal niche across Africa and Madagascar. No other reptile (aside from their avian cousins) dominates to such a degree. Walk through any montane forest in Africa or Madagascar with a spotlight at night, and you are sure to see a sleeping chameleon perched on a branch or twig. Although they are best known for their arboreal lifestyle, a number of species, indeed entire genera, are mainly terrestrial (Fig. 1.4 in color insert). While chameleons are primarily associated with forests, multiple species have capitalized on emerging habitats (grasslands and heathlands). In fact, in some places, they occupy the entire strata from ground to high canopy. Chapter 5, Ecology and Life History of Chameleons, summarizes the considerable body of studies on the interactions of chameleons with their environment. Here, we learn not only the more recognizable aspects of chameleon ecology, but are introduced to the lesser-known details. We are provided with an overview of their reproductive traits, which range from seasonal to year-round reproduction plus entire clades that are viviparous rather than oviparous. We find that although they are primarily insectivorous, their diet can be wide and may on occasion include their own kind. Even their foraging mode is thought to be exceptional among reptiles, being a combination of active foraging with sit-and-wait, termed "cruise foraging," which allows them to remain cryptic, while actively in pursuit of prey.

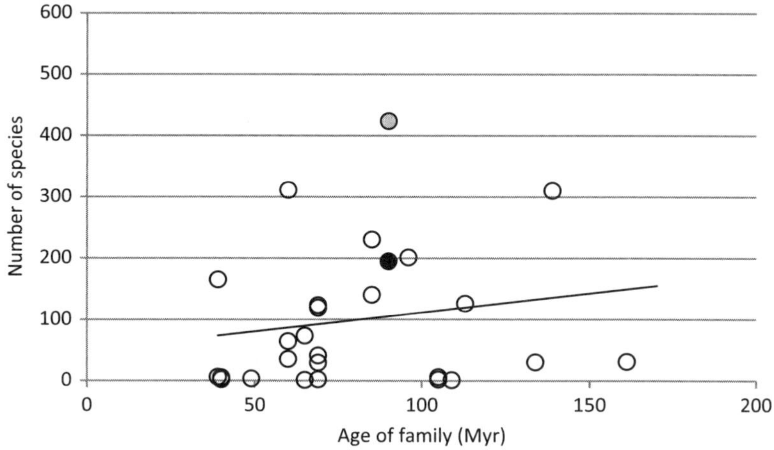

FIGURE 1.8. Age (millions of years) of most squamate families with number of species in that family. The Chamaeleonidae is shown as a black circle, and Agamidae as a gray circle. Scincidae, Gekkonidae, and Colubridae are not shown because of the large number of species in these families (1553, 917, and 1772, respectively). Based on data from Hedges and Vidal (2009) and Uetz (2012).

Despite their intensely cryptic lifestyle (Fig. 1.5 in color insert), chameleons have evolved some interesting behavioral traits. Indeed, chameleons are famous for their sometimes flamboyant ornamentation and their ability to change color. Chapter 6, Chameleon Behavior and Color Change, dispels the popular myths about chameleons and examines the how and why of their behavior. We are introduced to the physiological aspects of color change, and find that it is related to communication, camouflage, and thermoregulation (Fig. 1.6 in color insert). We also discover how ornaments are involved in sexual selection through display and aggression (Fig. 1.7 in color insert). Chameleon behavior is not limited to conspecific interactions, and includes some tricks for predator avoidance that go beyond crypsis, including free-falling from perches to avoid predation.

Moving from the individual level to the landscape level, chameleons have an interesting history and are an unusually diverse group of lizards. They are, in fact, a young clade of lizards, dating back only to the beginning of the Cenozoic, whereas most other lizard families are much older (Chapter 7, Evolution and Biogeography of Chameleons). Despite their young age, there are over 190 described species, and most experts agree that there are multiple species awaiting discovery and description (Chapter 8, Overview of the Systematics of the Chamaeleonidae). Actually, if the age of the clade is considered, the number of chameleon species is relatively high as compared with other many other squamate families (Fig. 1.8), which suggests a history of rapid lineage diversification. Chameleons are exceeded in this respect by their relatives, the Agamidae, but also Gekkonidae, Lacertidae, Scincidae, Colubrid snakes (Colubridae), and vipers (Viperidae). It appears that chameleon diversification

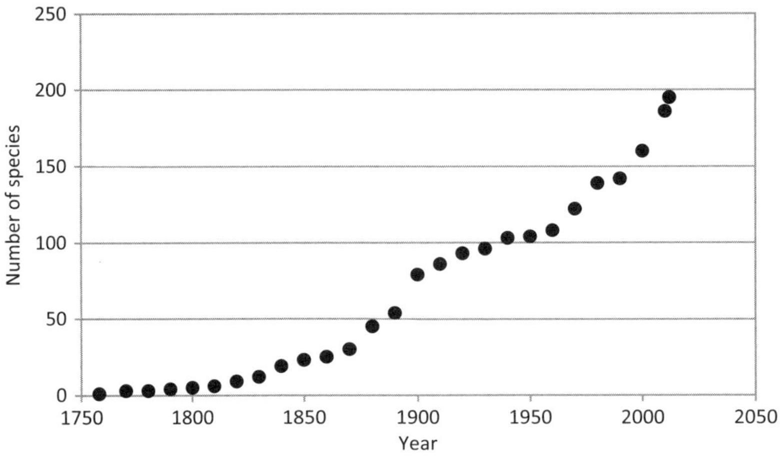

FIGURE 1.9. The cumulative number of chameleon species described, starting with the first description of *Chamaeleo chameleon* in 1758 by Linneaus.

is linked to their invasion of the arboreal habitat some 45 million years ago. Today, there are distinct centers of diversity, particularly in East Africa and Madagascar, that have elevated species richness and endemism. Although the fossil record is scarce, the clues provided by fossils that lead up to chameleons is rich, giving us insight to the early history of the group and ancestors. Chapter 9, Fossil History of Chameleons, takes us on a journey through time, beginning in the Cretaceous with the ancestors of chameleons, the priscagamids, and ends in the Holocene with fossils of extant species from Africa, Madagascar, Europe, and the Middle East.

Chapter 10, Chameleon Conservation, provides a comprehensive and up-to-date examination on the state of these lizards in our modern world. The conservation status and major threat categories are quantified and species are examined for their vulnerability. An alarming statistic that emerges, is that two thirds of chameleons already assessed by the International Union for Conservation of Nature (IUCN) are considered Threatened or Near Threatened, a figure that is much higher than that for other groups of reptiles. Of the tangible threats, habitat alteration appears to be the most prevalent, with a disproportionate impact on narrow-range endemics. Chameleons are also popular in the pet trade, and the more than 30 years of trade statistics have been summarized in this chapter. The majority of exported chameleons are harvested from wild populations, rather than through captive breeding programs, and the largest importer of chameleons is the United States, which is responsible for two thirds of the market.

The scientific literature available is surprisingly large and has filled this volume. Despite this, there is still much to learn about these intriguing animals. Even simple facts about their anatomy, physiology, and function remain unknown. Because of their cryptic nature, we know little of their daytime activities, and our impressions of their interactions with conspecifics and other animals are based on precious few studies. Aspects of their life history appear to be the most well known, yet for most species we have no understanding of their reproductive cycles, diets, home range sizes, and dispersal ability. Poorly explored forests in Africa and Madagascar reveal new species with every survey. The number of species described every year continues to rise (Fig. 1.9), and this is certainly not the end of a trend, given molecular phylogenies that show numerous undescribed lineages. There are huge gaps in the fossil record, particularly early in the history of this family, and new finds would certainly shed light on the morphology of early arboreal chameleon lineages. In exploring what we do know about chameleons, it seems we have succeeded in bringing to light the vast gap of what we do not know. But these are exciting times in the world of chameleons, and with the tools we have at our disposal today, it is clear that it will not take several more centuries to uncover many of these mysteries.

ACKNOWLEDGEMENTS

We would like to thank the authors of the these book chapters for contributing their time, expertise and energy toward making this book possible.

TWO

Chameleon Anatomy

CHRISTOPHER V. ANDERSON and TIMOTHY E. HIGHAM

The family Chamaeleonidae is a distinctive clade of squamate reptiles with a plethora of unusual structural adaptations. Chameleons exhibit numerous distinctive features, including a laterally compressed body, forcep-like feet with toes grouped in opposing bundles, prehensile tail, enlarged casque, independently rotating eyes, and long tongue capable of being projected from the mouth. While chameleons are unique animals, they are also extremely diverse, with species spanning an approximate 20-fold range in adult total length and a 2000-fold range in body mass. Moreover, chameleons exhibit an extensive range of ornamentation. In addition, chameleons live over incredible ranges of habitats and demonstrate an abundance of variation in their behavior and ecology (Chapters 5 and 6), features of which are often predicated on anatomical specializations.

As a result of their unique nature, people have been interested in the biology of chameleons for centuries. In order to understand many aspects of chameleon biology, however, understanding the associated morphological underpinnings can be of vital importance. Here we summarize what is known about the anatomy of chameleons, emphasizing the differences between chameleons and other reptiles and the differences among chameleons.

2.1 MUSCULOSKELETAL MORPHOLOGY

Axial

Cranial

The lateral compression of the chameleon in conjunction with their enlarged eyes and the formation of an enlarged casque have strong influences on the structure of the chameleon skull and the distribution and orientation of various cranial muscles. The extent to which these characteristics are developed, however, is also variable within the family, and thus examination of the anatomy of the skull and its musculature is not only of interest relative to other lizard groups, but also within the chameleons.

Skull and Teeth The extensive studies of the chameleon skull have included discussion and examination of the structure of the skull in *Archaius* (Hillenius, 1988; Rieppel and Crumly, 1997), *Bradypodion* (Fig. 2.1c,d,f) (Parker, 1881; Methuen and Hewitt, 1914; Brock, 1941; Engelbrecht, 1951; Rieppel, 1981; Hillenius, 1988), *Brookesia* (Fig. 2.1a,b) (Siebenrock, 1893; Methuen and Hewitt, 1914; Rieppel 1987; Rieppel and Crumly, 1997), *Calumma* (Methuen and Hewitt, 1914; Hillenius, 1988; Rieppel and Crumly, 1997), *Chamaeleo* (Parker, 1881; Siebenrock, 1893; Methuen and Hewitt, 1914; Prasad, 1954; Rieppel, 1987; Hillenius, 1988), *Furcifer* (Methuen and Hewitt, 1914; Hillenius, 1988; Rieppel and Crumly, 1997), *Kinyongia* (Hillenius, 1988), *Rhampholeon* (Werner, 1902b; Methuen and Hewitt, 1914; Frank, 1951; Rieppel, 1987; Hillenius, 1988), *Rieppeleon* (Rieppel, 1987; Hillenius, 1988), and *Trioceros* (Fig. 2.1e) (Rieppel, 1981, 1987, 1993; Hillenius, 1988). These studies have amassed a list of variations between the skulls of different genera and developmental stages. However, they have also resulted in varying interpretations of the skull bones, particularly of the temporal region. Here we summarize the morphology of the adult skull in chameleons following the terminology and interpretations of Rieppel (1981).

The premaxilla of chameleons is unpaired (fused) and lies medially between the maxillae (Fig. 2.1b,d) (Parker, 1881; Siebenrock, 1893; Werner, 1902b; Brock, 1941; Engelbrecht, 1951; Frank, 1951; Rieppel, 1981, 1987). The premaxilla in *Brookesia* (Siebenrock, 1893), *Chamaeleo* (Siebenrock, 1893), *Rieppeleon* (Rieppel; 1987), and *Rhampholeon* (Werner, 1902b; Frank, 1951) bear two vestigial teeth, whereas the premaxillae in *Trioceros* do not bear teeth (Rieppel, 1981). Engelbrecht (1951) reports that *Bradypodion pumilum* also lack premaxillary teeth. Rieppel (1981), on the other hand, observed indications of two vestigial teeth fused with the premaxilla, noting, however, that histological investigation is required to verify whether these are true teeth or paired bony projections on the transverse process of the premaxilla.

Ventrally, the vomerine (palatal) process of the premaxilla is reduced (Fig. 2.1f) (Romer, 1956), extending only a short distance posteriorly, and in *Bradypodion* (Fig. 2.1f) (Engelbrecht, 1951; Frank, 1951; Rieppel, 1981), it does not contact the vomer, as the palatal process of the maxillae meets behind the premaxilla. The maxillae do not meet behind the premaxilla in *Brookesia* (Siebenrock, 1893; Rieppel and Crumly, 1997), however. Typically, the vomers are fused (unpaired) in chameleons (Frank, 1951; Rieppel, 1981); however, in *Bradypodion pumilum,* the vomers are paired for most of their length and fused only anteriorly where they join with the maxillae (Engelbrecht, 1951; Rieppel, 1981). In *Archaius,* the vomer is paired (Rieppel and Crumly, 1997). At the posterior end, the vomer joins the palatines (Fig. 2.1f) (Engelbrecht, 1951; Rieppel, 1981). The palatines extend posteriorly and flare laterally to join the maxillaries (Fig. 2.1f) (Werner, 1902b; Engelbrecht, 1951; Rieppel, 1981).

Dorsally, the nasal process of the premaxilla extends posteriorly and fully separates the maxillae in most species (Fig. 2.1b,d) (Werner, 1902b; Engelbrecht, 1951; Frank, 1951;

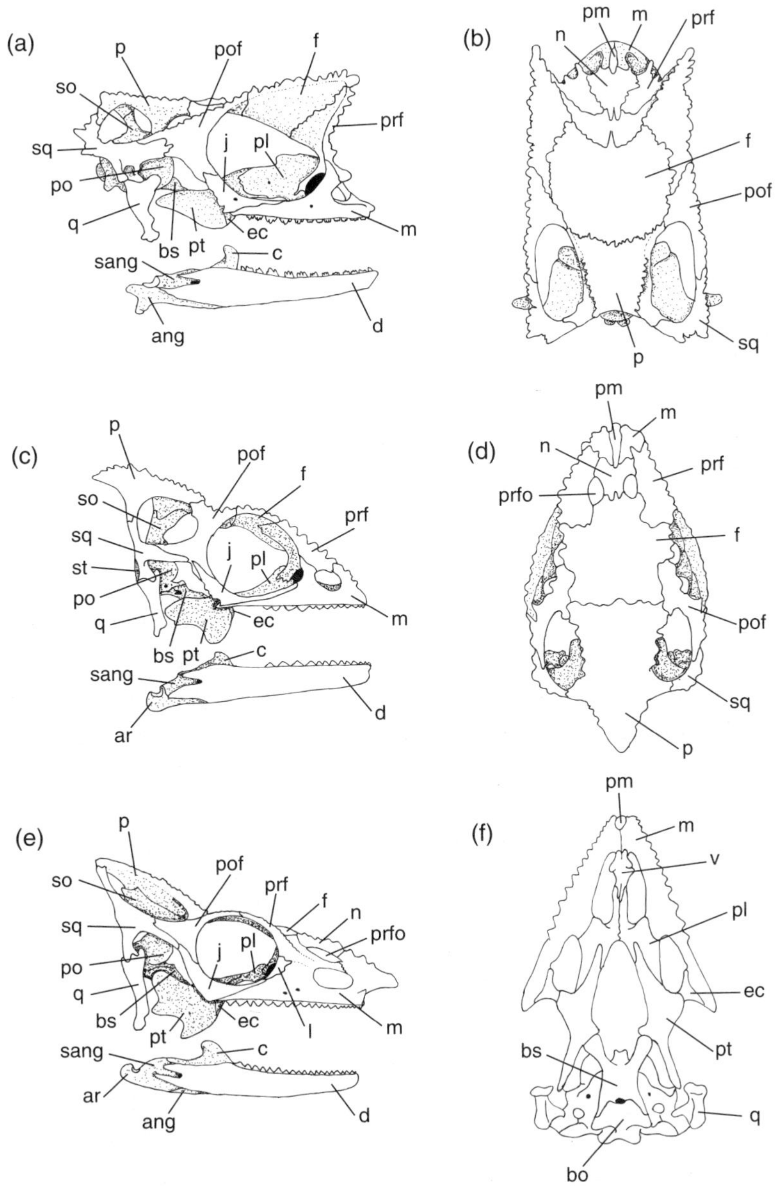

FIGURE 2.1. The skull of *Brookesia superciliaris* (a, b), *Bradypodion pumilum* (c, d, f), and *Trioceros melleri* (e) in lateral (a, c, e), dorsal (b, d), and ventral (f) views. (a, b) redrawn from Rieppel (1987) and (c, d, e, f) from Rieppel (1981).

LABELS: ang = angular; ar = articular; bo = basioccipital; bs = basisphenoid; c = coronoid; d = dentary; ec = ectopterygoid; f = frontal; j = jugal; m = maxilla; n = nasal; p = parietal; pl = palatine; pm = premaxilla; po = prootic; pof = postorbitofrontal; pf = prefrontal; prfo = prefrontal fontanelle; pt = pterygoid; q = quadrate; sang = surangular; so = supraoccipital; sq = squamosal; st = supratemporal; v = vomer.

Rieppel, 1981, 1987; Rieppel and Crumly, 1997). In *Bradypodion, Brookesia,* and *Chamaeleo,* the posterior edge of the nasal process of the premaxilla meets the fused nasals (Fig. 2.1b,d) (Seibenrock, 1893; Camp, 1923; Parker, 1942; Engelbrecht, 1951; Romer, 1956; Rieppel, 1981, 1987). However, the nasal process of the premaxilla separates the nasals and meets an anterior process of the frontal in *Rhampholeon* and *Rieppeleon* (Werner, 1902b; Parker, 1942; Frank, 1951, Rieppel, 1981, 1987). The nasals are paired and variably separated from the frontals by the premaxilla in *Calumma* and *Furcifer* (Rieppel and Crumly, 1997).

In Brookesia, the nasals circumscribe the dorsal margin of the nasal aperture (Fig. 2.1a,b) (Siebenrock, 1893; Engelbrecht, 1951; Romer, 1956; Rieppel, 1981). In *Bradypodion* (Fig. 2.1c,d) (Brock, 1941; Engelbrecht, 1951; Rieppel, 1981), *Chamaeleo* (Parker, 1881), and *Trioceros* (Fig. 2.1e) (Rieppel, 1981), the nasals do not participate in circumscribing the nasal aperture. The nasal aperture is bound by the maxillae on the anterior, ventral, and posterior edges (Siebenrock, 1893; Engelbrecht, 1951; Rieppel, 1981) and in *Bradypodion, Chamaeleo, Rieppeleon,* and *Trioceros,* the dorsal margin is bound by an anterior projection of the prefrontal lying lateral to the nasals (Fig. 2.1c–e) (Parker, 1881; Methuen and Hewitt, 1914; Engelbrecht, 1951; Rieppel, 1981, 1987). In *Bradypodion* and *Chamaeleo,* a prefrontal fontanelle is bound by the nasal, prefrontal, and frontal (Fig. 2.1d) (Parker, 1881; Engelbrecht, 1951; Rieppel, 1981). The prefrontal fontanelle is bound by the prefrontal, nasal, and maxilla in *Trioceros* (Fig. 2.1e) (Rieppel, 1981). In *Rhampholeon,* however, the prefrontal fontanelles are continuous with the nasal aperture and combined they are bound by the maxilla, prefrontal, nasal, frontal, and in some cases, the premaxilla (Werner, 1902b; Frank, 1951; Rieppel, 1981, 1987).

The prefrontal circumscribes the anterodorsal margin of the orbit in *Bradypodion* (Fig. 2.1c) (Parker, 1881; Engelbrecht, 1951; Rieppel, 1981), *Brookesia* (Fig. 2.1a) (Siebenrock, 1893; Romer, 1956; Rieppel, 1987), *Chamaeleo* (Parker, 1881), *Rhampholeon* (Werner, 1902b; Frank, 1951), *Rieppeleon* (Rieppel, 1987), and *Trioceros* (Fig. 2.1e) (Rieppel, 1981). The lacrimal is absent in *Bradypodion* (Brock, 1941; Engelbrecht, 1951; Rieppel, 1981), *Brookesia* (Siebenrock, 1893; Rieppel, 1987), *Calumma* (Methuen and Hewitt, 1913), *Rhampholeon* (Frank, 1951), and *Rieppeleon* (Rieppel, 1987), and the prefrontal joins the maxillae at the anterior edge of the orbit, allowing them to circumscribe the anteroventral margin of the orbit in these genera. In *Chamaeleo* (Parker, 1881; Camp, 1923) and *Trioceros* (Fig. 2.1e) (Rieppel, 1981), however, the lacrimal is present and joins the prefrontal at the anterior edge of the orbit, excluding the maxillae from involvement in circumscribing the orbit. Methuen and Hewitt (1914) note that the lacrimal is absent in *Furcifer lateralis,* whereas Rieppel and Crumly (1997) note that the lacrimal is usually observed in most *Furcifer* examined, including *F. lateralis* and with the exception of in *F. oustaleti.* The jugal joins with the lacrimal in *Chamaeleo* (Parker, 1881) and *Trioceros* (Fig. 2.1e) (Rieppel, 1981), and with the maxilla in *Bradypodion* (Fig. 2.1c) (Brock, 1941; Engelbrecht, 1951; Rieppel, 1981), *Brookesia* (Fig. 2.1a) (Siebenrock, 1893; Romer, 1956; Rieppel, 1987), *Calumma* (Methuen and Hewitt, 1914), *Furcifer* (Methuen and Hewitt, 1914), *Rhampholeon* (Werner, 1902b; Frank, 1951), and *Rieppeleon* (Rieppel, 1987); in both cases, they circumscribe the ventral and posteroventral edge of the orbit.

The frontal is fused and, when present, bears the pineal foramen (Romer, 1956; Rieppel, 1981). In *Archaius* (Rieppel and Crumly, 1997), *Bradypodion* (Fig. 2.1c,d) (Parker, 1881; Methuen and Hewitt, 1914; Engelbrecht, 1951; Rieppel, 1981), *Brookesia* (Fig. 2.1a,b) (Siebenrock, 1893; Romer, 1956; Rieppel, 1987), *Calumma* species other than *C. brevicorne* (Methuen and Hewitt, 1914; Rieppel and Crumly, 1997), *Furcifer bifidus* (Rieppel and Crumly, 1997), *Furcifer campani* (Rieppel and Crumly, 1997), and *Rhampholeon* (Werner, 1902b; Frank, 1951), the frontal circumscribes the dorsal margin of the orbit and joins with the prefrontal anteriorly and the postorbitofrontal posteriorly. In *Calumma brevicorne* (Methuen and Hewitt, 1914), *Chamaeleo* (Parker, 1881), *Furcifer* species other than *F. bifidus* and *F. campani* (Methuen and Hewitt, 1914; Rieppel and Crumly, 1997), and *Trioceros* (Fig. 2.1e) (Rieppel, 1981, 1993); however, the frontal is excluded from involvement in circumscribing the orbit by contact of the prefrontal with the postorbitofrontal. The postorbitofrontal joins the jugal at the posterior margin of the orbit (Fig. 2.1a,c,e) (Parker, 1881; Engelbrecht, 1951; Frank, 1951; Rieppel, 1981) and extends deep to meet the ectopterygoid (Rieppel, 1981).

In *Chamaeleo* (Parker, 1881), *Furcifer lateralis* (Methuen and Hewitt, 1914; Rieppel and Crumly, 1997), *F. pardalis* (Rieppel and Crumly, 1997), and *Trioceros* (Fig. 2.1e) (Rieppel, 1981), the dorsal tip of the jugal contacts the squamosal to form the upper temporal arch, with a posterior projection of the postorbitofrontal extending dorsal to the squamosal. In *Calumma* species other than *C. brevicorne* (Methuen and Hewitt, 1914; Rieppel and Crumly, 1997), *Bradypodion* (Fig. 2.1c) (Parker, 1881; Rieppel, 1981), *Brookesia* (Fig. 2.1a) (Siebenrock, 1893; Romer, 1856; Rieppel, 1987), *F. bifidus* (Rieppel and Crumly, 1997), *Rhampholeon* (Werner, 1902b; Frank, 1951), and *Rieppeleon* (Rieppel, 1987), the jugal and squamosal do not connect and the postorbitofrontal bridges the gap between them. Methuen and Hewitt (1914) note that the jugal and squamosal come into contact in *Calumma brevicorne* and *C. nasuta,* whereas Rieppel (1997) notes that the jugal may closely approach the squamosal in *C. nasuta* but does not touch it. A fontanelle in *C. brevicorne* may influence the possibility of contact between the jugal and squamosal. Further, Rieppel and Crumly (1997) note that contact between the jugal and squamosal is variable in *F. oustaleti* and *F. verrucosus.*

In *Archaius* (Hillenius, 1988), *Calumma* (Hillenius, 1988), *Chamaeleo* (Parker, 1881; Methuen and Hewitt, 1914; Hillenius, 1988), *Furcifer* (Hillenius, 1988), *Kinyongia* (Hillenius, 1988), *Rhampholeon* (Werner, 1902b; Frank, 1951; Rieppel, 1987), *Rieppeleon* (Rieppel, 1987), and *Trioceros* (Fig. 2.1e) (Rieppel, 1981; Hillenius, 1988), the parietal narrows posteriorly to form a sagittal crest, the parietal crest, extending posterodorsally to form the casque and meeting the supraoccipital ventrally. This posterior narrowing is slower, forming a more trigonal shape, in *Calumma* and *Rhampholeon* (Hillenius, 1988; Rieppel and Crumly, 1997), with some *Calumma* species having a broadening again posteriorly (Rieppel and Crumly, 1997). In *Archaius* (Rieppel and Crumly, 1997), *Calumma* (Hillenius, 1988), *Chamaeleo* (Parker, 1881; Methuen and Hewitt, 1914), *Furcifer* (Hillenius, 1988), *Rhampholeon* (Werner, 1902b; Frank, 1951; Rieppel, 1987), *Rieppeleon brevicaudatus* (Rieppel, 1987), and *Trioceros* (Fig. 2.1e) (Rieppel, 1981), a dorsal process of the squamosal meets the posterodorsal tip of the parietal crest, whereas in

Rieppeleon brachyurus and *Rieppeleon kerstenii*, the dorsal process is reduced and no longer meets the parietal (Rieppel, 1981). In *Bradypodion* (Fig. 2.1d) (Parker, 1881; Methuen and Hewitt, 1914; Engelbrecht, 1951; Rieppel, 1981; Hillenius, 1988) and *Brookesia* (Fig. 2.1b) (Siebenrock, 1893; Rieppel, 1987), the parietal forms a tapered plate extending posterodorsally to form the casque. A sagittal crest is formed on the ventral surface of the parietal, which meets the supraoccipital (Fig. 2.1a,c,e) (Rieppel, 1981) and a lateroventral processes extends off the posterolateral edge of the parietal to meet the dorsal process of the squamosal (Fig. 2.1a,c,e) (Parker, 1881; Methuen and Hewitt, 1914; Brock, 1941; Engelbrecht, 1951; Rieppel, 1981, 1987). A small supratemporal lies medial to the squamosal, wedged between the otic capsule wall and the head of the quadrate and squamosal (Fig. 2.1c) (Brock, 1941; Engelbrecht, 1951; Rieppel, 1981). The supratemporal is absent in *Rieppeleon* (Rieppel, 1987).

The lateral head of the quadrate's cephalic condyle articulates with the anterior surface of the squamosal's posteroventral process (Fig. 2.1a,c,e) (Rieppel, 1981). The pterygoid joins posteriorly with the palatine, laterally with the ectopterygoid, and posteromedially with the basisphenoid (Fig. 2.1f) (Engelbrecht, 1951; Rieppel, 1981). The pterygoid extends posterolaterally from the junction with the basisphenoid toward the quadrate, expanding into a wing-shaped structure in the process, but does not reach the quadrate, forming only a ligamentous connection with it (Fig. 2.1a,c,e,f) (Romer, 1956; Rieppel, 1981). The basisphenoid joins with the basioccipital at its posterior edge, and the occipital condyle is at the posterior edge of the basioccipital (Fig. 2.1f) (Parker, 1881; Werner, 1902b; Rieppel, 1981). In *Calumma*, the occipital condyle can be formed by the exoccipital with only participation of the basioccipital (Rieppel, 1987). The occipital condyle articulates with the proatlas of the vertebral column (Hoffstetter and Gasc, 1969).

The anterodorsal edge of the squamosal, posterodorsal edge of the postorbitofrontal, and the ventrolateral edge of the parietal circumscribes the upper temporal fossa (Fig. 2.1a–e) (Parker, 1881; Werner, 1902b; Engelbrecht, 1951; Frank, 1951; Rieppel, 1981, 1987). The anterior edge of the quadrate, ventral edge of the squamosal, posterior edge of the jugal, and in *Chamaeleo* and *Trioceros*, the posteroventral edge of the postorbitofrontal, circumscribe the posttemporal fossa (Fig. 2.1a–e) (Parker, 1881; Siebenrock, 1893; Werner, 1902b; Engelbrecht, 1951; Frank, 1951; Rieppel, 1981, 1987).

The dentaries are the sole tooth-bearing bones of the lower jaw; they join at a symphysis anteromedially (Fig. 2.1a,c,e) (Parker, 1881; Werner, 1902b; Engelbrecht, 1951; Frank, 1951; Rieppel, 1981, 1987). The coronoid attaches to the dentary medially with a dorsal coronoid process extending beyond the dorsal edge of the dentary (Fig. 2.1a,c,e) (Parker, 1881; Werner, 1902b; Engelbrecht, 1951; Frank, 1951; Rieppel, 1981, 1987). In *Calumma, Chamaeleo,* and *Trioceros*, the tooth row extends posteriorly beyond the anterior edge of the coronoid process (Fig. 2.1a) (Rieppel and Crumly, 1997). The angular attaches to the ventromedial aspect of the dentary (Fig. 2.1a,e) (Parker, 1881; Engelbrecht, 1951; Frank, 1951; Rieppel, 1981, 1987). The surangular joins to the posterior aspect of the coronoid and medial aspect of the dentary (Fig. 2.1a,c,e) (Parker, 1881; Engelbrecht, 1951; Rieppel, 1981, 1987).

The articular joins at the posterior edge of the surangular (Fig. 2.1c,e) (Parker, 1881; Engelbrecht, 1951; Frank, 1951; Rieppel, 1981, 1987). The retroarticular process in chameleons is reduced (Romer, 1956).

Chameleons, as do Agamidae and Leiolepididae, possess an acrodont dentition (Camp, 1923; Romer, 1956; Schwenk, 2000). Acrodont teeth are ankylosed to the apical surface of the upper and lower jaw, are added posteriorly to the tooth row during growth, and are worn throughout life and not replaced (Schwenk, 2000). Because teeth are not replaced, the stability of tooth position allows for strong occlusion patterns (Camp, 1923; Schwenk, 2000).

Cranial Musculature The musculature of the skull can be divided into a few broad complexes, the jaw abductor muscles, and the complex jaw adductor musculature, which is very well developed in chameleons, and finally the constrictor dorsalis musculature. Muscles of the throat, buccal cavity and hyobranchial apparatus are presented in the "Hyobranchial" section below and those of the eye in the section titled "Eye," which focuses on the eye as a whole.

JAW ABDUCTOR MUSCULATURE The Musculus (M.) depressor mandibulae complex in chameleons has not been described in detail (Haas, 1973). It is noted, however, to consist of an internus and externus division (Engelbrecht, 1951). Combined, it originates on the ascending process of the squamosal (Meyers and Clarke, 1998), through the posterior surface of the lateral ridge on the quadrate (Frank, 1951; Meyers and Clarke, 1998), and inserts into the posterior end of the mandible (Mivart, 1870; Meyers and Clarke, 1998) on the articular bone (Meyers and Clarke, 1998).

In some species, however, some of the posterior superficial fibers, called the M. depressor mandibulae pars auricularis, insert onto the lateral surface of the occipital lobes (Meyers and Clarke, 1998). The M. depressor mandibulae pars auricularis function to abduct the occipital lobes during display, which are then passively adducted by recoil of the skin and connective tissue on the lobes' medial surface (Meyers and Clarke, 1998). Utilization of a portion of the M. depressor mandibulae for this function, however, also results in movement of the occipital lobes during feeding (C.V. Anderson, personal observation).

JAW ADDUCTOR MUSCULATURE The quadratomaxillary ligament (the zygomatic ligament of Mivart, 1870, and Ogilvie, 1966, and the ligamentum jugomandibulare of Meyers and Clarke, 1998) lies beneath the skin on the side of the head, posterior to the corner of the mouth, and spans between the bones along the posteroventral margin of the orbit and the ventral end of the suspensorium (Mivart, 1870; Poglayen-Neuwall, 1954; Ogilvie, 1966; Rieppel, 1981; Meyers and Clarke, 1998). Deep to this ligament lies the rictal plate (Poglayen-Neuwall, 1954), which serves as a site of muscle attachment for some of the jaw adductor muscles (Schwenk, 2000). Finally, the jaw adductor tendon, called the "bodenaponeurosis" or "basal aponeurosis," which is a large aponeurotic plate attached to the lower jaw, spans between the lower jaw and some of the jaw adductor muscles in multiple sheets or septa (Poglayen-Neuwall, 1954; Rieppel, 1981, 1987). The lateral septum of the bodenaponeurosis is a narrow dorsal projection that extends dorsally from the coronoid process (Rieppel, 1981, 1987). The posterior sheet of the bodenaponeurosis extends dorsally from the posterior to the coronoid process (Rieppel, 1981). The anterior sheet of the bodenaponeurosis is narrow and fan-shaped; it also extends dorsally well into the upper temporal opening from the coronoid process (Rieppel, 1981).

The most superficial of the jaw adductor muscles is the M. levator anguli oris, which consists of an anterior and a posterior division (Rieppel, 1981, 1987). These divisions are weakly separated in some taxa (e.g., *Bradypodion pumilum* [Rieppel, 1981]; *Brookesia superciliaris* [Rieppel, 1987]), whereas in others they are more distinctly separated (e.g., *Trioceros melleri* [Rieppel, 1981]). The M. levator anguli oris anterior originates on the upper temporal arch and the M. levator anguli oris posterior originates on the quadrate (Rieppel, 1981, 1987), with both inserting on the rictal plate at the corner of the mouth (Schwenk, 2000). The M. tensor anguli oris is absent in chameleons (Rieppel, 1981).

Beneath the M. levator anguli oris and rictal plate is the M. adductor mandibulae externus superficialis (Rieppel, 1981). It originates on the medial surface of the upper temporal arch and inserts on the dorsolateral surface of the lower jaw (Rieppel, 1981). Superficially, the fibers of the M. adductor mandibulae externus superficialis extend dorsal to posterodorsally but the deeper fibers transition to more oblique angles approaching the more sharply posterodorsal angle of the fibers of the M. adductor mandibulae externus medialis, which lies beneath it (Rieppel, 1981).

The M. adductor mandibulae externus medialis lies deep to the M. adductor mandibulae externus superficialis and originates and inserts broadly (Rieppel, 1981). The anteriormost fibers originate on the dorsal and dorsoventral edge of the upper temporal fossa on the parietal and squamosal bones, and insert on the dorsal part of the lateral septum of the bodenaponeurosis (Rieppel, 1981, 1987). More medial fibers originate on the medial surface of the posterior side of the upper temporal arch on the squamosal bone and from the cephalic condyle of the quadrate and then insert on the posteroventral portion of the lateral septum of the bodenaponeurosis (Rieppel, 1981, 1987). The posteriormost fibers originate on the lateral surface of the quadrate and insert on the posterior sheet of the bodenaponeurosis (Rieppel, 1981) or on the surangular of the lower jaw (Rieppel, 1987). Some deep fibers of the M. adductor mandibulae externus medialis originate on the parietal and ascending process of the squamosal and insert on the lateral surface of the anterior sheet of the bodenaponeurosis (Rieppel, 1981).

The M. adductor mandibulae externus profundus is divided into three heads, or portions, and overall is enlarged because of the formation of the casque (Rieppel, 1981). The posteroventralmost portion, the so-called 3a-head, corresponds to fibers originating on the anteromedial surface of the quadrate and insert on the dorsomedial surface of the jaw, deep to the posterior sheet of the bodenaponeurosis, but in chameleons cannot be clearly defined (Rieppel, 1981). The pattern of origin and insertion of the anterodorsalmost portion, the so-called 3b-head, varies from one casque structure to another. In *Trioceros,* which have a strong sagittal crest on the parietal bone, the 3b-head originates on the sagittal crest deep to the M. adductor mandibulae externus medialis and inserts on the medial aspect of the anterior sheet of the bodenaponeurosis (Rieppel, 1981). In *Bradypodion,* which have a broad parietal bone with a ventrolateral process, the 3b-head originates on the lower surface of the parietal and on to the ventrolateral process and inserts on the medial surface of the anterior sheet of the bodenaponeurosis (Rieppel, 1981). In both cases, the origin extends anteriorly over the insertion of the M. pseudotemporalis superficialis (Rieppel, 1981) and in some cases onto the posterior edge of the postorbital, where it meets with the parietal (Engelbrecht, 1951). Between the 3a- and 3b-heads lies the so-called 3c-head of the M. adductor mandibulae externus profundus (Rieppel, 1981). The 3c-head originates on the lateral and ventral aspect of the prootic (Rieppel, 1981, 1987), the covering of the surface of the otic capsule wall (Brock, 1941; Engelbrecht, 1951), and the anterior and anterodorsal aspect of the paroccipital process of the back of the skull (Rieppel, 1981). It inserts on the medial surface on the basal portion of the bodenaponeurosis and on the medial surface of the coronoid process itself (Rieppel, 1981).

The M. adductor posterior lies deep to the M. adductor mandibulae externus profundus and is rather large in chameleons (Rieppel, 1981). It originates on the medial edge of the quadrate; on the membrane between the quadrate, prootic, and pterygoid; and on the dorsolateral part of the pterygoid wing (Haas, 1973; Rieppel, 1981). The M. adductor posterior inserts on the medial aspect of the surangular of the lower jaw (Rieppel, 1981).

Although regarded by some as not differentiated (Brock, 1941), the M. pseudotemporalis consists of a superficialis and profundus division and is also deep to the M. adductor mandibulae externus profundus (Haas, 1973; Rieppel, 1981, 1987). The M. pseudotemporalis superficialis originates on the anteromedial aspect of the casque and inserts on the dorsal portion of the tendinous raphe extending dorsally from the coronoid process deep to the bodenaponeurosis (Rieppel, 1981). The M. pseudotemporalis profundus originates on the anterior edge of the prootic and on the membranous sidewall of the braincase anterior to it (Rieppel, 1981). Anterior fibers of the M. pseudotemporalis profundus insert on the posterior base of the tendon, whereas the deeper and more posterior fibers of the M. pseudotemporalis superficialis insert on the medial aspect of the lower jaw direction just posteroventral to the coronoid process (Rieppel, 1981, 1987).

The M. pterygoideus consists of a superficial (ventral) and deep (dorsal) head (Rieppel, 1981). The superficial head originates on the ventral edge and ventromedial aspect of the pterygoid wing and inserts on the lower edge and ventrolateral surface of the lower jaw (Rieppel, 1981). The deep head originates on the lateral aspect of the posteroventral portion of the pterygoid wing and inserts on the medial surface of the lower jaw, just ventral and anteroventral to the jaw joint (Rieppel, 1981).

CONSTRICTOR DORSALIS MUSCULATURE The muscles of the constrictor internus dorsalis complex, which are typically involved in cranial kinesis (Schwenk, 2000), are highly reduced in chameleons (Brock, 1941; Engelbrecht, 1951; Frank, 1951; Haas, 1973; Rieppel, 1981). This is in large part a result of the akinetic structure of the chameleon skull (Haas, 1973).

Whereas the M. levator pterygoidei has been reported in young *Chamaeleo* (Lakjer, 1926; Rieppel, 1981), other research has failed to identify it (Lubosch, 1933; Brock, 1941; Engelbrecht, 1951; Frank, 1951; Poglayen-Neuwall, 1954; Rieppel, 1981, 1987).

The M. protractor pterygoidei is strongly developed in some taxa but only weekly developed in others (Frank, 1951; Haas, 1973; Rieppel, 1981, 1987). It originates on the basipterygoid process (Frank, 1951; Poglayen-Neuwall, 1954; Haas, 1973; Rieppel, 1981, 1987) and inserts on the medial to dorsomedial aspect of the pterygoid wing (Haas, 1973; Rieppel, 1981, 1987). In *Bradypodion,* it is also noted to insert on the ligament connecting the pterygoid wing and the quadrate, thus acting as a quadrate protractor in these taxa (Rieppel, 1981).

The M. levator bulbi ventralis is also lacking in chameleons (Poglayen-Neuwall, 1954; Haas, 1973). The M. levator bulbi dorsalis, however, is present and originates on the prootic wing and inserts on the ventral portion of the eye (Poglayen-Neuwall, 1954; Haas, 1973).

Hyobranchial

One of the more highly specialized features in chameleons is the tongue. In order to achieve ballistic tongue projection, the tongue apparatus has undergone a series of anatomical changes from their agamid-like ancestors. Interest in the tongue of chameleons has resulted in a wide range of studies on its structure and function over the years. These have subsequently resulted in a wide range of varying interpretations and names of the tongue's structures, particularly muscular structures. These name synonyms and the names we've adopted are summarized in Table 2.1 and described in the following sections.

Tongue Skeleton The chameleon hyobranchial apparatus is comprised of a reduced basihyoid, an elongate lingual process, and two pairs of cornua (Fig. 2.2a,b) (Bell, 1989; Herrel et al., 2001b; Meyers et al., 2002). The hyobranchial apparatus is suspended in the region of the neck and throat by muscle connection between it and the lower jaw, sternum, and pectoral girdle (Zoond, 1933; Wainwright et al., 1991). The elongate lingual process lies medially and extends anteriorly into the buccal cavity (Houston, 1828).

The elongate lingual process, called the "entoglossal process," is parallel-sided over most of its length, with a tapered anterior tip (Fig. 2.2a,b) (Gnanamuthu, 1930; van Leeuwen, 1997; Wainwright and Bennett, 1992b; Herrel et al., 2001b, 2009; de Groot and van Leeuwen, 2004). The degree of tapering reported in the literature varies from the anterior 10% (Wainwright and Bennett, 1992b) to 1 to 1.5% (Herrel et al., 2001b). Histological sections of the entoglossal process indicate that it is cartilaginous, with hyaline cartilage along its body and a thick layer of dense fibrocartilage near the tip (Herrel et al., 2001b). Some degree of calcification of the entoglossal process is evident, however, as the entoglossal process on cleared and stained specimens stains for bone (Herrel et al., 2001b).

The anterior pair of cornua consists of the ceratohyalia, which are shorter than the posterior pair of cornua (Fig. 2.2a,b) (Gnanamuthu, 1930; Bell, 1989; Wainwright et al., 1991; Herrel et al., 2001b, 2009; Meyers et al., 2002) and are completely cartilaginous (Wainwright et al., 1991; Herrel et al., 2001b; Meyers et al., 2002). Each ceratohyal is divided into two parts, with the proximal part being more robust and the distal part being more flexible (Herrel et al., 2001b). The two parts of the ceratohyals articulate with a synovial joint and the proximal part articulates on the anterior dorsal side of the basihyoid with a U-shaped synovial joint (Herrel et al., 2001b). From the basihyoid, the ceratohyals extend anterodorsally when the tongue apparatus is in its rest position and viewed laterally (Houston, 1828; Bell, 1989;

TABLE 2.1 Muscle Synonomies of Chameleon Hyobranchial Musculature

This Review	Published Synonyms	Sources
Musculi (Mm.) mandibulohyoideus medialis[a], lateralis 1[b] et lateralis 2[c]	Mm. geniohyoid "internal"[a] et "external"[b,c]	Houston, 1828; Dewevre, 1895
	M. geniohyoid[a]	Mivart, 1870; Zoond, 1933
	M. ceratomandibular[b,c]	Mivart, 1870
	Mm. geniohyoideus medialis[a] et lateralis[b,c]	Lubosch, 1932; Altevogt and Altevogt, 1954; Altevogt, 1977; Wainwright et al., 1991; So et al., 1992
	M. geniohyoideus	Brücke, 1852a; Wainwright and Bennett, 1992a
	M. geniohyoideus[a]	Kathariner, 1894; Germershausen, 1913
	M. genio-ceratoideus[b,c]	Kathariner, 1894; Germershausen, 1913
	M. mandibulohyoideus	Meyers and Nishikawa, 2000; Herrel et al., 2009
	Mm. mandibulohyoideus 1[b], 2[a], et 3[c]	Herrel et al., 2001b
	Mm. mandibulohyoideus 1[c], 2[a], et 3[b]	Meyers et al., 2002
M. omohyoideus	M. omohyoid	Houston, 1828; Mivart, 1870
	M. scapula-hyoidien	Dewevre, 1895
	M. omohyoideus	Kathariner, 1894; Germershausen, 1913; Lubosch, 1932; Herrel et al., 2001b; Meyers et al., 2002
Mm. sternohyoideus superficialis[a] et profundus[b]	M. sternohyoideus[a]	Houston, 1828; Mivart, 1870; Kathariner, 1894; Germershausen, 1913; Gnanamuthu, 1930, 1937; Zoond, 1933; Altevogt and Altevogt, 1954; Altevogt, 1977; Wainwright et al., 1991; So et al., 1992; Wainwright and Bennett, 1992a; Meyers et al., 2000; Herrel et al., 2001b, 2009
	M. sternoceratoideus[b]	Houston, 1828; Kathariner, 1894; Germershausen, 1913
	M. sternothyroideus[b]	Mivart, 1870; Gnanamuthu, 1930, 1937; Zoond, 1933; Wainwright et al., 1991; So et al., 1992; Wainwright and Bennett, 1992a; Meyers et al., 2000; Herrel et al., 2001b

(Continued)

TABLE 2.1 *(Continued)*

This Review	Published Synonyms	Sources
	Mm. sterno-hyoiden "anterieur"[a] et "postero-lateral"[b]	Dewevre, 1895
	M. sternohyoidei	Lubosch, 1932
	Mm. sternohyoideus superficialis[a] et profundus[b]	Meyers et al., 2002
Mm. genioglossus anterior[a] et posterior[b]	M. genioglossus	Mivart, 1870; Kathariner, 1894; Gnanamuthu, 1930; 1937; Bell, 1989
	M. génio-périglosse	Dewevre, 1895
	Mm. genioglossus anterior[a] et posterior[b]	Herrel et al., 2001b
	Mm. genioglossus medialis[b] et lateralis[a]	Meyers et al., 2002
M. constrictor colli	M. constrictor colli	Herrel et al., 2001b
Mm. intermandibularis anterior[a] et posterior[b]	M. mylohyoideus	Houston, 1828; Brücke, 1852a; Dewevre, 1895; Gnanamuthu, 1930
	Mm. mylohyoideus anterior[a] and posterior[b]	Mivart, 1870; Kathariner, 1894
	M. intermaxillaris[a]	Germershausen, 1913
	M. mylo-hyoideus posterior[b]	Germershausen, 1913
	Mm. intermandibularis anterior[a] et posterior[b]	Herrel et al., 2001b; Meyers et al., 2002
M. branchiohyoideus	M. ceratohyoideus	Mivart, 1870; Gnanamuthu, 1930
	M. branchiohyoideus	Herrel et al., 2001b
M. hyoglossus	M. hyoglossus	Houston, 1828; Brücke, 1852a; Kathariner, 1894; Gnanamuthu, 1930, 1937; Lubosch, 1932; Zoond, 1933; Altevogt and Altevogt, 1954; Altevogt, 1977; Bell, 1989; So et al., 1992; Wainwright and Bennett, 1992a; Herrel et al., 2000, 2001a,b, 2002, 2009; Meyers and Nishikawa, 2000, 2002; de Groot and van Leeuwen, 2004
	M. glosso-hyoidiens	Dewevre, 1895
	"glossohyal muscle"	Gans, 1967
	"retractor muscle"	Wainwright et al., 1991
	"hyoglossal muscle"	Wainwright and Bennett, 1992a

TABLE 2.1 *(Continued)*

This Review	Published Synonyms	Sources
M. accelerator linguae	"Annular muscle"	Houston, 1828
	"ring muscle"	Gnanamuthu, 1930, 1937; Zoond, 1933
	"accelerator muscle"	Gans, 1967; Wainwright et al., 1991; So et al., 1992; Wainwright and Bennett, 1992a,b; van Leeuwen, 1997; Meyers et al., 2002; de Groot and van Leeuwen, 2004
	M. accelerator	Kathariner, 1894; Lubosch, 1932; Altevogt and Altevogt, 1954; Altevogt, 1977; Herrel et al., 2000, 2001a,b, 2009; Meyers and Nishikawa, 2000
	M. accelerator linguae	Brücke, 1852a; Kathariner, 1894; Sewertzoff, 1923; Altevogt and Altevogt, 1954; Altevogt, 1977; Schwenk and Bell, 1988; Bell, 1989
M. "retractor pouch"	M. longitudinales linguae adductoris	Brücke, 1852a; Gnanamuthu, 1930, 1937; Zoond, 1933; Bell, 1989
	M. hyoglossus superficialis	Kathariner, 1894
	M. pouch retractor	Herrel et al., 2000
	M. "retractor pouch"	Herrel et al., 2001b
M. longitudinalis linguae ventralis	M. submucosus	Brücke, 1852a
	M. hyoglossus profundus	Kathariner, 1894
	M. longitudinalis linguae extensoris	Gnanamuthu, 1930, 1937; Zoond, 1933; Bell, 1989
	M. longitudinalis linguae ventralis	Herrel et al., 2001b
M. pulvinaris	M. pulvinar	Brücke, 1852; Lubosch, 1932
	"Ringmuskel für den Fangnapf"	Altevogt, 1977
	M. pulvinaris	Bell, 1989; Herrel et al., 2001b
Mm. transversalis linguae anterior[a] et posterior[b]	M. lateralis linguae	Brücke, 1852a
	M. transversalis linguae[a]	Gnanamuthu, 1930, 1937
	M. lateralis linguae[b]	Gnanamuthu, 1930
	M. transversalis linguae externi[a]	Bell, 1989
	M. superficialis linguae[b]	Bell, 1989
	Mm. transversalis linguae anterior[a] and posterior[b]	Herrel et al., 2001b

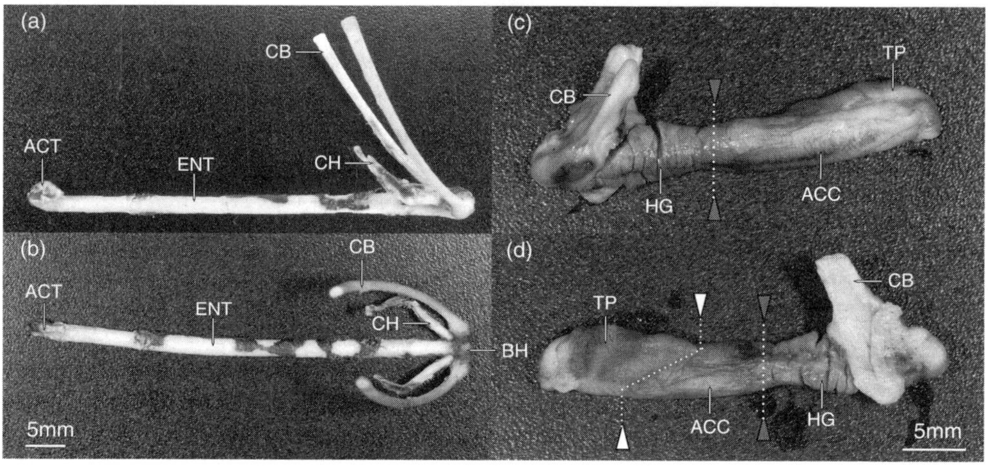

FIGURE 2.2. Skeletal and muscular components of the chameleon tongue apparatus. Lateral (a) and dorsal (b) views of the skeletal elements of the tongue of *C. p. parsonii*. Ventrolateral (c) and dorsolateral (d) views of the muscular elements of the tongue of *T. johnstoni* at rest. Anterior end of elements at left in (a), (b), and (d), and at right in (c). Scale bar at bottom left applies to (a) and (b), and that at bottom right to (c) and (d). Modified from Anderson et al. (2012).

LABELS: ACC = M. accelerator linguae; ACT = articulating cartilaginous tip; BH = basihyoid; CB = ceratobranchial; CH = ceratohyal; ENT = entoglossal process; HG = M. hyoglossus; TP = tongue pad. Dotted lines between gray triangles in (c) and (d) indicate division between HG and ACC. Dotted line between white triangles in (d) indicate posterior limits of the TP.

Herrel et al., 2001b). When viewed in a transverse plane, the ceratohyals form a U-shape. As the entoglossal process is pulled forward during tongue protrusion, the ceratohyals rotate and are pointed upward (Herrel et al., 2001b). In some species, the distal part of the ceratohyal has a flat triangular piece of cartilage attached to it (Gnanamuthu, 1930; Herrel et al., 2001b).

The posterior pair of cornua is the ceratobranchials, which are ossified and longer than the ceratohyals (Fig. 2.2a,b) (Gnanamuthu, 1930; Bell, 1989; Herrel et al., 2001b; Meyers et al., 2002). The ceratobranchials articulate with the posterior side of the basihyoid with a saddle-shaped synovial joint (Herrel et al., 2001b; Meyers et al., 2002). At rest, the ceratobranchials extend anterodorsally (Gnanamuthu, 1930; Bell, 1989) to dorsally in a nearly perpendicular direction to the long axis of the hyobranchial apparatus when viewed laterally (Herrel et al., 2001b; Meyers et al., 2002). When viewed in a transverse plane, the ceratobranchials form a U-shape. During protrusion of the tongue, the ceratobranchials rotate and are folded backward (Wainwright et al., 1991; Herrel et al., 2009).

Hyobranchial Musculature The hyobranchial apparatus is suspended in the throat by muscles that originate outside the hyobranchial apparatus and insert on the hyobranchial skeleton (Bell, 1989; Wainwright et al., 1991). These muscles serve to draw the hyobranchial apparatus forward and back during tongue protrusion and hyobranchial retraction (Gnanamuthu, 1930; Bell, 1990; Herrel et al., 2009).

The paired M. mandibulohyoideus consists of three distinct divisions (Gnanamuthu, 1930; Herrel et al., 2001b). The M. mandibulohyoideus medialis originates near the symphysis of the lower jaw via a short aponeurosis and inserts on the ventral surface of the basihyoid (Wainwright and Bennett, 1992a; Herrel et al., 2001b). The M. mandibulohyoideus lateralis 1 originates lateral to the symphysis of the lower jaw and inserts on the tip of the ceratohyal (Herrel et al., 2001b). The M. mandibulohyoideus lateralis 2 originates on the jaw between the M. mandibulohyoideus medialis and M. mandibulohyoideus lateralis 1, is attached to the M. mandibulohyoideus lateralis 1 for most of the latter's length, and inserts on the distal third of the ceratobranchial (Herrel et al., 2001b). Together, the M. mandibulohyoideus serves to draw the hyobranchial apparatus anteriorly during tongue protrusion and protraction and is active during prey transport (Brücke, 1852a; Dewevre, 1895; Gnanamuthu, 1930; Zoond, 1933; Wainwright et al., 1991; Wainwright and Bennett, 1992a; Meyers and Nishikawa, 2000). The two divisions of the M. mandibulohyoideus lateralis may also serve to facilitate articulation of the cornua with the basihyoid by drawing the tips of the cornua forward as the M. sternohyoideus draws the basihyoid back during hyobranchial retraction (Dewevre, 1895; Gnanamuthu, 1930).

The paired M. omohyoideus originates on the anterior, ventral side of the scapula and inserts on the posterior side of the lateral aspect of the basihyoid (Mivart, 1870; Gnanamuthu, 1930, 1937; Meyers et al., 2002). From the basihyoid, however, it extends dorsally to wrap around the M. sternothyroideus before returning ventrally and curving under the M. episternocleidomastoideus toward the scapula (Herrel et al., 2001b; Meyers et al., 2002). It serves to draw the basihyoid upward (Mivart, 1870; Gnanamuthu, 1930).

The paired M. sternohyoideus consists of a superficialis and a profundus division (Meyers et al., 2002). The M. sternohyoideus superficialis originates on the posteroventral surface of the xiphisternum (xiphoid process) and inserts on the ventral side of the basihyoid (Gnanamuthu, 1930; Herrel et al., 2001b; Meyers et al., 2002). It serves to draw the basihyoid posteriorly during hyobranchial retraction (Gnanamuthu, 1930, 1937; Zoond, 1933; Wainwright and Bennett, 1992a). The M. sternohyoideus profundus consists of two divisions (Herrel et al., 2001b). The anterior division originates on the midbody connective-tissue band anterior to the xiphisternum and inserts on the posterior tip of the ceratobranchial (Herrel et al., 2001b). The posterior division also originates on the midbody connective-tissue band but immediately anterior to the xiphisternum and inserts onto the posterior side of the dorsal half of the ceratobranchial (Herrel et al., 2001b). Together, they serve to draw the distal end of the ceratobranchials in a posteroventral direction during tongue protrusion (Gnanamuthu, 1930, 1937; Wainwright and Bennett, 1992a).

Within the throat and buccal cavity, support and movement of the hyobranchial apparatus is facilitated by intermandibular musculature (Gnanamuthu, 1930; Herrel et al., 2001b). These muscles originate on the skull and mandible and generally serve to elevate the throat and gular regions, and in doing so elevate the hyobranchial apparatus within the throat and buccal cavity (Gnanamuthu, 1930).

The paired M. genioglossus consists of an anterior and a posterior division (Herrel et al., 2001b). They originate on the inner surface of the mandible and insert on the buccal-floor epithelium (Gnanamuthu, 1930; Bell, 1989; Herrel et al., 2001b), with the anterior portion inserting via a tendon (Herrel et al., 2001b). The anterior division inserts at the level of the basihyoid, whereas the posterior division inserts on an aponeurosis at the floor of the throat (Herrel et al., 2001b). The M. genioglossus forms a pouch around the tongue inside the mouth and when contracted form paddle-like lips on either side of the tongue (Gnanamuthu, 1930, 1937; Bell, 1989).

The paired M. constrictor colli originates on the dorsal nuchal/cervical fascia and inserts on the midventral fascia (Herrel et al., 2001b; Meyers et al., 2002). From its origin, it extends ventrally and then posterior to the lower jaw and curves medially toward the midventral fascia (Herrel et al., 2001b). It serves to elevate the throat (Gnanamuthu, 1930).

The paired M. intermandibularis consists of two divisions, an anterior and a posterior one (Gnanamuthu, 1930; Herrel et al., 2001b), with the anterior division being further divided into a principalis and profundus sheet by some researchers (Gnanamuthu, 1930, 1937). They originate broadly along the inner surface of the mandible and lower jaw, with the posterior division originating via a short aponeurosis (Herrel et al., 2001b). The anterior division inserts on the midventral fascia, which is attached to the jaw symphysis, and the posterior division inserts on the midventral fascia via an aponeurosis (Herrel et al., 2001b; Meyers et al., 2002). The M. intermandibularis anterior principalis runs anteromedially toward its insertion, whereas the M. intermandibularis anterior profundus runs posteromedially toward its insertion (Gnanamuthu, 1930). Together they serve to elevate the floor of the mouth.

Within the hyobranchial apparatus itself, the paired M. branchiohyoideus spans between the posterior edge of the distal third of the ceratohyal and the anterior side of the distal quarter of the ceratobranchial (Gnanamuthu, 1930; Herrel et al., 2001b). It enables movement of the cornua with respect to each other, as the aforementioned muscles that insert on them modulated their position.

The paired M. hyoglossus originates on the medial surface of the ceratobranchial along its entire length and inserts under the strong outer layer of connective tissue on the lateral aspect of the M. accelerator linguae at approximately a quarter of its length (Fig. 2.2c,d) (Herrel et al., 2001b; Meyers et al., 2002). The muscle is bulky near its origin and quickly narrows as it runs ventrally to the proximal end of the ceratobranchial (Gnanamuthu, 1930; Bell, 1989; Herrel et al., 2001b). It then passes under the ceratohyal by its articulation with the basihyoid (Gnanamuthu, 1930; Bell, 1989; Herrel et al., 2001b). At rest, the M. hyoglossus is heavily pleated around the posterior end of the entoglossal process until it reaches the posterior edge of the M. accelerator linguae (Fig. 2.2c,d) (Herrel et al., 2001b; Meyers et al., 2002). The M. hyoglossus is surrounded by a sheath of epimysium between the M. accelerator linguae and the base of the entoglossal process (Meyers et al., 2002). Fully elongated, the M. hyoglossus extends up to 600% of its resting length (Herrel et al., 2001a, 2002). This extreme shortening capability is the result of supercontracting muscle fibers with perforated Z discs, which allow filaments within each muscle sarcomere to extend through the Z discs and into adjacent sarcomeres (Rice, 1973; Bell, 1989; Schwenk, 2000; Herrel et al., 2001a, 2002). The M. hyoglossus serves to retract the M. accelerator linguae back onto the entoglossal process following tongue projection (Altevogt and Altevogt, 1954; Gans, 1967; Bell, 1989; Wainwright and Bennett, 1992a).

The M. accelerator linguae surrounds the entoglossal process at rest and is surrounded by an inner and outer tendinous connective-tissue sheath (Fig. 2.2c,d) (Gnanamuthu, 1930; Bell, 1989; Herrel et al., 2001b). The posterior three quarters form a muscular tube around the entoglossal process (Gnanamuthu, 1930; Bell, 1989; Herrel et al., 2001b), with muscle fibers extending radially between the inner and outer tendinous sheaths in a cross-helical fashion (Gnanamuthu, 1930; Gans, 1967; Bell, 1989; van Leeuwen, 1997; Herrel et al., 2001b; de Groot and van Leeuwen, 2004). The anterior quarter of the M. accelerator linguae divides into a dorsal and a ventral projection (Gnanamuthu, 1930; Bell, 1989; Herrel et al., 2001b). The dorsal bundle extends to the posterior edge of the tongue pouch, whereas the ventral bundle extends all the way to the tip of the tongue (Herrel et al., 2001b). The dorsal bundle is continuous with the posterior three quarters of the M. accelerator linguae, with muscle fibers oriented perpendicularly to the long axis of the hyobranchial apparatus (Herrel et al., 2001b). The ventral bundle has similarly oriented muscle fibers and is continuous with the posterior three quarters up to the approximate location where the dorsal bundle ends, at which point a vertical connective-tissue septum separates the remaining length of the ventral projection of the M. accelerator linguae (Herrel et al., 2001b). The M. accelerator linguae serves to push the tongue off the entoglossal process (Gans, 1967; Altevogt, 1977; Bell, 1989; Wainwright and Bennett, 1992a,b) and load elastic elements involved in tongue projection (de Groot and van Leeuwen, 2004).

The paired M. "retractor pouch" originates on the dorsolateral side of the M. accelerator linguae on the posterior third of its length and inserts medially on the inner side of the membrana grandulosa of the tongue pad, which is invaginated at rest (Herrel et al., 2001b). It serves to draw the center of the tongue pad posteriorly during prey prehension (Herrel et al., 2000), thus invaginating the membrane grandulosa of the tongue pad.

The paired M. longitudinalis linguae ventralis originates immediately posterior to the bifurcated tongue tip on the internal surface of the tongue pad and inserts on the lateral side of the anteroventral, noncircular portion of the M. accelerator linguae (Bell, 1989; Herrel et al., 2001b). Whereas some studies suggest that this muscle's action involves extension of the tongue (Gnanamuthu, 1930), it appears clear that it is not an extensor (Herrel et al., 2001b) but likely serves to draw the ventral aspect of the tongue pad back, possibly drawing the bifurcated tongue tip ventrally in the process.

The paired M. pulvinaris is restricted to the tongue pad, where it develops at its posterior end, extends anteriorly, and ends immediately anterior to the tongue pouch (Brücke, 1852a; Bell, 1989; Herrel et al., 2001b).

The M. transversalis linguae consists of an anterior and a posterior division. The paired M. transversalis linguae anterior originates on the dorsal aspect of the anterior, noncircular portion of the M. accelerator linguae and inserts immediately anterior to the pouch on the inner surface of the tongue pad (Herrel et al., 2001b). The paired M. transversalis linguae posterior originates on the dorsolateral surface of the M. accelerator linguae on its posterior end and inserts posterior to the pouch on the medial inner surface of the tongue pad (Herrel et al., 2001b).

Located between the entoglossal process and the tongue muscles that surround it is an assortment of connective tissue (Gnanamuthu, 1930; Zoond, 1933; Gans, 1967; Bell, 1989; Herrel et al., 2001b; de Groot and van Leeuwen, 2004) that is comprised of a series of nested intralingual sheaths (Gnanamuthu, 1930; Bell, 1989; de Groot and van Leeuwen, 2004). At the anterior tip of the entoglossal process, a short articulating cartilaginous projection, which is folded back on the entoglossal process at rest, is found (Herrel et al., 2001b). This articulating cartilaginous tip connects the layer of longitudinal collagen fibers surrounding the entoglossal process, the perichondrium, to the innermost intralingual collagen sheath between the entoglossal process and the M. accelerator linguae (de Groot and van Leeuwen, 2004). The innermost intralingual sheaths are longer than more peripheral sheaths, as the inner sheaths attach more proximally on the hyobranchial apparatus than the outer sheaths (de Groot and van Leeuwen, 2004). The innermost sheath is attached at its posterior end to the fascia of the M. hyoglossus, near the articulation of the entoglossal process and the certobranchials, and each subsequent sheath attaches slightly anterior to the previous sheath (de Groot and van Leeuwen, 2004). More peripheral sheaths eventually are connected to the inner fascia of the M. accelerator linguae (de Groot and van Leeuwen, 2004). The inner sheaths are connected to each other only via their attachment to the M. hyoglossus and are able to slide past each other in a telescoping fashion as the tongue extends (de Groot and van Leeuwen, 2004). In addition to their structural connection at their attachment, the more peripheral sheaths are also interconnected by collagenous trabeculae (de Groot and van Leeuwen, 2004). The collagen fibers of the peripheral sheaths, which lie medial to the M. accelerator linguae, are arranged in a cross-helical pattern (de Groot and van Leeuwen, 2004). The fibers of the inner sheaths are also arranged in a cross-helical pattern in the portions of these sheaths that lie medial to the M. accelerator linguae; however, the fibers in the portions that are posterior to the M. accelerator linguae run parallel to the long axis of the entoglossal process (de Groot and van Leeuwen, 2004).

Trunk and Tail

Chameleons are adapted to be able to produce a large amount of dorsoventral flexion. This is particularly true in their highly prehensile tail, which they are able to curl tightly under their body. Various works have been done on the vertebral column and tail of the chameleon (e.g. Siebenrock, 1893; Camp, 1923; Ali, 1948; Romer, 1956; Etheridge, 1967; Hoffstetter and Gasc, 1969; Zippel et al., 1999) but surprisingly little has been done on the trunk musculature (e.g., Mivart, 1870; Sathe, 1959).

Vertebral Column and Ribs The vertebral column of chameleons has been examined in only a handful of taxa. These studies have found the vertebral column within the family to be variable in a number of regards and to possess a number of functional specializations as compared with other Saurians.

One of the most variable features of the chameleon vertebral column is the number of vertebrae. The number of presacral (cervical, thoracic, and lumbar) vertebrae is known to range from 14 (Bergmann and Irschick, 2011) to 23 (Hoffstetter and Gasc, 1969). Whereas having fewer

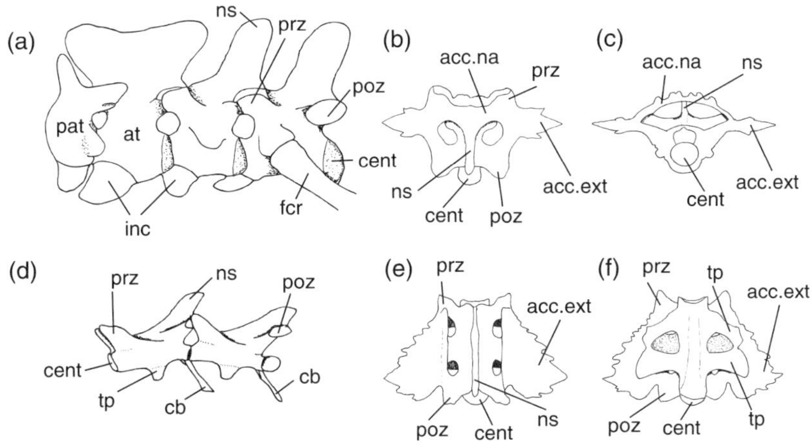

FIGURE 2.3. Vertebral elements of a generalized *Bradypodion* (a), *Brookesia superciliaris* (b, c, e, f) and *Chamaeleo zeylanicus* (d). Depicted are a lateral view of the first four cervical vertebrae (a), a dorsal (b) and transverse (c) view of a trunk (thoracic/lumbar) vertebrae, a lateral view of two proximal caudal vertebrae (d), and a dorsal (e) and ventral (f) view of the sacral vertebrae (fused into a synsacrum in *Brookesia*). (a) redrawn from Raw (1976; originally based on Hoffstetter and Gasc, 1969), (b, c, e, f) from Siebenrock (1893), and (d) from Ali (1948).

LABELS: acc.ext = accessory extension; acc.na = accessory neural arch; at = atlas; cb = chevron bone (hemal arch); cent = centrum; fcr = first cervical rib; inc = intercentra; ns = neural spine; pat = proatlas; poz = postzygapophysis; prz = prezygapophysis; tp = transverse process.

than 23 presacral vertebrae is known only within Saurians that occur in the suborder Iguania, this range represents a reduction in the typical number of presacral vertebrae seen in the Iguanidae and Agamidae (Hoffstetter and Gasc, 1969). Further, with 14 presacral vertebrae, *Brookesia superciliaris* has among the lowest number of trunk vertebrae of all squamate reptiles (Bergmann and Irschick, 2011). Posteriorly, there are two sacral vertebrae in all taxa (Hoffstetter and Gasc, 1969) and caudal vertebrae are known to range from 17 (Nečas, 2004) to 62 (Etheridge, 1967) with smaller, more terrestrial genera typically having fewer caudal vertebrae then larger, more arboreal genera (Etheridge, 1967; Nečas, 2004; Boistel et al., 2010).

All vertebrae have large procoelous centrums (Camp, 1923; Hoffstetter and Gasc, 1969; Raw, 1976) with elongated, cylindrical centra (Camp, 1923; Romer, 1956) and intercentra confined only to the cervical region (Fig. 2.3a) (Hoffstetter and Gasc, 1969). The neural spine is generally quite tall and typically extends posterodorsally with a posterior incline and terminates with a straight, axe-shaped dorsal edge (Fig. 2.3a,d) (Hoffstetter and Gasc, 1969). The neural spine can be elongated considerably in certain regions of the spine in some species, such at *Trioceros cristatus* or *T. montium* (Case, 1909). In these cases, the distal ends of the neural spines are connected by strong connective tissue threads and covered with a skin membrane forming a strongly elevated crest (Case, 1909). Zygosphenes and zygantra are absent from the vertebra (Raw, 1976). The articular facets of the prezygapophyses and postzygapophyses are steep, resulting in near-vertical articulation, thus allowing for increased dorsoventral flexion (Hoffstetter and Gasc, 1969). Precaudal vertebrae lack ventral hypapophyses in some species (Raw, 1976); however, three or four cervical intercentra are typical (Fig. 2.3a) (Hoffstetter and Gasc, 1969).

All chameleons have five cervical vertebrae (Siebenrock, 1893; Werner, 1902b; Hoffstetter and Gasc, 1969; Raw, 1976; Nečas, 2004). However, some researchers (Camp, 1923; Romer, 1956) have indicated the presence of only three, because the last two cervical vertebrae bear long cervical ribs (Fig. 2.3a) (Siebenrock, 1893; Werner, 1902b; Hoffstetter and Gasc, 1969; Raw, 1976; Nečas, 2004). These cervical ribs do not fuse to the sternum, however (Siebenrock, 1893; Hoffstetter and Gasc, 1969). The first two cervical vertebrae are the proatlas and atlas, respectively (Fig. 2.3a) (Hoffstetter and Gasc, 1969; Raw, 1976), with the proatlas appearing among Saurians only in the chameleons (Hoffstetter and Gasc, 1969). The three or four intercentra are always separated and maintain an intervertebral position on the ventral aspect of the cervical region of the vertebral column (Fig. 2.3a) (Hoffstetter and Gasc, 1969).

Given a constant number of cervical vertebrae (Hoffstetter and Gasc, 1969; Raw, 1976; Nečas, 2004), the combined number of thoracic and lumbar vertebrae varies from 9 (Bergmann and Irschick, 2011) to 18 (Hoffstetter and Gasc, 1969). Thoracic vertebrae have both sternal and parasternal ribs (Hoffstetter and Gasc, 1969), and the first two lumbar vertebrae typically have reduced lumbar ribs (Hoffstetter and Gasc, 1969; Raw, 1976).

The ribs in chameleons are unicipital, with a single articulation between the rib and the vertebra (Hoffstetter and Gasc, 1969) on the lateral margin of the vertebra (Rieppel, 1993). The ribs generally have two proximodistal segments, a bony vertebrocostal and a cartilaginous sternocostal segment, although a third cartilaginous intercostal segment between the two aforementioned segments is sporadically seen within the Chamaeleonidae (Hoffstetter and Gasc, 1969). Three to four sternal ribs, which are joined to the sternum or mesosternum, are seen (Methuen and Hewitt, 1914; Hoffstetter and Gasc, 1969). Parasternal ribs have their distal cartilaginous segments fuse on the midventral line, forming a parasternum posterior to the sternum (Camp, 1923; Hoffstetter and Gasc, 1969) and range in number from 5 to 11 (Sathe, 1959; Hoffstetter and Gasc, 1969). Two reduced lumbar ribs are observed on the anteriormost lumbar vertebrae (Hoffstetter and Gasc, 1969; Raw, 1967).

The thoracic and lumbar vertebrae in *Brookesia* are somewhat different from those of other chameleons. Their thoracic and lumbar vertebrae have a bony arch between the prezygapophyses and postzygapophyses on one side of a single vertebrae and an accessory neural arch extending from this arch to the ridge of the neural spine (Fig. 2.3b,c) (Siebenrock, 1893; Parker, 1942; Rieppel, 1987). These bony shields (Romer, 1956) result in channels on either side of the neural spine, which have muscles running within them (Fig. 2.3c) (Siebenrock, 1893; Parker, 1942), and a more rounded dorsal crest with a less distinct ridge. In addition, some *Brookesia* have accessory extensions projecting laterally off the arch between the prezygapophyses and postzygapophyses, corresponding to their laterovertebral spines (Fig. 2.3b,c,e,f) (Siebenrock, 1893; Parker, 1942).

The sacrum is typical in possessing two vertebrae (Werner, 1902b; Hoffstetter and Gasc, 1969; Raw, 1976). These vertebrae bear wing-like transverse processes or sacral pleurapophyses (Hoffstetter and Gasc, 1969; Raw, 1976), which are made up of fused sacral ribs (Hoffstetter and Gasc, 1969). The sacral vertebrae in *Brookesia* are fused to form a synsacrum (Fig. 2.3e,f) (Siebenrock, 1893; Klaver, 1979; Nečas, 2004).

The number of caudal vertebrae varies considerably between species (Etheridge, 1967; Nečas, 2004; Boistel et al., 2010). The transverse processes are dorsoventrally compressed and project ventrally rather than laterally in *Chamaeleo* (Fig. 2.3d) (Ali, 1948); however, in *Furcifer*, there is a transition from the transverse processes projecting ventrolaterally on the proximal portion of the tail to more laterally on the more distal portions of the tail (Zippel et al., 1999). Intervertebral chevron bones form the hemal arch (Fig. 2.3d) (Ali, 1948; Romer, 1956; Etheridge, 1967; Hoffstetter and Gasc, 1969) and become smaller in size distally, disappearing toward the end (Etheridge, 1967; Hoffstetter and Gasc, 1969). These chevron bones start on the first caudal vertebra in *Brookesia*; however, in other chameleons they do not begin immediately and result in up to four proximal caudal vertebrae lacking a chevron bone (called "pygal vertebrae") (Hoffstetter and Gasc, 1969). No caudal autotomy and no autotomy planes exist in chameleons (Romer, 1956; Etheridge, 1967; Hoffstetter and Gasc, 1969).

A number of adaptations for increased dorsoventral flexion of the tail are seen in the morphology of the caudal vertebrae. The surface of the anterior and posterior vertebral centrum are not evenly rounded, with a reduced ventral lip on the concave surface of the anterior centrum and a stronger sloped dorsal half of the convex surface of the posterior centrum (Ali, 1948). This pattern becomes more prominent distally and allows for increased dorsoventral articulation between adjacent caudal vertebrae (Ali, 1948). In addition, the prezygapophysis and postzygapophysis of the caudal vertebrae are elongated (Fig. 2.3d) (Ali, 1948; Zippel et al., 1999). The steep slope of the facets on the prezygapophysis and postzygapophysis (Ali, 1948; Hoffstetter and Gasc, 1969), allow for dorsoventral movement, while restricting lateral movements, and their length allows the interlocking prezygapophysis and postzygapophyses to remain in close contact even when the tail is fully coiled (Ali, 1948).

The form of the caudal vertebrae differs once again in *Brookesia*. Like the thoracic and lumbar vertebrae, the more proximal caudal vertebrae of *Brookesia* have a bony arch extending from the ridge of the neural spine to an arch between the prezygapophysis and postzygapophyses (Boistel et al., 2010). They also have an additional arch extending from the arch between the zygapophyses and the ridge of the transverse processes, which extends ventrolaterally in *Brookesia* (Boistel et al., 2010). More terminal caudal vertebrae, however, lack this bony shield, likely resulting in increased vertebral mobility of the distal portion of the tail (Boistel et al., 2010).

Trunk Musculature The trunk musculature in reptiles is broadly arranged into epaxial and hypaxial musculature based on innervation from either the dorsal or the ventral branch of the spinal nerves, respectively, rather than on topographic criteria, as in fishes (Gasc, 1981). In general, however, the trunk musculature in chameleons has not been thoroughly examined (see Mivart, 1870, and Sathe, 1959) and is only superficially discussed here as a result.

In general, epaxial muscles of reptiles are divided into medial, central, and lateral columns, consisting of the M. transversospinalis group, M. longissimus group, and M. iliocostalis group, respectively (Gasc, 1981). The epaxial musculature in chameleons is highly reduced (Gasc, 1981), largely because movements of the girdles substitute for bending of the body column (Peterson, 1973; Gasc, 1981). Mivart (1870) refers to upper and inferior portions of the longissimus dorsi, presumably referring to the M. transversospinalis and M. longissimus, respectively, which extend onto the tail (see "Tail Musculature," below). Muscles that could be associated with the M. iliocostalis group are not clearly described by Mivart (1870) and the arrangement in other Saurians is diverse (Gasc, 1981).

Hypaxial musculature in reptiles is typically divided into medial, lateral, and subvertebral layers (Gasc, 1981). The medial layer includes the M. transversus, M. obliquus internus, M. intercostalis internus, and M. rectus (Mivart, 1870; Gasc, 1981). The lateral layer consists of the M. intercostalis externus and M. obliquus externus (Mivart, 1870; Gasc, 1981). Finally, the subvertebral layer is generally restricted to the neck in Saurians (Gasc, 1981), and muscles that could be associated with this layer are not clearly described by Mivart (1870).

Tail Musculature The caudal muscles are primarily organized into four longitudinal muscle bundle pairs (Ali, 1948; Zippel et al., 1999). Two of these pairs lie dorsal to the axis of rotation of the vertebrae and represent the epaxial musculature of the tail (M. transversospinalis and M. longissimus), whereas the other two pairs lie ventral to the axis of rotation and represent the hypaxial musculature (M. iliocaudalis and M. inferocaudalis) (Ali, 1948; Zippel et al., 1999). One to two tendinous bands originate from each of these muscles every vertebral length, creating a segmented pattern to the tail musculature (Zippel et al., 1999). These tendinous bands insert onto one or more distal or proximal vertebral processes (Ali, 1948; Zippel et al., 1999). Activity of the epaxial musculature is responsible for extending the tail, whereas activity of the hypaxial musculature is responsible for curling it (Ali, 1948; Zippel et al., 1999).

The M. transversospinalis occupies the space between the neural spine and the zygapophyses on each side of the caudal vertebrae (Ali, 1948; Zippel et al., 1999). Each segment of this muscle gives rise to a single tendinous band, about halfway between two successive neural spines, which extend posteroventrally (Zippel et al., 1999). Approximately halfway between the posterior of the two aforementioned neural spines and the next most distal neural spine, these tendons bifurcate, with one branch continuing to run posteroventrally to insert on the next postzygapophysis and the second branch running posterodorsally to insert on the next neural spine before continuing posteroventrally (Zippel et al., 1999). A division of the M. transversospinalis, called the "M. interspinalis," originates on a neural spine and inserts on the next most distal neural spine via a tendon, which continues posteroventrally (Zippel et al., 1999).

The M. longissimus occupies the space between the zygapophyses and the transverse process on each side of the caudal vertebrae (Ali, 1948; Zippel et al., 1999). Each segment gives rise to a single, broad tendon posteroventrally to a zygapophysis (Zippel et al., 1999). This tendinous band runs anteriorly past the next most proximal zygapophysis, where it extends superficially from beneath the previous muscle segment's tendon (Zippel et al., 1999). The tendon continues anteriorly, where it is supplanted by the next tendon, to which it continues to run dorsal, eventually inserting on a prezygapophysis a few vertebrae anterior to where its associated muscle gave rise to it (Zippel et al., 1999). While the tendon is running superficially to the previous muscle segment's tendon and before it is supplanted by the next muscle segment's tendon, a branch splits off the tendon and runs anteroventrally and inserts onto circumferential connective tissue dorsal to the transverse processes (Zippel et al., 1999).

The M. iliocaudalis occupies the space below the transverse processes on each side of the caudal vertebrae but also extends between and slightly above them as well (Ali, 1948; Zippel et al., 1999). This muscle is composed of distinct dorsalis and ventralis divisions (Ali, 1948; Zippel et al., 1999), but because of the transition of the transverse process from a ventrolateral projection proximally to a primarily lateral projection more distally in some species, their positions relative to this process may vary along the length of the tail (Zippel et al., 1999). On the proximal portion of the tail, the tendinous band of the M. iliocaudalis dorsalis originates above the transverse process but transitions to originating below the transverse process more distally (Zippel et al., 1999). This tendon runs posteriorly and spans at least one vertebra before inserting on the tip of the transverse process on a more posterior vertebrae on the proximal end of the tail or on the circumferential connective tissue more distally (Zippel et al., 1999). As it runs across the transverse process of the vertebrae proximal to its insertion, a branch breaks off of this tendon and extends posterodorsally (Zippel et al., 1999). The insertion of this branch, however, varies along the length of the tail (Zippel et al., 1999). On the proximal portion of the tail, the tendinous band of the M. iliocaudalis ventralis originates anteroventral to the transverse process but transitions to originating in the cleft separating the M. iliocaudalis and M. inferocaudalis more distally (Zippel et al., 1999). Although Ali (1948) finds that these tendons run anteriorly, Zippel et al. (1999) find that they run posteriorly and emerge from within the muscle at the next transverse process. Zippel et al. (1999) go on to describe that approximately halfway to the next transverse process, the tendon bifurcates, with the dorsal branch extending beyond that of the next transverse process and inserting on the next more distal one; the ventral branch joins with the superficial circumferential connective tissue as it goes deep between the M. iliocaudalis ventralis and M. inferocaudalis, presumably then inserting on the hemal arch (Zippel et al., 1999).

The M. inferocaudalis occupies the space along the ventral side of the caudal vertebrae and is separated at the midline by a vertical septum (Ali, 1948). The tendinous bands from this muscle originate along this septum and run posteriorly into the cleft between the two sides of this muscle, where they presumably insert on the hemal arches (Zippel et al., 1999). The number of vertebrae these bands span appears to increase distally (Zippel et al., 1999). A branch off the more proximal tendons extends dorsally along the surface of the muscle and joins with the superficial circumferential connective tissue as it runs deep between the M. iliocaudalis ventralis and M. inferocaudalis (Zippel et al., 1999). In *Brookesia*, in which the proximal portion of the tail is largely immobile in many species, these ventromedial tendons are more strongly developed distally where tail mobility is increased (Boistel et al., 2010).

Appendicular

As with other aspects of chameleon biology, the appendicular musculoskeletal system is highly specialized. This is likely influenced by the arboreal habitat in which most (but not all) chameleons live. In particular, chameleons tend to use perches of relatively small diameter, and there is often considerable perch discontinuity (Peterson, 1984). In addition to the constraints imposed by an arboreal habitat (see Chapter 4), chameleons are cryptic and move very slowly in their natural habitat (Hopkins and Tolley, 2011). Despite the unique morphological and behavioral attributes, a paucity of information exists on the locomotor system of chameleons. This is in contrast to our understanding of general appendicular morphology in other lizards (Jackson, 1973; Losos, 1990; Garland and Losos, 1994; Aerts et al., 2000; Melville and Swain, 2000; Johnson and Russell, 2009; Higham and Russell, 2010). The terminology used in this section follows that of Russell and Bauer (2008) and is somewhat different from the older terminology used by Mivart (1870). This section is not meant to be exhaustive. Rather, we will focus on the skeletal elements and muscles that differ in anatomy from other lizards, such as *Iguana iguana* and *Agama agama*.

Pectoral Girdle and Forelimb

Skeletal Elements The shoulder region is probably one of the most noted features of the chameleon locomotor apparatus (Peterson, 1984). When considering the anatomy of the pectoral girdle, a common theme is the increased girdle mobility (Peterson, 1984). In addition, the girdle of chameleons is more laterally compressed, which has traditionally been linked to a relatively upright posture as compared with other lizards. However, see Chapter 4 for a detailed discussion regarding posture in chameleons.

The breast–shoulder apparatus in chameleons differs in key respects from that of other lizards. For example, the two halves of the sternum form an acute angle opposite the posterior end of the coracosternal joint, and the sternum is compressed into a V-shape (Russell and Bauer, 2008). The midventral edge of the sternum is sharp and keeled, and the M. sternohyoideus and the M. pectoralis attach here. Although the presternum is often perforated with fontanelles in lizards, this is the derived state (Lecuru, 1968a). Interestingly, chameleons were noted by Lecuru (1968a) as having an imperforate presternum, whereas others have noted the presence of a sternal fontanelle, such as in the genus *Bradypodion* (Skinner, 1959). The presence of a sternal fontanelle in other chameleons was also noted by (Peterson, 1973).

Another key difference between chameleons and other lizards is the way in which the sternum articulates with the coracoid (coracosternal articulation). In most lizards, this articulation lies in the horizontal plane (Russell and Bauer, 2008). In chameleons, however, this articulation is turned dorsally (Werner, 1902b). In this case the glenoid is located considerably dorsal to the coracosternal articulations. This ultimately leads to a more depressed posture of the limb and passive closing (at least partially) of the coracosternal articulation (Russell and Bauer, 2008).

The predominant feature of the scapulacoracoid that has distinguished different groups of lizards is the fenestration pattern (Russell and Bauer, 2008). It was proposed by Lecuru (1968b) that there are six types of lacertilian scapulacoracoid, based primarily on the pattern of fenestrae. In this scheme, chameleons share a similar type with some geckos, characterized by an emarginated scapula and an unfenestrated coracoid separated by a scapulocoracoid emargination (Lecuru, 1968b).

The clavicle apparently appears early during development in *Bradypodion* and is then reabsorbed and replaced (in terms of location) by the sternocoracoid ligament (Skinner, 1959). It has consequently been suggested that this ligament is homologous with the clavicle. Interestingly, in a developmental study of *Trioceros hoehnelii*, there was no indication of a clavicle at any stage (Rieppel, 1993). The interclavicle is also lacking in chameleons (Peterson, 1973; Russell and Bauer, 2008). However, the longitudinal arm of the sternocoracoid ligament is homologous with the bony interclavicle (Peterson, 1973).

Another ligament, the scapulosternal, is important for preventing anterior and lateral displacement of the girdle (Peterson, 1973). Although terrestrial lizards have a coracoidal arm of this ligament, which limits displacement in the coracosternal joint, chameleons lack this arm. This permits increased movement of the coracosternal joint during locomotion.

The humerus lies distal to the pectoral girdle and articulates with the glenoid (glenohumeral joint). In lacertilians, the glenohumeral joint is relatively flexible, approximating a ball-and-socket joint (Haines, 1952; Russell and Bauer, 2008). In chameleons, the main articulation (there is a small second articulation on the lateral surface of the scapulocoracoid) faces posteriorly on the girdle (Peterson, 1973). In general, the ligaments of the articulation tend to be looser and are fewer in number relative to other lizards. This likely contributes to the increased range of movement of the humerus. The articular surface itself is relatively larger in chameleons, as compared with generalized nonarboreal lizards.

The humerus of chameleons has a number of attributes that differ from other terrestrial lizards. First, the humerus tends to be longer, there is reduced torsion, and the bone is straighter (Peterson, 1973). In a study of eight species of lizard comprising both arboreal and terrestrial forms, including *Anolis* (5 species), *Dipsosaurus, Chamaeleo,* and *Agama*, chameleons exhibited the longest standardized humerus length. In addition, *Chamaeleo* exhibited 22 degrees of long-axis torsion, which was considerably lower than that of other terrestrial genera such as *Dipsosaurus* (44 degrees) and *Agama* (28 degrees) (Peterson, 1973). Other differences between *Chamaeleo* and terrestrial lizards include a narrower humerus, and muscle attachments that are located more proximally. Finally, the humerus is longer in terrestrial chameleons than in arboreal ones (Bickel and Losos, 2002).

Chameleons have extremely mobile forelimbs that emphasize an increased range of motion associated with moving in an arboreal habitat. Their limb motion tends to be more in a parasagittal plane than that of other lizards given the relatively upright posture. This also results in a reduced amount of long-axis humeral rotation. The glenohumeral articulation is thus modified to enhance motion via expansion of the articular surfaces. This allows the humerus to slide laterally during protraction. The lateral orientation of the articular surface also enhances excursion into the anterior quadrants of the glenoid (Russell and Bauer, 2008). Together, these morphological specializations allow up to 150 degrees of movement in the horizontal plane (Peterson, 1973, 1984).

The wrist in chameleons is highly modified over that of other lizards, and this is associated with their specialized pattern of locomotion. In both anatomical and developmental studies, it is clear that fusion of elements in the carpus is prevalent among chameleons (Gasc, 1963; Rieppel, 1993). However, several aspects of this fusion have been the source of debate, with studies presenting varying conclusions (for a discussion, see Russell and Bauer, 2008).

In terms of function, the proximal carpal row is aligned functionally with the antebrachium. In this case, the wrist joint is a pivot between the proximal and distal rows of carpals. This joint, which involves articulation between the ulnare and the large element of the distal carpal row, has been interpreted as being mechanically equivalent to a ball and socket joint (Gasc, 1963).

The metacarpals of chameleons are extremely different from those of other lizards. The metacarpals are divided into two bundles that articulate with the largest element of the distal carpal row (Gasc, 1963), where the first three digits form one bundle (mesial) and the fourth and fifth form another (lateral). These two groups of digits form the grasping mechanism of the forelimb.

Muscular Elements

AXIAL MUSCULATURE ACTING ON THE PECTORAL GIRDLE The M. episternocleidomastoideus has been a challenging muscle for anatomists studying lizards, given that not all of the skeletal elements are actually associated with this muscle in all species. The association of the M. episternocleidomastoideus with the M. trapezius has been discussed previously, and it has been suggested that this muscle is actually part of the M. trapezius (Jollie, 1962). This muscle originates at the posterior aspect of the ascending process of the parietal and the posterolateral margin of the paroccipital process of the exoccipital and inserts onto the anterolateral borders of the sternum (Mivart, 1870; Skinner, 1959).

The M. trapezius is small and thin in chameleons, relative to other lizards. The fibers insert along the anterior margin of the dorsal part of the scapula (Peterson, 1973). As for the origin, cervical fibers are absent, which differs from other lizards. Instead, the origin is from the first three thoracic vertebrae. The clavotrapezius is absent in chameleons.

The M. levator scapulae originates from the transverse processes of the first cervical vertebra (atlas). The insertion is entirely marginal and lies dorsal to the acromial region (Skinner, 1959). There is an additional origin of this muscle in chameleons. It is from the basioccipital condyle of the skull, which is tendinous and shared with the cervical axial muscles (Mivart, 1870). This muscle is typically associated with lateral undulation in terrestrial lizards (Peterson, 1973). Given the reduced lateral undulation in chameleons (Peterson, 1984), and the lack of clavicular attachment, this muscle brings about scapular rotation in the parasagittal plane.

The M. serratus anterior is primarily involved in suspending the body from the pectoral girdle. Lizards typically have both dorsal and ventral portions, including multiple bellies within each portion (Russell and Bauer, 2008). This muscle is reduced in chameleons, which have only two dorsal bellies and a single ventral belly (Mivart, 1870; Furbringer, 1900; Skinner, 1959). In addition to this reduction, the fibers of this muscle are all in line with the M. levator scapulae (Peterson, 1973). Finally, the bellies of this muscle in chameleons are longer than in other lizards, and this is thought to assist in the displacement of the girdle on the body wall (Peterson, 1973).

SHOULDER MUSCULATURE The shoulder musculature is typically important for protraction and retraction of the humerus in lizards and also plays an important role in stabilizing the shoulder joint. As described below, the shoulder musculature of chameleons is drastically different from that of other lizards because of, or associated with, increased mobility of the forearm. The M. sternocoracohumeralis is comparable to the M. clavodeltoideus in lizards other than chameleons. However, it maintains a different name because of the altered origin and the lack of a clavicle in chameleons (Peterson, 1973). There is also considerable variation in the morphology of this muscle among lizards. In chameleons, a small M. sternohumeralis belly originates from the superficial surface of the L. scapulosternale anterior. This is near the junction of the transverse with the longitudinal arm of the ligament. This region is analogous with the interclavicle–clavicle joint region (Peterson, 1973). The M. coracohumeralis portion exhibits a dorsolateral origin, at the level of the glenoid. In chameleons, as compared with other lizards, the M. sternocoracohumeralis is relatively small. The M. sternohumeralis fibers and the ventral M. coracohumeralis fibers form a bipinnate tendon that inserts along the proximal part of the dorsolateral aspect of the deltopectoral crest of the humerus (Peterson, 1973).

The M. supracoracoideus has been noted to be very different in chameleons as compared with other lizards (Mivart, 1870; Furbringer, 1900; Ribbing, 1938; Skinner, 1959; Gasc, 1963; Peterson, 1973). This muscle is divided into two discrete portions, originating from the lateral surface of the coracoid and the ventral scapula. The dorsal limit of the origin is the acromion. In chameleons, this muscle inserts along the anterior face of the lateral tuberosity between the glenohumeral joint capsule medially and the insertion of the M. pectoralis laterally. This muscle pulls the head of the humerus forward, protracts the humerus, and stabilizes the glenohumeral joint (Peterson, 1973).

The M. medial suprascapularis was first named by Peterson (1973) and is present only in chameleons. This muscle is found on the anteromedial surface of the scapular blade and is deep to the M. levator scapulae insertion. This muscle lies anterior to the origin of the M. subscapularis (Peterson, 1973). The M. suprascapularis medialis inserts on the proximal humerus. The origin includes the anteromedial margin of the scapula and suprascapular cartilage and a sheet of dense fascia, which separates the muscle belly from the M. subscapularis. Peterson (1973) noted the unique arrangement of fibers in this muscle. For example, the dorsalmost origin is fleshy or has short, fine tendons arising from the fascial sheath. Within a few millimeters of the dorsal limit of the origin, there exists a central tendon within the muscle belly. Interestingly, as muscle fibers stem from the scapular margin, they coil anteriorly, then medially, and ultimately posteriorly and deep into the belly, where they meet the tendon (Peterson, 1973). It is thought that the M. medial suprascapularis is derived from the M. supracoracoideus complex, and that it shares the actions of this complex. In addition, the evolutionary origin of this muscle suggests that it is related to the adaptation for protraction and a greater range of forelimb movement in chameleons (Peterson, 1973).

The M. biceps originates between the origins of the M. supracoracoideus and the M. coracobrachialis brevis, and near the ventral border of the coracoid (Peterson, 1973). The tendon of chameleons, which is small and round, originates more dorsally and occurs at the level of the inferior glenoid buttress. The M. biceps fuses with the M. brachialis over the distal third of the arm and then inserts on the proximal portion of the radius and ulna (Peterson, 1973).

The M. pectoralis is the largest muscle in the shoulder area; it covers the entire ventral aspect. In chameleons, this muscle originates from a sternal keel in the midline and the lateral surface of the sternum posterior to the first sternocostal articulation (Peterson, 1973). The insertion of the M. pectoralis is the deltopectoral crest, but is less tendinous and involves a smaller humeral area in chameleons. Relative to other lizards, the insertion onto the humerus is more proximal, which permits a greater range of motion (greater arc) of the humerus. This muscle will retract the humerus.

The M. latissimus dorsi is biarticular, spanning both the coracosternal and glenohumeral articulations. It is essentially a flat sheet that originates from an aponeurosis near the dorsal midline over the level of cervical vertebra 5 to thoracic vertebra 5 (Peterson, 1973). The origin in chameleons also incorporates the third, fourth, and fifth thoracic ribs. The M. latissimus dorsi inserts onto the proximal portion of the humerus. Like the M. pectoralis, this muscle is also a humeral retractor (within the parasagittal plane) with little ability to rotate the humerus.

In lizards, the M. triceps complex is typically comprised of four bellies, two of which originate from the shaft of the humerus and two of which originate from the primary girdle (Russell and Bauer, 2008). They all have a common tendinous insertion on the ulna. However, chameleons exhibit a three-headed condition, missing the coracoid arm of the sternoscapular ligament and the M. coracotriceps. It is thought that the absence of this ligament and the M. coracotriceps permits an increased range of motion and forward reach in chameleons. In other species of lizard, this ligament and muscle impose limitations.

LOWER FORELIMB MUSCULATURE Much of what is known about forelimb musculature in chameleons is related to the muscles acting at the girdle. This is likely due to the extreme motion at the level of the girdle during locomotion. The function of the lower forelimb (antebrachium) has not received as much attention, and future work will help illuminate the functional consequences of the specialized morphology of chameleons.

The M. extensor digitorum longus typically occupies the anterior area of the forearm. The origin, via a short tendon, is just dorsal to the radial condyle of the humerus (Russell and Bauer, 2008). In chameleons, there are two bellies. The first runs along the ulna and inserts close to the proximal ends of the fourth and fifth metacarpals. The second belly exhibits a tendinous insertion onto the third metacarpal. Although the exact function of this muscle could be related to flexing the carpus dorsally during the swing phase or pulling the antebrachium forward over the manus, more work is needed. The fact that the number of insertion points in chameleons is reduced from three (typical lizard) to two suggests a relation to the pincer-like nature of the manus and the specialized locomotor behavior.

The M. flexor digitorum longus of chameleons exhibits substantial differences from that of other lizards, but appears to share some similarities with *Gekko* (Russell and Bauer, 2008). This muscle is divided into four heads. Two of them originate from the humerus and the other two originate from the ulna. The posterior head that originates from the humerus inserts on digit five only. The other head, originating from the humerus, serves the other digits. The ulnar heads follow an insertion pattern similar to that of the humeral heads, with the anterior deep head inserting on digits one to four. The posterior deep head inserts onto digit five.

Pelvic Girdle and Hindlimb

Skeletal Elements The pelvic girdle is comprised of the dorsal ilia, anteroventral pubes, and the posteroventral ischia (Russell and Bauer, 2008). On each side, the three components share a common suture, which is centered on the acetabulum. In the midventral line, the pubes and ischia also share a suture. The epipubis is an ossified structure that is between and anterior to the pubes in chameleons.

The femur articulates with the acetabulum via an oval and gently curving condyle. For most lizards, an internal trochanter lies anterior and somewhat ventral to the condyle. However, the internal trochanter has been reduced to a ridge in chameleons (Cope, 1892). The distal portion of the femur has rarely been examined in lizards, but it is clear that chameleons exhibit differences that are related to their locomotor mode. The lateral distal condyle in lizards is typically larger than the mesial condyle, but chameleons do not follow this pattern (Russell and Bauer, 2008). Instead the lateral and mesial condyles are comparable in size. In addition, the demarcation of the patellar surface is absent in chameleons.

The tarsus is made up of a proximal row, including the astragalocalcaneum, and a distal row, which is functionally a part of the pes (Russell and Bauer, 2008). The mesotarsal joint (ankle joint) is located between the crus (lower limb) and the proximal tarsal row. The astragalus and calcaneum of lizards typically develop as independent condensations, with the astragalus ossifying first. The tarsus of chameleons, however, is unique in that it originates from a single large cartilage distal to the tibia and fibula (Russell and Bauer, 2008).

With the exception of chameleons, all lizards that have been examined exhibit a flattened astragalocalcaneum. (For a more detailed description for lizards other than chameleons, see Russell and Bauer [2008].) In chameleons, it is curved and depressed to form a ventrally directed concavity. Tendons run within this concavity, and the astragalocalcaneum takes on the role of a pulley. However, chameleons do not exhibit this modification.

In chameleons, a globular fourth tarsal alone articulates with all of the metatarsals. As with the metacarpals, the metatarsals are grouped into two bundles. However, unlike the forelimb, the first and second are grouped together while the third, fourth, and fifth are grouped together (Rieppel, 1993).

Muscular Elements The musculature of the hindlimb of the chameleon was first determined by Mivart (1870), using the Parson's chameleon (*Calumma parsonii*). A more recent study of hindlimb muscle anatomy examined the veiled chameleon, *Chamaeleo calyptratus* (Higham and Jayne, 2004a).

The M. caudofemoralis is a robust muscle originating from the transverse processes of the four most proximal caudal vertebrae and inserting to both the greater trochanter of the femur and the proximal portion of the fibula via an auxiliary tendon (Fig. 2.4 in the color insert). This muscle typically slows femur protraction during late swing in lizards. However, this does not seem to be the case in chameleons, which is likely due to their slow locomotor speeds (Higham and Jayne, 2004a). Instead, the M. caudofemoralis likely flexes the knee during early stance.

The M. iliofibularis originates via a tendon from the posterior and lateral margin of the ilium and inserts on the fibula distal to the insertion of the M. caudofemoralis auxiliary tendon (Fig. 2.4). Like other lizards, activity in the M. iliofibularis is predominantly during swing (Higham and Jayne, 2004a).

In chameleons, the M. iliotibialis originates via a tendon from the posterior portion of the ilium just dorsal to the origin of the M. iliofibularis and inserts to the proximal tibia via the connective tissue on the anterior face of the knee.

The M. flexor tibialis externus originates from the ilioischiadic tendinous arch, runs along the posterior and ventral portion of the thigh and sends a long tendon, running along the posterior edge of the lower leg, to the plantar ossicle (Fig. 2.4). In addition, the M. flexor tibialis externus sends a shorter tendon that crosses the M. iliofibularis and inserts on the fibula just proximal to the insertion of the M. iliofibularis (Fig. 2.4).

The M. puboischiotibialis is on the ventral surface of the thigh and originates from the puboischiatic symphysis (midventral line) and inserts on the proximal portion of the tibia (Fig. 2.4). This muscle likely contributes to knee flexion and perhaps maintains the horizontal orientation of the femur (Higham and Jayne, 2004a).

The M. gastrocnemius originates from both the distal part of the femur and the posterior aspect of the tibia and runs along the posterior edge of the lower leg where it inserts on the plantar ossicle (Fig. 2.4). This is a stance phase muscle and is primarily involved in ankle extension.

The M. extensor digitorum longus originates from the distal portion of the femur and from the posterior portion of the fibula and inserts onto both the fourth and fifth digit (Fig. 2.4). However, others have suggested that this muscle inserts only onto the third metatarsal (Mivart, 1870). This muscle, according to the muscle-activation patterns, is predominantly a stance-phase muscle, with a smaller burst occurring during the swing phase (Higham and Jayne, 2004a).

The M. peroneus originates from the proximal portion of the anterior face of the fibula and from the proximal portion of the posterior tibia and inserts on the proximal and dorsal portion of the fifth metatarsal (Fig. 2.4 in the color insert). This muscle is typically active during the first half of stance and is likely responsible for knee flexion (Higham and Jayne, 2004a).

The M. tibialis anterior originates from the proximal portion of the tibia and inserts onto the proximal portion of the first metatarsal (Fig. 2.4). When measured under *in vivo* conditions, this muscle exhibits variable activity and is often active for a large portion of the stride (Higham and Jayne, 2004a).

2.2 EXTERNAL MORPHOLOGY AND INTEGUMENT

In addition to functions of protection, water balance, grasping and substrate interaction, etc., the external integument in chameleons also contains an assortment of signaling capabilities. These signals range from color and pattern changes, the mechanistic basis of which range from sexual selection to species-recognition characteristics (Chapter 6). The structure of many of these external morphological characteristics is therefore important to much of the broader biology of chameleons.

Scalation

Whereas superficial ossifications in the form of bony shield arches and accessory extensions are found in *Brookesia* above portions of the vertebral column (Siebenrock, 1893; Romer, 1956; Boistel et al., 2010), body osteoderms, as in other iguanian lizards, are absent in chameleons (Romer, 1956). Further, whereas some species exhibit small patches of bare skin (e.g., *Bradypodion damaranum*), the majority of the skin and external surface in chameleons is covered with keratinous and generally nonoverlapping scales (Nečas, 2004; Tilbury, 2010). These scales come in various sizes, shapes, and arrangements and are often the basis of some of the larger ornamentations.

Scale Types and Scalation Patterns

Broadly, the scalation of chameleons is characterized by the consistency of the size and shape of the scales. When the scales appear to be of much the same size and shape, the animal is said to exhibit homogeneous scalation. When the scales appear to be of highly variable size and shape, the animal is said to exhibit strongly heterogeneous scalation. In species with heterogeneous scalation, these scales can be distributed seemingly randomly or can be organized into distinct patterns, such as rows of enlarged scales or circular rosettes of scales on the flanks. Largely homogeneous or heterogeneous scalation patterns, however, can involve a variety of scale types.

While all scale-type designations are intended to be descriptive of the shape of the different scales, scale shapes are often grouped differently (e.g., Nečas, 2004; Tilbury, 2010). Because there is no single accepted set of scale types for chameleons, our chosen set of scale types may vary from other sources; however, examination of their respective descriptions should help rectify inconsistencies. Here we divide the scale types in chameleons into conical, granular, labial, keeled, tubercular, lenticular, plate-like, and stellate and polygonal scales.

Conical scales are elongate, lanceolate, or cone-shaped (Nečas, 2004; Tilbury, 2010). They are typically found the dorsal and gular crests but can also be found on the flanks, throat, tail, and head and rostral processes in some species (Nečas, 2004).

Granular scales are small, bumpy, and granular-shaped. A more or less homogeneous arrangement of granular scales, such as in *Chamaeleo senegalensis* or *C. laevigatus,* is seen when these scales are spread across almost the entire body or in large patches with occasional interspersed larger scales. Arrangements of these scales can also span the spectrum to arrangements in which these granular scales are seemingly found only interstitially between larger scale types (Tilbury, 2010).

Labial scales are semicircular scales found around the mouth. They are found in a single row around the mouth in all chameleons and form what appear to be lips.

Keeled scales exhibit a ridge down the middle of the scale coming to a point. They are uncommon in chameleons but are observed in the caudal scales of some *Brookesia* (Müller and Hildenhagen, 2009).

Tubercular scales are scales that form a rounded eminence or projection from the surface. These scales are typically found on the cranial crests (Nečas, 2004); however, some authors group lenticular, plate-like, and stellate scales as forms of tubercular scales (Tilbury, 2010).

Lenticular scales are rounded, circular scales that are taller in their center than on their periphery and are often lumped together with tubercular scales. They are often found on the flanks but can also be found on the limbs, tail, throat, and head (Nečas, 2004). They are often interspersed among smaller granular scales and can be enlarged to varying degrees, even within a single individual.

Plate-like scales are rounded, flat scales and are often considered a type of tubercular scale. They are frequently found on the flanks but are also seen on the casque, rostral protuberances, occipital lobes, and extremities of some species (Nečas, 2004). They are often interspersed among smaller grandular and lenticular scales and can be of varying size, even within a single individual.

Stellate and polygonal scales are scales typically found on the flanks that have irregular-shaped sides. In some *Brookesia, Rhampholeon,* and *Rieppeleon* species, the scalation of the body consists of heterogeneous, interlocking, star-shaped, or stellate, scales (Nečas, 2004; Tilbury, 2010). Some other chameleon species have body scalation consisting of heterogeneous polygon-shaped scales (Tilbury, 2010).

Feet

The scales on the palms and soles of the feet in chameleons are generally rounded to give a cobblestoned or smooth appearance (Mariaux and Tilbury, 2006; Tilbury, 2010). In *Brookesia* and *Rieppeleon,* however, the scales on the feet are sharply pointed or spinous with acuminate spines (Mariaux and Tilbury, 2006; Tilbury, 2010). In *Rhampholeon,* one to three spinous projections, called "accessory plantar spines," are found at the base of each claw (Mariaux and Tilbury, 2006; Tilbury, 2010).

Dermal Pits

In many chameleon species dermal invaginations are found at the base of the limbs (Mariaux and Tilbury, 2006; Tilbury, 2010), which frequently contain mites (Tilbury, 2010). They take the form of axillary pits on the posteroventral base of the forelimbs and the inguinal pits on the anteroventral base of the hindlimbs. At least one of these sets of pits are found in most *Calumma, Furcifer, Rhampholeon,* and *Rieppeleon* species, but in some species their presence is inconsistent between individuals (Tilbury, 2010).

Microstructure

Scanning electron microscope examination of the scales on the subdigital and subcaudal surfaces in chameleons shows a complex microstructure in many species. This can include complex arrangements of adhesive bristles or setae (Schleich and Kästle, 1979, 1985; Canham, 1999; Müller and Hildenhagen, 2009), rectangular to hexagonal honeycomb shapes, or thorny points (Müller and Hildenhagen, 2009). The length, shape, and combination of different microstructures in these regions varies between genera and species (Schleich and Kästle, 1979; Canham, 1999; Müller and Hildenhagen, 2009). For instance, the setae in some species appear rounded, whereas in other species they appear to come to a point or even appear heterogeneous in length and shape (Schleich and Kästle, 1979; Canham, 1999; Müller and Hildenhagen, 2009). There can also be variation in the length and shape of setae within individual pads (Müller and Hildenhagen, 2009). Finally, within the subcaudal region, there also appears to be differentiation between different areas, with a scansorial pad displaying distinct features relative to adjacent portions in some taxa (Schleich and Kästle, 1979).

Although most genera exhibit some form of setae and honeycomb-shaped surfaces on their subdigital and subcaudal scales, there are also some distinct differences. The more agama-like thorny points, for instance, are only found in *Brookesia* and *Rieppeleon* (Müller and Hildenhagen, 2009). Further, *Brookesia* lack adhesive bristles altogether, having only a thorny point and/or rounded honeycomb structure (Müller and Hildenhagen, 2009).

Claws

Chameleons have a claw projecting from each toe or two to three claws on each set of fused digits for a total of five claws per foot. All species of *Archaius, Bradypodion, Brookesia, Calumma, Chamaeleo, Furcifer, Kinyongia, Nadzikambia,* and *Trioceros* have a simple claw, whereas all species of *Rhampholeon (Rhinodigitum), Rhampholeon (Bicuspis),* and *Rieppeleon* have bicuspid claws with the formation of a secondary point approximately midway along the main claw Klaver, 1979; Nečas and Schmidt, 2004; Mariaux and Tilbury, 2006; Tilbury, 2010). These claws are strongly bicuspid in *Rhampholeon (Rhinodigitum)* and *Rhampholeon (Bicuspis)* but only weakly bicuspid in *Rieppeleon,* although *Rieppeleon kerstenii* may have strongly bicuspid rear feet (Tilbury, 2010). Within *Rhampholeon (Rhampholeon),* on the other hand, only *Rhampholeon (Rhampholeon) spectrum* has bicuspid claws (Müller and Hildenhagen, 2009), whereas all others have simple claws (Tilbury, 2010).

Ornamentation

Chameleons exhibit a vast assortment of ornamentation based on their skeletal, dermal, and other structures. This ornamentation takes the form of various crests, cranial protuberances, fan-like elongations on the vertebral column, occipital lobes, and intricate arrangements of their scalation.

Crests

The bones of the skull form a number of sharp angles and ridges on the head, which are often subsequently adorned with tubercular scales of varying sizes. The degree to which these crests are developed and their shape are often associated with species recognition and sexual-selection characteristics.

The paired lateral crest extends anteriorly from the apex of the casque, over the orbits, and to above the mouth tip, where they fuse (Nečas, 2004). Each lateral crest is divided into three parts: the rostral crest, orbital crest, and lateral crest proper (Nečas, 2004). The rostral crests extend from the anterodorsal margin of the orbit forward to just above the mouth tip, where they join. They are formed by the prefrontals, maxillae, and premaxilla. The ocular crests are constrained to the upper margin of the orbits and are formed by either the prefrontals and postorbitofrontals, or the prefrontals, frontals, and postorbitofrontals, depending on whether or not the prefrontals and postorbitofrontals join. The lateral crest proper extends from the posterior margin of the orbit to the apex of the casque and is formed by the postorbitofrontal and squamosal portion of the upper temporal arch and ascending process of the squamosal.

In species with a narrow parietal bone, the parietal crest lies medially and extends posterodorsally from immediately posterior to the eyes to the apex of the casque (Nečas, 2004) and is formed by the elevated ridge of the parietal bone. The parietal crest can be flat or highly concave and can be quite tall in some species. In species with a broad parietal bone, such as *Bradypodion* and *Brookesia,* the lateral aspects of the parietal bone form a pair of ridges lateral to the midline and medial to the lateral crests on each side called the "parasagittal crests" (Raxworthy, 1991).

In some species, an additional crest, called the "temporal crest," extends anteroventrally from the lateral crest posterior to the eye (Nečas, 2004). This crest is formed by the postorbitofrontal and in some cases the dorsal projection of the jugal.

Along the spine from behind the skull backward, a medial ridge is present in most chameleons. A dorsal crest is said to be present when a series of enlarged, often conical, scales are present along this ridge. The dorsal crest, however, can be limited to only a few conical scales immediately behind the head or consist of a large number of conical scales extending down the back and even onto the tail. In *Brookesia* species, a dorsal crest is not seen; however, many species have accessory extensions projecting laterally off the vertebrae to form laterovertebral spines (Siebenrock, 1893).

A ridge of enlarged scales is also frequently found along the midline on the ventral side of the body, running from the symphysis of the lower jaw posterior to the cloaca. From the jaw symphysis to the anterior edge of the sternum, this ridge is called the "gular crest," whereas from the sternum back to the cloaca it is called the "ventral crest." In most species, the gular crest is formed by a single medial row of enlarged scales, whereas in *Trioceros tempeli* it is formed by two rows of enlarged scales and in *Trioceros affinis* it is formed by paired dermal ridges.

Cranial Protuberances

Perhaps the most notable ornamentations in chameleons are the variety of cranial protuberances adorned by many members of the family. These protuberances are highly variable in their form and function but include keratin-covered annulated horns, bony projections, and soft dermal lobes.

Among the most recognized of these cranial protuberances are the true or annulated horns. These horns have a bony base, are elongated and narrow, and are covered by an annulated keratin sheath formed by a single hypertrophied scale (Nečas, 2004). These horns are typically located preorbitally or rostrally along the lateral crests; however, *Trioceros melleri* has an unusual structure, in which a single annulated horn is located on the end of a medial bony rostral projection separate from the rostral crest portion of the lateral crest (Rieppel, 1981). When present, preorbital annular horns are paired with a single horn projecting from the anterodorsal aspect of each ocular crest. Rostral horns, on the other hand, can be either one, two, four, or six in number, and project side by side from one another on the anterior portion of the rostral crest.

Additional types of cranial protuberances are false or bony horns (Fig. 2.5). These horns are formed by projections of the cranial bones with a layer of scale-covered skin over them (Nečas, 2004). These scales are typically enlarged tubercular or plate-like scales. Often the false horns are paired and laterally compressed extensions of the rostral crest projecting forward beyond the tip of the jaw and formed by modified prefrontal and maxillary bones (Fig. 2.5) (Rieppel and Crumly, 1997). In some species, these paired false horns have become medially fused to each other, giving the appearance of a single laterally compressed paddle. False horns can also take the form of smaller elevated points along the lateral crests, such as the superior nasal cones and superior ocular cones in *Brookesia* (Raxworthy, 1991), or the single elevated rostral cones of species like *Trioceros hoehnelii*.

The cranial protuberances of some species, however, are entirely flexible. These soft or dermal horns lack a bony base and are made of soft, scale-covered skin (Nečas, 2004). These dermal horns are typically covered by granular scales or other slightly enlarged scales that remain soft and pliant. They can be found preocularly or rostrally. Preocular dermal horns are typically paired, and rostral dermal horns can be either bulbous or laterally compressed.

Finally, semipliant horns are seen in a couple species and appear as intermediaries between false and dermal horns (Nečas, 2004). These horns have a bony base and a flexible tip (Nečas, 2004). The rigidity of this tip varies, possibly because of a fibrous or cartilaginous tissue structure in the distal portions.

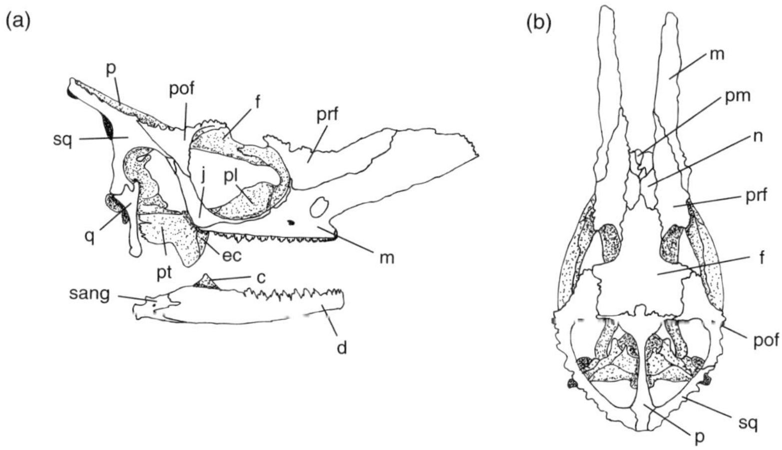

FIGURE 2.5. The skull of a male *Furcifer bifidus* in lateral (a), and dorsal (b) views showing formation of false horn. Redrawn from Rieppel and Crumly (1997).

LABELS: ang = angular; ar = articular; bo = basioccipital; bs = basisphenoid; c = coronoid; d = dentary; ec = ectopterygoid; f = frontal; j = jugal; m = maxilla; n = nasal; p = parietal; pl = palatine; pm = premaxilla; po = prootic; pof = postorbitofrontal; pf = prefrontal; prfo = prefrontal fontanelle; pt = pterygoid; q = quadrate; sang = surangular; so = supraoccipital; sq = squamosal; st = supratemporal; v = vomer.

Sails

In some species, a tall sail along the vertebral column of the back or proximal portion of the tail is seen. This sail is formed by elongated neural spines of the vertebrae with strong connective-tissue threads between their distal ends and a skin covering (Case, 1909). These elongations form a fan-like sailfin on the proximal portion of the tail in some male West African *Trioceros* species, an elevated sail-like dorsal ridge in *Trioceros cristatus* and to a lesser extent *T. deremensis*, and a crenulated dorsal crest in *T. melleri*.

Occipital Lobes

At the posterior margin of the head, many chameleon species have posteriorly oriented skin flaps called "occipital lobes." These lobes can vary from quite narrow strips to large ear-like lobes. In some species, these occipital lobes have a connective-tissue structure and attachment to the squamosal bone, giving the lobes a semirigid structure (Meyers and Clarke, 1998). This connective-tissue "skeleton" is covered with mostly plate-like scales and in some species has an insertion by the M. depressor mandibulae pars auricularis, enabling the lobes to be erected during display (Meyers and Clarke, 1998).

Tarsal Spurs

Whereas most species lack them, some species and some sexes of certain species exhibit a short posterior projection from their hindfoot called the "tarsal spur," which is a bony extension of the tarsal bone covered in skin and scales (Tilbury, 2010). Of the species that do exhibit them, in most it is more strongly developed in males; however, in some species they are present in both males and females.

2.3 SENSORY STRUCTURES

Chameleons are known to have an increased dependence on visual cues relative to their other senses. As a result, chameleons' eyes have become highly developed. The remaining sensory structures, on the other hand, have become reduced or even vestigial in some cases.

Eye

The eye of chameleons is their most developed sensory organ, with higher image magnification than any other vertebrate eye when scaled to the same size (Ott and Schaeffel, 1995). The eyes are notably enlarged, are placed laterally on the head, bulge almost entirely out of the orbit, and move independently of each other. They are surrounded exteriorly by scale-covered eyelids, which are fused to the sclera of the eye and have only a small center opening for the pupil. This arrangement allows for an impressive oculomotor range, which exceeds both 180 degrees horizontally and 90 degrees vertically (Sándor et al., 2001). Moreover, they are likely the only reptiles to achieve binocular fixation with a central fovea (Underwood, 1970).

Their oculomotor range is enabled by four rectus muscles and two oblique muscles (Leblanc, 1924, 1925). The four rectus muscles have a fascicular origin posteroventrally on the medial side of the orbit on the interorbital membrane (Leblanc, 1925). The two oblique muscles, on the other hand, originate on the anteromedial aspect of the orbit at the junction of the palatine and prefrontal (Leblanc, 1925).

The M. rectus superior is very broad and extends anterolaterally to inserts on the sclera on the dorsal surface of the eye just behind the cornea (Leblanc, 1925). It serves to elevate the cornea and rotate the dorsal surface of the eye posteroventrally (Leblanc, 1924, 1925). The M. rectus medialis extends horizontally behind the eye and then turns laterally to insert on the sclera on the anterior surface of the eye behind the cornea (Leblanc, 1925). It serves to draw the cornea anteromedially (Leblanc, 1924, 1925). The M. rectus inferior has two bundles (Leblanc, 1924, 1925) that extend ventrolaterally and insert on the sclera behind the cornea on the ventral side of the eye and just ventromedially to the insertion of the M. rectus medialis (Leblanc, 1925). They serve to draw the cornea ventromedially (Leblanc, 1924, 1925). Finally, the M. rectus lateralis has two bundles that extend laterally to slightly dorsolaterally (Leblanc, 1924, 1925). The upper bundle inserts on the sclera behind the cornea on the posterior side of the eye, whereas the lower bundle inserts on the anteroventral side of the conjunctival sac (Leblanc, 1925). They serve to draw the cornea posteromedially and draw the conjunctival sac over the Harderian gland (Leblanc, 1924, 1925).

The M. obliquus superior extends posteriorly in a dorsolateral direction and inserts broadly onto the sclera of the dorsal portion of the eye immediately behind and below the M. rectus superior (Leblanc, 1925). It serves to rotate the dorsal surface of the eye anteroventrally and thus is an antagonist to the M. rectus superior (Leblanc, 1925). The M. obliquus inferior extends horizontally and slightly laterally to insert on the sclera on the ventral side of the eye, perpendicular to the insertion of the M. rectus inferior, which inserts along the edge of the cornea (Leblanc, 1925). It serves to rotate the ventral surface of the eye anterodorsally and thus acts with the M. rectus superior as an antagonist to the M. obliquus superior (Leblanc, 1925).

The scleral cartilage (ring) is present and in *Chamaeleo* is formed by 11 scleral ossicles, creating a conical form (Gugg, 1939; Underwood, 1970). It is confined to the orbital hemisphere in the scleral layer of eye, with the cornea extending out of center (Leblanc, 1925; Underwood, 1970; Pettigrew et al., 1999). This scleral ossicle is coated with fine muscle fibers from the M. depressor palpebralis inferior of the eyelid just below the surface of the skin (Leblanc, 1924, 1925). This eyelid depressor muscle extends from the rim of the eyelid ventromedially around the eye in a thin sheet to the ventral and medial aspect of the orbit, where it originates on the palatine and interorbital membrane (Leblanc, 1925). This muscle serves to draw the rim of the eyelid and scleral ossicle ventrally to cover and protect the eye (Leblanc, 1924, 1925), as seen when chameleons rub their eyes during cleaning. The M. levator bulbi is absent in chameleons (Underwood, 1970).

Chameleons are unique among vertebrates in having a negatively powered lens (Ott and Schaeffel, 1995), thus reducing the contribution of the lens and increasing the contribution of the cornea to the total optical power of the eye (Ott and Schaeffel, 1995; Pettigrew et al., 1999). This serves to elongate the focal length of the eye and create a large retinal image (Ott and Schaeffel, 1995; Ott, 2001). Because the crystalline lens is relatively thick, with its lateral and medial surfaces being relatively flat, the internal isoindical shells of the lens are concave in shape in order to establish this negative refractive power (Ott and Schaeffel, 1995). The cornea is small (Underwood, 1970) and has a very small radius of curvature (Ott and Schaeffel, 1995), indicating that the cornea extends abruptly outward. Corneal curvature, however, is modulated for corneal accommodation by the M. cornealis, which inserts directly onto the corneal stroma (Pettigrew et al., 1999).

Finally, chameleons have extremely high visual resolution. They have a deep-pit fovea, with the retina being thick at its center and declining in thickness at its periphery (Ott and Schaeffel, 1995; Pettigrew et al., 1999). This retina has a dense photoreceptor package, with an estimated 756,000 cones/mm^2 (Harkness, 1977; Ott, 2001). This estimate is the higher than in all other lizards, and whereas some researchers have indicated that this estimate is likely high, it is within the range found in humans and birds of prey (Harkness, 1977).

Parietal Organ and Pineal Gland

The function of the parietal organ and pineal gland in chameleons is not clear, and it is thought to be rudimentary in mature chameleons (Nečas, 2004). The pineal gland in chameleons is located dorsal to the midbrain and cerebellum (Schmidt, 1909; Quay, 1979). Overall, it is tubular in shape, first extending posterodorsally and then bending at nearly a right angle into a strongly inclined anterodorsal extension toward the roof of the skull, terminating in a long, thin tip (Schmidt, 1909).

When present, the parietal organ, or pineal eye, is dorsoventrally compressed and has either a round or slight sagittally elongated shape (Schmidt, 1909). It lies just under the skin and considerably anterior to the pineal gland, or pineal organ (Schmidt, 1909; Quay, 1979), in or above the pineal (parietal) foramen of the frontal bone (Romer, 1956; Rieppel, 1981). The location of this foramen in the frontal bone represents a forward shift from ancestral forms (Trost, 1956).

The parietal organ is connected to the pineal organ by the parietal eye nerve (Quay, 1979). A spot associated with the presence of the parietal organ, called the "parietal spot," is visible in *Bradypodion, Brookesia, Chamaeleo,* and *Furcifer,* but absent in *Rhampholeon, Rieppeleon,* and *Trioceros* (Schmidt, 1909; Gundy and Wurst, 1976). The diameter of the parietal organ relative to the diameter of the pineal foramen in chameleons is known to vary from half its size to nearly twice its size (Edinger, 1955), so estimating development of the parietal organ based on the size of the pineal foramen is difficult.

Ear

The ear in chameleons is greatly reduced. There is no external ear opening or tympanic membrane (Brock, 1941; Engelbrecht, 1951; Frank, 1951; Wever, 1968, 1973; Wever and Werner, 1970), and the traditional round window is absent or extreme reduced (Wever, 1968, 1969a, 1973). Some species further lack a tympanic cavity (Engelbrecht, 1951; Frank, 1951; Simonetta, 1957; Toerien, 1963; Wever, 1968), and the columella is often reduced or modified to a level that it is regarded as nonfunctional (Toerien, 1963; Wever, 1968, 1969b). Further, the extracolumella exhibits various modifications and is noted to terminate on various tissues, affecting potential conductance (Wever, 1968, 1969b; Wever and Werner, 1970). Whereas these reductions have not resulted in the loss of ability to detect airborne sound, their hearing is greatly reduced (Wever, 1968, 1969a,b, 1973; Wever and Werner, 1970).

Whereas *Rhampholeon* lack a tympanic cavity (Frank, 1951) and *Bradypodion* is said to either lack (Engelbrecht, 1951; Simonetta, 1957) or possess a vestigial tympanic membrane (Brock, 1941), in *Chamaeleo* and *Trioceros* the tympanic cavity is well defined and encloses the middle ear (Wever, 1968, 1969b). In these taxa, the tympanic cavity is separated from the pharyngeal region by a membrane, although a small oval-shaped opening corresponding to the Eustachian tube is found (Wever, 1968).

The stapedial footplate of the osseous columella rests in the oval window at the floor of the otic capsule (Fig. 2.6) (Toerien, 1963; Wever, 1968, 1969a,b). In *Chamaeleo* and *Trioceros* the footplate is large and nearly fills the oval window (Fig. 2.6) (Wever, 1968, 1969b), whereas in *Bradypodion* the footplate is small and does not fit closely within the oval window (Toerien, 1963) and in *Rhampholeon* the footplate is extremely small or vestigial (Frank, 1951; Toerien, 1963).

In *Bradypodion* (Engelbrecht, 1951; Toerien, 1963) and *Rhampholeon* (Frank, 1951) the columella is poorly developed and may not form a connection with the quadrate (Toerien, 1963). When it does, a cartilaginous extracolumella lies at the distal end of the columella (Brock, 1941; Wever, 1968, 1969b). In *Chamaeleo,* the extracolumella has anterior and posterior processes (Fig. 2.6a) (Wever, 1968), whereas the anterior process is lacking in *Trioceros* (Fig. 2.6b) (Wever, 1969b). The anterior process of *Chamaeleo* extends along a membrane between the

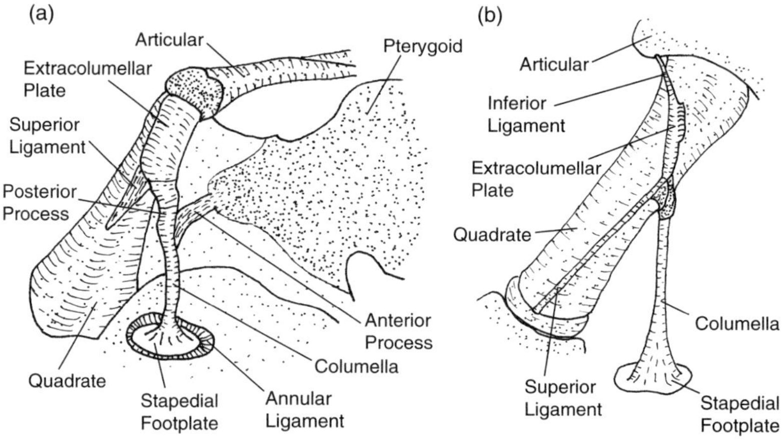

FIGURE 2.6. Drawings of the inner aspects of the right ears of *C. senegalensis* (a) and *T. hoehnelii* (b) from a ventral, medial, and slightly anterior direction. (a) redrawn from Wever (1968) and (b) from Wever (1969b).

quadrate and pterygoid to the thin edge of the pterygoid wing, where it forms a ligamentous attachment (Fig. 2.6a) (Wever, 1968). The posterior process of the extracolumella extends to the ventral part of the quadrate, forming a flat plate in the process (Fig. 2.6) (Wever, 1968, 1969b), which is smaller in *Trioceros* (Wever, 1969b). Dorsal (superior) and ventral (inferior) ligaments extend from the extracolumellar plate, with the dorsal ligament extending from the proximal end of the plate along the quadrate toward the squamosal (Fig. 2.6) (Wever, 1968, 1969b), and the ventral ligament extending from the distal end of the plate to the articulation between the quadrate and articular in *Chamaeleo* (Fig. 2.6a) (Wever, 1968), and to the posterior end of the articular in *Trioceros* (Fig. 2.6b) (Wever, 1969b).

In *Chamaeleo*, it is hypothesized that the pterygoid wing and the membrane extending to the quadrate, combined with the columellar system acts as a substitute tympanic membrane by serving as a conductive mechanism for airborne sound (Fig. 2.6a) (Wever, 1968). The lack of ligamentous connection between the columellar system and the pterygoid in *Trioceros*, however, results in a lack of a tympanic membrane substitute (Wever, 1969b).

Whereas a traditional round window is lacking in chameleons (Wever, 1968, 1969a, 1973), a substitute for it and its pressure discharge mechanism during oscillation of the oval window is known in *Chamaeleo* and *Trioceros* (Wever, 1968, 1969a, 1973). This substitute is present in the form of a fluid-filled path extending from the posterior wall of the scala tympani of the ear, posteriorly into the exoccipital bone and then laterally through the foramen of the glossopharyngeal nerve and into the tympanic cavity (Wever, 1968, 1969a).

The vestibular system of the inner ear has been examined in only a limited number of taxa. It is characterized by three well-developed semicircular canals with the curves of the posterior and anterior canals extending ventrally and the curve of the horizontal canal extending medially (Boistel et al., 2010). These semicircular canals are relatively flattened and oblong in shape in *Brookesia*, whereas in *Archaius*, they are more rounded (Boistel et al., 2010). Curiously, the horizontal canal in chameleons is only oriented horizontally when the head it elevated (Boistel et al., 2010).

Tongue Pad and Taste Buds

The bulbous portion of the tongue in chameleons that is projected from the mouth can be divided into the tongue tip, the foretongue, and the hindtongue (Herrel et al, 2001b). The tongue tip is composed of the bifurcated anteroventral end of the tongue and the area adjacent and posterior to it (Herrel et al., 2001b). The foretongue consists of the portion of the tongue pad that is invaginated to create a lingual pocket, or dimple, with an upper and lower lobe (Herrel et al., 2000, 2001b) and is often called the "membrana glandulosa" (Bell, 1989). The hindtongue consists of the epithelium surrounding the M. accelerator linguae posterior to the tongue pad (Herrel et al., 2001b).

The tongue tip is bifurcated, with paired ventral plicae. This region is comprised of dense, closely packed papillae that show little to no visible microstructure (Herrel et al., 2001b). At the bifurcated tip, these papillae appear to be randomly oriented; however, posteriorly toward the foretongue, they are arranged in transverse rows (Herrel et al., 2001b). Taste buds are present on the tongue tip, but they are not abundant (Schwenk, 1985; Herrel et al., 2001b).

The foretongue or membrana glandulosa consists of densely packed reticular papillae oriented in transverse rows and exhibiting a prominent microstructure (Herrel et al., 2001b). Extending posteriorly, the density of these papillae decreases (Herrel et al., 2001b). This region is rich in epithelial-gland cells producing serous and mucous secretions (Bell, 1989; Herrel et al., 2000; Schwenk, 2000). Free plumose cells are known to be scattered occasionally on the edges of the foretongue in some species (*Trioceros melleri*; Herrel et al., 2001b); however, they are reported to be numerous in the lingual pouch of chameleons (Schwenk, 1983, 2000). Further, studies have found this region to lack taste buds in some species (*Trioceros melleri*; Herrel et al., 2001b), whereas others have indicated that they are present, although not abundant, in other species (*Trioceros jacksonii*; Schwenk, 1985).

The hindtongue lacks papillary structures and consists instead of a smooth epithelium around the M. accelerator linguae (Herrel et al., 2001b). Still, a prominent microstructure can be observed (Herrel et al., 2001b). Some studies have located taste buds in this region at higher concentrations than in the anterior regions (Herrel et al., 2001b), whereas other studies have found this region to be devoid of taste buds (Schwenk, 1985).

Overall, chameleons possess fewer gustatory receptors than other Iguanian lizards (Schwenk, 1985; Herrel et al., 2001b). More broadly, however, taste buds are said to always be numerous in the oral epithelium of lizards, with the exception of varanids, which lack taste buds altogether, and chameleons, which lack them on the oral epithelium (Schwenk, 1985).

Nasal Capsule and Nasal Cavity

Overall, the nasal capsule and nasal cavity is of reduced size, having been shortened and compressed in the process of being pushed anterodorsally because of the enlarged eye and tongue (Brock, 1941; Malan, 1945; Engelbrecht, 1951; Frank, 1951; Visser, 1972; Slaby, 1984). The reduction and poor development of a number of features of the nasal capsule and nasal cavities, in addition to the olfactory nerves and olfactory nerve branches, has generally resulted in chameleons being considered microsmatic at best (Haas, 1937).

The cartilaginous nasal capsule is highly complex and differs considerably from that in ancestral lineages (Haas, 1937; Malan, 1945; Engelbrecht, 1951; Slaby, 1984; Hallermann, 1994). Its roof and sidewalls are quite complete, whereas the floor is relatively incomplete (Haas, 1937; Engelbrecht, 1951). The interpretation of the formation or the floor, however, is the subject of a variety of interpretations, particularly with regard to the presence or absence of paraseptal cartilages (Haas, 1937; Brock, 1941; Malan, 1945; Engelbrecht, 1951; Slaby, 1984). Discussion of the specific fine structure of the nasal capsule is not discussed here but can be reviewed elsewhere (Haas, 1937; Brock, 1941; Malan, 1945; Engelbrecht, 1951; Frank, 1951; Visser, 1972; Slaby, 1984; Hallermann, 1994).

The nostrils in chameleons are positioned laterally and enter the elongate and large diameter nasal vestibules at an oblique anterior direction (Engelbrecht, 1951; Parsons, 1970; Visser, 1972). The vestibular wall is composed of erectile muscular tissue with a layer of keratinized epithelium covering it (Malan, 1945; Engelbrecht, 1951; Frank, 1951), whereas the rest of the nasal cavities are lined with ciliated epithelium (Engelbrecht, 1951). The vestibules open laterally at their posterior end into the olfactory chamber located beneath it via a wide slit (Malan, 1945; Engelbrecht, 1951; Parsons, 1970). This slit is elongated in *Bradypodion* but shorter in *Chamaeleo*, forming a blind cavity posterior to the opening to the olfactory chamber (Malan, 1945; Haas, 1937; Parsons, 1970). The olfactory chamber is small, the most reduced of any reptile, and the olfactory epithelium is highly reduced (Haas, 1937; Malan, 1945; Frank, 1951; Parsons, 1970). The nasal conchae, or turbinates, are reduced to a rudimentary flat ledge (Haas, 1937) or absent altogether (Hallermann, 1994). The choanae lie directly beneath the opening between the vestibules and olfactory chamber in *Bradypodion* (Malan 1945; Engelbrecht, 1951). Inspired air is thus able to travel from the vestibules directly into and through the choanae in *Bradypodion* (Malan, 1945; Engelbrecht, 1951), whereas air must travel a more elaborate route in *Chamaeleo* through the olfactory chamber and into the choanae (Haas, 1937; Malan, 1945). Inspired air then travels from the choanae into the oral cavity (Engelbrecht, 1951). The paired choanal grooves in the palate of the oral cavity are deep and bordered by choanal folds, which are supported by the ectochoanal cartilages, which are in turn supported by the medial process of the maxillae (Engelbrecht, 1951; Frank, 1951).

Vomeronasal Organ

The predominating theory on the presence and development of a vomeronasal, or Jacobson's, organ in chameleons is based on that of Haas (1947), who described the presence of a "reduced" and "functionless" vomeronasal organ in *Chamaeleo chamaeleon*. Based on this study, many report chameleons in general to possess a rudimentary or vestigial vomeronasal organ (Nečas, 2004; Gehring and Lutzmann, 2011); however, some others simply state that the vomeronasal organ is absent in chameleons (Døving and Trotier, 1998). In reality, there is no standard condition within the family (Parsons, 1970). For instance, the vomeronasal organ has been reported to be completely absent in some taxa (Slaby, 1984), including in *Trioceros hoehnelii* (Malan, 1945) and *Rhampholeon platyceps* (Frank, 1951), whereas it is regarded as rather rudimentary in *C. dilepis* (Born, 1879) and *C. chamaeleon* (Born, 1887; Haas, 1947) and well developed in *Bradypodion pumilum* (Malan, 1945; Engelbrecht, 1951; Visser, 1972) and *B. ventrale* (Brock, 1941). This lack of ubiquity within the family is not typically discussed, because while we know very little about the structure of the vomeronasal organ in different chameleons, we know even less about its functionality.

When present, the paired vomeronasal organs are located in the roof of the mouth anterior to the nostrils (Brock, 1941; Malan, 1945; Haas, 1947; Engelbrecht, 1951; Visser, 1972). Their openings into the oral cavity lie between the anterior tip of the vomer and maxillae (Brock, 1941; Haas, 1947; Engelbrecht, 1951), and the vomeronasals are separated from the choanae by Fuchs secondary palate rather than opening into them (Malan, 1945; ngelbrecht, 1951). Their height is reduced, and whereas the vomeronasal organ in most lizards lies beneath the nasal vestibules, they lie medial to them in chameleons (Brock, 1941; Malan, 1945; Engelbrecht, 1951; Visser, 1972). This more anterior and dorsal positioning is thought to be due to the need to accommodate the large eyes and tongue (Malan, 1945; Engelbrecht, 1951; Visser, 1972; Slaby, 1984).

The vomeronasal organs are covered dorsally and laterally by two cartilaginous plates (Brock, 1941; Malan, 1945; Engelbrecht, 1951), likely derived from the roofing cartilage, forming their own cartilaginous roof (Malan, 1945; Engelbrecht, 1951). Coverage by these cartilaginous plates is interrupted dorsolaterally by a fontanelle (Brock, 1941; Malan, 1945; Engelbrecht, 1951), which is not covered by the septomaxillary as in other lizards, as it is absent in chameleons (Malan, 1945; Engelbrecht, 1951; Frank, 1951; Visser, 1972; Hallermann, 1994). The ventral edges of the lateral cartilaginous plates are bent medially, forming a floor for the lateral portions of each vomeronasal organ (Brock, 1941; Malan, 1945; Engelbrecht, 1951). The vomeronasal organs are lined with ciliated epithelium (Engelbrecht, 1951). There is no connection between the ductus nasolacrimalis and the vomeronasal organs in chameleons (Malan, 1945; Engelbrecht, 1951).

Brain and Nervous System

The neurology of the chameleon has been studied by a number of researchers over the years. Here we very briefly comment on a couple of general trends as compared with other reptilian brains that apply more broadly to trends seen in other aspects of chameleon anatomy and ecology.

The cerebellum in chameleons is highly developed; it is long, narrow, and curved forward in shape, possibly because of its function in maintaining equilibrium, which is important in arboreal animals (Shanklin, 1930). Further, whereas the olfactory bulbs are typically large in reptiles, in chameleons they are minute, and the peduncles are very slender, adding further support to the notion that chameleons are microsmatic (Shanklin, 1930; Goldby and Gamble, 1957). Similarly, the main vomeronasal-recipient structure, the nucleus sphaericus, is reduced in size and devoid of a cortical-like arrangement (Senn and

Northcutt, 1973; Northcutt, 1978). The basal optic-root ganglion is well developed and likely correlated with the wide range of eye movements in chameleons (Shanklin, 1930). Finally, the hypoglossal nucleus is highly differentiated, likely in association with the complex and highly evolved tongue and its complex projection mechanism (Shanklin, 1930).

Further information on the neurology of the chameleon brain can be found in Shanklin (1930). Other studies of the chameleon brain and nervous system have focused on the cerebral tube (Bergquist, 1952), neopallium (Dart, 1934), wall of the forebrain (Källén, 1951a,b), motor pathways of the eye (Stefanelli, 1941), and the nucleus opticus tegmenti (Shanklin, 1933).

2.4 VISCERAL SYSTEMS

In general, relatively little is known about the visceral systems in chameleons. Of note, however, are the lung and hemipenal morphology, which has been extensively examined for taxonomic purposes. Here we briefly describe the anatomy of these and other visceral systems.

Circulatory

Overall, the circulatory system of chameleons has not been well studied. The pathways and branching patterns of aspects of the arterial (Rathke, 1857; Mackay, 1886; Beddard, 1904; Adams, 1953, 1957) and venous (Beddard, 1904; Bruner, 1907) systems have been described in detail elsewhere and are not discussed here. Instead, a brief summary is provided on the aspects of the anatomy of the three-chambered heart of chameleons, which has received only minimal attention from researchers.

Internally, the ventricle of the heart is known to have seven apical chambers, as is typical of most reptilian hearts, but little else is known of the internal structure (Farrell et al., 1998). Externally, the sinus venosus is well developed, with visible swelling at the confluence of the postcaval and right precaval veins (Kashyap, 1960; Farrell et al., 1998). The terminal portion of the left precaval vein is also swollen but has a considerable constriction at its junction with the aforementioned confluence (Kashyap, 1960; Farrell et al., 1998). The right and left atria are of approximately equal size and an atrial diverticulum is present between the paired carotid arteries (Kashyap, 1960; Farrell et al., 1998). Whereas in most reptiles the conus arteriosus has been absorbed into the ventricle, traces of a vestigial conus arteriosus are visible at the base of the arterial trunk in chameleons (Kashyap, 1960; Farrell et al., 1998). The apex of the heart is attached to the pericardium by a gubernaculum cordis and the apical two thirds of the ventricle is attached to the pericardium by a mesocardial membrane (Kashyap, 1960; Farrell et al., 1998).

Respiratory

The lungs in chameleons are highly variable and can be extremely elaborate. Their structure has been extensively studied for use as a taxonomic marker (e.g., Klaver, 1973, 1977, 1979, 1981) as the configuration of the pulmonary septa are conserved within groups of related species (Klaver and Böhme, 1986).

The larynx is formed, as in other reptiles, by the cricoid cartilage and arytenoid cartilages (Germershausen, 1913). In some species, an inflatable sac, called the "gular pouch," is connected with the ventral wall of the trachea just behind the larynx (Germershausen, 1913; Klaver, 1981; Klaver and Böhme, 1986).

The lungs in chameleons occupy a large portion of the body cavity, with lung volumes that are among the largest for their size of any reptile (Perry, 1998). The luminal walls of the lungs have numerous terminal air sacs for gas exchange, called "edicula," which are at least as wide as they are deep (Perry, 1998) and supported by a trabeculated smooth-muscle network (Klaver, 1981; Perry, 1998; Tilbury, 2010). The lungs can be simple and sac-like or can have internal septa that project into the lumen of the lung in one of five patterns (Klaver, 1981; Klaver and Böhme, 1986; Tilbury, 2010). Further, diverticula of differing shape, size, position, and number can project off the ventral and terminal aspects of the lungs in some species (Beddard, 1907; Methuen and Hewitt, 1914; Klaver, 1973, 1977, 1979, 1981; Klaver and Böhme, 1986; Tilbury, 2010).

A nonseptate condition is seen in all *Brookesia, Rhampholeon,* and *Rieppeleon* species (Klaver, 1979; Klaver and Böhme, 1986; Tilbury, 2010), except *Rhampholeon spinosus* (Klaver, 1981). In this condition, the lung lumen forms a simple sac devoid of any septae (Klaver, 1979; Klaver and Böhme, 1986; Tilbury, 2010).

The first septation condition lacks long longitudinal septa, but the lungs are clearly divided, with the dorsal, cranial and ventral walls having varying numbers of small to moderately sized septa (Klaver, 1973, 1977, 1981; Klaver and Böhme, 1986; Tilbury, 2010). This pattern is seen in *Rhampholeon spinosus* (Klaver, 1981) and members of the *Bradypodion, Calumma, Furcifer, Kinyongia,* and *Nadzikambia* genera (Klaver, 1973, 1977, 1981; Klaver and Böhme, 1986; Tilbury, 2010).

The remaining four types of divisions are characterized by large longitudinal septa running posteriorly through the lumen from the orifice of the bronchus (Klaver and Böhme, 1986). One of these types, as seen in *Chamaeleo* species, has two septa that end freely in the lumen (Klaver, 1973, 1977; Klaver and Böhme, 1986; Tilbury, 2010). The other three types are seen in the genus *Trioceros* and have one, two, and three septa that connect to the ventral wall at their distal end, completely subdividing the lumen into chambers (Klaver, 19,73 1977, 1981; Klaver and Böhme, 1986; Tilbury, 2010).

Digestive

A limited number of studies have discussed the anatomy of the digestive system in chameleons, and most of this is related to the folding relief of the gastrointestinal tract. Therefore, here we only briefly describe some of the structure of the digestive system in chameleons.

Whereas in most lizards the esophagus has smooth-surfaced longitudinal folds of relatively consistent diameter, the esophageal folds in chameleons are rough-surfaced and of varying diameter (Parsons and Cameron, 1977). The liver in chameleons is typically brownish gray in color and has two lobes, with the left lobe being larger and having the greenish-colored gallbladder positioned on its dorsolateral edge (Beddard, 1907; Nečas, 2004).

The pancreas is yellowish in color (Nečas, 2004) and bilobed, although these lobes are not always distinct, forming instead a single curved, elongated mass (Beddard, 1907). Part of the pancreas lies on the ventral side of the stomach, between the stomach and duodenum, with an additional portion extending toward the dorsal side of the stomach and back toward its posterior end (Beddard, 1907). The spleen is purplish red in color and located just ventral to the stomach (Nečas, 2004).

Longitudinal folds in the stomach are of varying diameter and are not parallel, with both wavy and straight portions (Parsons and Cameron, 1977). The wall of the stomach between the longitudinal folds has a fine pebble-like surface (Parsons and Cameron, 1977). The tunica muscularis of the stomach is smooth muscle with an inner circular and outer longitudinal layer (Luppa, 1977). The muscular layer is of reduced thickness toward the pylorus of chameleons (Luppa, 1977).

The intestinal tract is short and poorly differentiated (Nečas, 2004). Longitudinal folds of the duodenum have an irregular pattern with tall, thin folds that can appear membranous (Parsons and Cameron, 1977). Their borders are crenulated and the edges bear projections (Parsons and Cameron, 1977). The wall of the duodenum between the folds is very rough and has occasional fine longitudinal ridges (Parsons and Cameron, 1977). The rest of the small intestine has thicker folds with borders that are even more irregular (Parsons and Cameron, 1977).

The colon has very large, thick, transverse folds that are separated by deep clefts (Parsons and Cameron, 1977). These folds have smaller, randomly arranged, longitudinal folds running along their surface (Parsons and Cameron, 1977). The large folds are very rough, with grooves and small projections similar to villi (Parsons and Cameron, 1977). Intestinal glands (glands of Lieberkühn) are reported in the colon of chameleons (Luppa, 1977). The cloaca also is reported to have simple tubular (unbranched) glands, which are independent of one another (Luppa, 1977).

Urogenital

Most of our knowledge of the chameleon urogenital system stems from the use of the male reproductive parts as taxonomic markers (e.g., Klaver and Böhme, 1986). Here we only briefly discuss the structure of other urogenital structures and focus on the hemipenes, because of their importance in species differentiation and taxonomy.

The kidneys are located in the posterodorsal portion of the body cavity along the spine (Nečas, 2004) and are elongate pear-shaped to uniformly elongate (Fox, 1977). A urinary bladder is present in chameleons and opens ventrally into the cloaca (Fox, 1977). The urinary bladder may be used for water storage (Burrage, 1973).

In females, the oviducts and eggs occupy a large portion of the body cavity when a clutch is being developed (Nečas, 2004). In males, the testes are black and the seminal vesicles have a tubular arrangement (Fox, 1977).

Male chameleons, like other squamates, have a paired intromittent organ called the "hemipenes." The hemipenes are held inside the body in an inverted position while at rest.

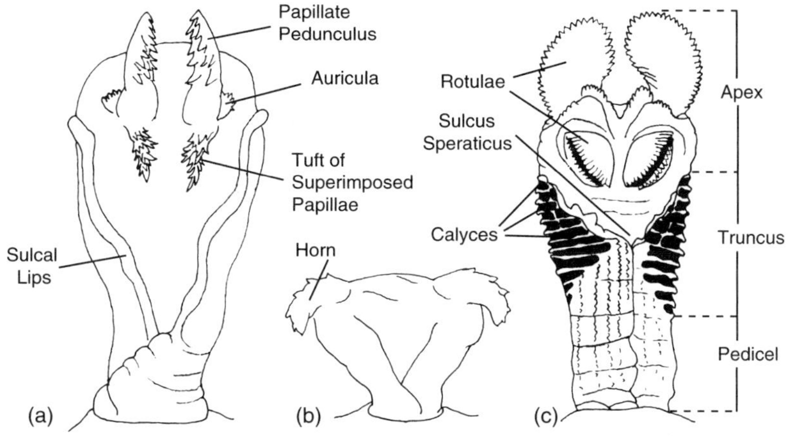

FIGURE 2.7. Schematic sulcal views of hemipenis morphology for *F. lateralis* (a), *R. platyceps* (b), and *C. calyptratus* (c). Redrawn from Klaver and Böhme (1986).

It is held in a pocket posterior to the vent in the base of the tail, often forming a hemipenal bulge, which can be useful in determining the sex of individuals.

Each hemipenis in chameleons has either a strong clavate shape, in the case of *Brookesia, Rhampholeon,* and *Rieppeleon* species (Fig. 2.7b), or weakly clavate to subcylindrical shape, as in other genera (Fig. 2.7a,c), when everted (Klaver and Böhme, 1986). Overall the hemipenis can be divided into three regions: the pedicle, the truncus, and the apex (Fig. 2.7c) (Klaver and Böhme, 1986). The pedicle is the proximal base of the hemipenis, the truncus is the medial portion, and the apex is the distal tip (Fig. 2.7c) (Klaver and Böhme, 1986).

The pedicle of the hemipenis has a relatively smooth surface (Klaver and Böhme, 1986). The truncus can either be calyculate, with reticulated honeycomb-like pits, called "calyces," ornamenting its surface (Fig. 2.7c), or acalyculate, with a smooth surface, making differentiation between the pedicle and truncus difficult (Fig. 2.7b) (Klaver and Böhme, 1986).

A channel-shaped groove, called the "sulcus spermaticus," bordered by sulcal lips, runs along the hemipenal surface of the pedicle and truncus for sperm transport during copulation (Fig. 2.7) (Klaver and Böhme, 1986; Nečas, 2004). The sulcus spermaticus is smooth, whereas the sulcal lips can be smooth or have ridge traces from the surrounding calyces (Klaver and Böhme, 1986). The sulcal lips may exhibit a capitate state, where they diverge distally to form a clear ridge boundary between the truncus and apex, or be noncapitate (Klaver and Böhme, 1986).

The apex is simple to slightly bilobed at its distal end and is often elaborately ornamented, with ornamentation being arranged bilaterally (Klaver and Böhme, 1986). Ornamentation may include papillae, pedunculi, auriculae, rotulae, horns, and crests (Klaver and Böhme, 1986). Papillae are fleshy and flexible projections that vary in size and shape and can be single, paired, scattered, arranged in rows, or concentrated in papillary fields (Fig. 2.7a) (Klaver and Böhme, 1986; Nečas, 2004). Pedunculi are thick stalks protruding over the distal end of the sulcus spermaticus and can be papillate themselves (Fig. 2.7a) (Klaver and Böhme, 1986). Auriculae, on the other hand, are curved dentriculate ridges that occur on the asulcal side of the apex— that is, the opposite side of the hemipenis from where the sulcus spermaticus occurs (Fig. 2.7a) (Klaver and Böhme, 1986; Nečas, 2004). Rotulae are similar to auriculae but are more developed and semicircular discs with a denticulate or serrated outer margin (Fig. 2.7c) (Klaver and Böhme, 1986). Horns, as seen in many *Rhampholeon* species, are broad, rotund projections that taper toward their distal ends and curve toward the sulcal side of the apex (Fig. 2.7b) (Klaver and Böhme, 1986). Finally, crests, as seen in some *Brookesia* species, are papillate or dentriculate crests or crested lobes on the apex of the hemipenis (Klaver and Böhme, 1986). Interestingly, the development of these apical structures appears to be related to seasonal and hormonal factors, and specimens may exhibit intraspecific variation depending on reproductive state or the time of year (Klaver and Böhme, 1986; Tilbury, 2010).

Endocrine and Exocrine

Our knowledge of the anatomy of endocrine and exocrine structures in chameleons is extremely limited. Endocrine glands in chameleons have been examined only to a limited extent, and the description of their morphology is extremely superficial or limited to the broader context of larger groups of lizards (e.g., Lynn and Walsh, 1957; Gabe and Martoja, 1961; Bockman, 1970; Gabe, 1970; Girons, 1970; Lynn, 1970), and is therefore not discussed here. Examinations of exocrine structures in chameleons are similarly limited; however, a unique, suspected holocrine gland is known in some chameleons.

Whereas chameleons lack the femoral glands common to many other lizards (Camp, 1923), some do have a structure that is thought to be similar to the femoral gland in lizards and analogous to the sebaceous gland of mammals (Ogilvie, 1966). This structure, called the "temporal gland," is a dermal pouch in the temporal region of the head that excretes decaying cornified skin cells (Ogilvie, 1966). When present, it is located between the superficial muscles of the temporal region of the skull and the external layer of skin, anterior to the M. depressor mandibulae (Ogilvie, 1966). Its base lies beneath the quadratomaxillary ligament, and the pouch opens into the commissure of the jaws when the lower jaw is depressed (Ogilvie, 1966). It is believed that this pouch may have arisen as a result of an increased area of skin present at the angle of the jaw (Ogilvie, 1966).

The development of the temporal gland is highly variable between chameleon species and genera, with some of the most developed examples occurring in *Trioceros*, whereas *Rieppeleon* have only a small temporal pouch that is difficult to detect under a microscope, and *Rhampholeon* are believed to lack the pouch altogether (Ogilvie, 1966). Overall, the pouch has been observed to varying degrees of development in *Bradypodion, Chamaeleo, Kinyongia, Rieppeleon,* and *Trioceros*, but it is absent in *Calumma, Furcifer,* and *Rhampholeon* (Ogilvie, 1966).

In addition, some chameleons are known to excrete salt from nasal salt glands (Burrage, 1973). The structure of the nasal salt glands has not been examined in chameleons specifically; however, in lizards the salt gland is formed by the modified lateral nasal gland (Dunson, 1976; Hazard, 2004) and consists of branching secretory tubules projecting radially around a central duct (Burrage, 1973; Dunson, 1976) that opens into the nasal vestibule (Peaker and Linzell, 1975). These glands produce a brine of potassium, sodium, and chloride that is exuded from the nostrils and dries, forming deposits around the nares (Burrage, 1973).

While the multitude of unique features of chameleons has resulted in many researchers examining various aspects of chameleon anatomy over the years, a considerable gap in our knowledge remains. Future work will likely reveal morphological differences between species and genera of chameleons, especially those that live in different types of habitats.

Not all chameleons, for example, are arboreal, although terrestrial chameleons still appear to maintain many of the same morphologies as their arboreal relatives. Key questions regarding commonalities and divergence between disparate groups of chameleons remain, however. As noted by Tolley and Burger (2007), terrestrial chameleons tend to be small, and they typically exhibit relatively short tails. How internal morphology relates to a terrestrial lifestyle in chameleons remains relatively unknown.

Further, a great deal of behavioral variation exists between different lineages within the family. Many of these behavioral differences may have underlying morphological variations associated with them. Behavioral observations of tongue-touch behavior in various species (Ogilvie, 1966; Gehring and Lutzmann, 2011; C.V. Anderson, personal observation) suggest a need for more in-depth examination of the morphological variation and functionality of the vomeronasal organ, for example.

ACKNOWLEDGMENTS

Support during the writing of this chapter was provided by a Fred L. and Helen M. Tharp Endowed Scholarship (to C.V.A.). We thank Jack Conrad, Stephen Deban, and Anthony Herrel for consultation and extremely helpful comments on earlier drafts of this chapter.

THREE

Chameleon Physiology

ANTHONY HERREL

3.1 NEUROPHYSIOLOGY

Sensory Physiology

The sensory system of chameleons is highly tuned toward visual stimuli. Not only do chameleons rely on visual signals in a social context, they are also visual predators (Gans, 1967). As such, a large body of literature has been devoted to the functioning and anatomy of the visual system. For example, it has been shown that chameleons have a negatively powered lens (Land, 1995; Ott and Schaeffel, 1995) and use accommodation cues to judge distance (Harkness, 1977), features convergent with the sandlance (Pettigrew et al., 1999). The auditory system has also been studied in some detail and shows lower hearing sensitivity as compared with that of other lizards (Wever, 1968, 1969a,b). Little is known about the physiology of the olfactory (smell), vomerolfactory, and gustatory (taste) systems in chameleons beyond anatomical descriptions (Schwenk, 1985; Halpern, 1992). In general, the reduced development of these systems is thought to be associated with reduced functionality, which is attributed to the highly visual nature of chameleons (Chapter 2).

Visual System

Chameleons rely heavily on the visual system for a wide variety of behaviors ranging from foraging and prey capture to social signaling and predator detection. Many of the unusual features of the eyes (e.g., their ability to move the eyes independently over 180 degrees horizontally and over 90 degrees vertically; Sándor et al., 2001), and of the visual system in general, have evolved in tight relationship with their unique ballistic tongues used during prey capture (Chapters 2, 4). Although initially it was believed that chameleons used stereopsis (i.e., binocular vision) to judge distance (Duke-Elder, 1957), it was later demonstrated that prey distance is estimated using accommodation cues (i.e., changing of lens shape to maintain focus on an object as its distance varies; Ott and Schaeffel, 1995). Consequently, chameleons are capable of successfully capturing prey with one eye occluded (Kirmse et al., 1994). The accommodation range is large in chameleons, and the focusing precision is better than in other vertebrates (Ott and Schaeffel, 1995; Ott et al., 1998). Surprisingly, chameleons are similar to nonhuman primates in having a vestibuloocular reflex response that optimally stabilizes the entire retinal image (Haker et al., 2003). Although chameleons use independent eye movements to scan their environment directly preceding prey capture, chameleons use binocular fixation at the moment of prey capture (Flanders, 1985; Ott, 2001).

In addition to functional and physiological studies, a series of papers has investigated the neuroanatomy of the visual system in chameleons (Bennis et al., 2001, 2005). The neuronal organization of the visual system in chameleons suggests several features unique to chameleons. First, the ipsilateral retinofugal projection is absent; second, an additional hypothalamic visual center is present; and third, the nucleus opticus tegmenti is unusually large as compared with other lizards (Bennis et al., 1994; 1996). Studies on the optic nerve and retina show that chameleons have a complex retina that is in some ways similar to the mammalian one (El Hassni et al., 1997; Bennis et al., 2005). Moreover, the retina possesses four distinct types of single cones (Bowmaker et al., 2005), rendering the visual system of chameleons sensitive to visible light and the near ultraviolet with a cutoff at about 350 nm (Hunt et al., 2001; Bowmaker et al., 2005). This suggests that signaling in the ultraviolet range may make up an important component of the behavioral repertoire in chameleons (Stuart-Fox and Moussalli, 2008; Chapter 6).

Auditory System

Several studies have been devoted to the auditory system of chameleons (Wever, 1968, 1969a,b). Chameleons are unusual in having no external ear opening and no visible tympanic membrane. Moreover, the round window of the cochlea is lacking (Wever, 1968). Consequently, chameleons have a relatively poor auditory sensitivity as compared with other lizards (Wever, 1968, 1969a). Interestingly, different auditory sensitivities were recorded for species of the genus *Chamaeleo* versus the genus *Trioceros*, with *T. hoehnelii* and *T. jacksonii* having poorer auditory performance than all species of *Chamaeleo* (Wever, 1968, 1969a). Moreover, severing of the columella decreased hearing performance in species of the genus *Chamaeleo* but not *Trioceros*. This was explained by a difference in the sound-reception mechanisms in the two genera. Whereas aerial sounds are received by the pterygoid plate embedded in the tissues at the side of the head in species of *Chamaeleo* (Wever, 1968, 1969b), this system is dysfunctional in *Trioceros* because of the lack of the anterior process of the extracolumella (Wever, 1969a). Although at the time it was unknown that these species with different morphologies belonged to different genera (Townsend and Larson, 2002; Chapter 7), this observation suggests a strong phylogenetic effect. Unfortunately, given the lack of data for other genera, the basal condition of the group cannot be determined. Moreover, whether the difference in anatomy and hearing sensitivity is related to differences in habitat use (with *Trioceros* typically being forest dwellers) remains unknown.

Sleep

Although in general, relatively little is known about sleep in reptiles, one study has investigated sleep in chameleons (Tauber et al., 1966). One of the big debates in the literature is whether rapid eye movement (REM) sleep is present in reptiles other than birds (Tauber et al., 1968; Siegel, 2008). Although the evidence is equivocal, most studies tend to suggest that, although reptiles do sleep, no true REM sleep can be demonstrated based on recordings of brain activity (Ayala-Guerrero and Mexicano, 2008; Siegel, 2008). Yet, Tauber and colleagues (1966) demonstrated an intriguing pattern of eye movements in chameleons during physiological sleep states. Specifically, they demonstrated that typically brief periods of REM are present while sleeping. Unilateral eye movements with the eyelid opened are also observed without any changes in sleep posture. However, their electroencephalographic recordings from the telencephalon suggested no true REM sleep is present. Given the radically different anatomy of the lizard brain, recordings from the brain stem (where the structures homologous to those recorded in mammals and birds reside) would be needed to confirm this. In conclusion, neurophysiological studies in reptiles are woefully lacking behind those of other vertebrates and much remains to be investigated.

3.2 MUSCLE PHYSIOLOGY

Studies on chameleon muscle physiology are rather scarce and can be subdivided into two groups: those that investigate the physiological basis of the slow locomotion in chameleons, and those interested in the structure and physiology of the tongue muscles. Chameleons are well known for being slow, with sprint speeds being about 10 times slower than those of closely related agamid lizards of similar size (Abu-Ghalyun et al., 1988; Losos et al., 1993; Herrel et al., 2011; see also, Chapter 4). To better understand the physiological basis of this slow locomotor behavior, several studies have examined the histochemical, ultrastructural, and metabolic profiles of chameleon muscle (Guppy and Davison, 1982; Abu-Ghalyun, 1990; Mutungi, 1992). These studies show that chameleons possess, on average, more slow fibers in the arm and leg muscles as compared with other lizards (Abu-Ghalyun, 1990; Mutungi, 1992). Moreover, fast lizards such as skinks have higher metabolic potential in skeletal and cardiac muscle than chameleons, yet both skinks and chameleons have mammalian-level metabolic capacities, as indicated by levels of citrate synthase and β-hydroxybutaryl-coenzyme A dehydrogenase in the heart (Guppy and Davison, 1982). The slow locomotion of chameleons thus likely resides in the contractile capacities of its locomotor muscles (Abu-Ghalyun et al., 1988) coupled with the changes in limb posture (Chapter 4), and lower overall muscle mass as compared with that of other lizards of similar size (A. Herrel, personal observation). The contractile properties of the iliofibularis muscle show generally slower (up to four times slower) contraction velocities as compared with similarly sized agamids (Abu-Ghalyun et al., 1988).

In contrast to the limb muscles, the tongue muscles of chameleons are anything but slow. Indeed, histochemical profiling revealed only fast glycolytic and fast oxidative glycolytic fibers in the tongue muscles (Herrel et al., 2001b). Consistent with this observation, muscle innervations were always of the "en plaque" type (Herrel et al., 2001b). Muscle-physiology experiments showed that the tongue retractors in chameleons are somewhat slower than those in closely related agamid lizards (time to peak tension, 42 vs. 23 ms; time to half relaxation, 35 vs. 23 ms), yet they produce high forces for their cross-sectional area (Herrel et al., 2001a, 2002; Anderson and Deban, 2013). Chameleons have unusual tongue-retractor muscles that are of the supercontractile type (Rice, 1973; Herrel et al., 2001a , 2002, 2009). These muscles are characterized by perforated Z discs and high, nearly invariant force-generation capacity over a wide range of muscle lengths (Herrel et al., 2001a, 2002). Despite the thermal independence of overall tongue projection (Anderson and Deban, 2010), the physiology of the accelerator muscle responsible for tongue projection is highly temperature-dependent (Anderson and Deban, 2013). Thus, the reduced thermal sensitivity of tongue-projection performance appears to be the result of the morphologic arrangement and the incorporation of elastic elements into the projection mechanism rather than physiological specialization of the associated muscles (Anderson and Deban, 2013).

3.3 METABOLISM, SALT, AND WATER BALANCE

Surprisingly few studies have investigated the metabolism of chameleons (Burrage, 1973; Wheeler, 1984; Zari, 1993). Given their slow sit-and-wait or cruise foraging behavior and generally cryptic lifestyle (Butler, 2005; Chapter 5), chameleons could be expected to have low standard metabolic rates. Yet, metabolic rates of chameleons are similar to those measured for other lizards and are highly dependent on body mass (Burrage, 1973; Bennett and Dawson, 1976; Andrews and Pough, 1985; Zari, 1993). Moreover, standard metabolic rate is highly temperature-dependent, with Q_{10} values of 3.38 to 3.81 in the range from 20 to 30°C for *C. calyptratus* (Zari, 1993) but with lower Q_{10} values for *C. namaquensis* and *B. pumilum* (1.26 to 2.91) (Burrage, 1973). Interestingly, for the desert-dwelling species (*C. calyptratus* and *C. namaquensis*) Q_{10} values decrease above 30 to 35°C, suggesting an adaptation to their hot desert environment (Burrage, 1973; Zari, 1993). One study investigated the water relationships and dehydration rates of chameleons (Burrage, 1973). Differences between species were observed in dehydration rates with the desert-dwelling *C. namaquensis* dehydrating much more slowly than *B. pumilum* (note that the *B. pumilum* included in the study by Burrage, 1973, are likely individuals belonging to both *B. pumilum* and *B. occidentale*). Furthermore, the *C. namaquensis* excreted salt around the nares suggesting the presence of salt glands (Burrage, 1973). This has been suggested to be an adaptation of desert-dwelling lizards that feed on halophytic plants (Norris and Dawson, 1964) and/or the presence of cloacal water reabsorption (Schmidt-Nielsen, 1963).

3.4 TEMPERATURE

Chameleons are found among widely varying thermal regimes and climatic conditions ranging from hot and dry desert habitats, through tropical rainforests and Mediterranean climates, to high-altitude environments (Burrage, 1973; Hebrard et al., 1982; Reilly, 1982; Bennett, 2004; Chapter 5). Although chameleons are often described as being thermoconformers (e.g., Dimaki et al., 2000), more recent studies show that chameleons carefully regulate body temperatures using behavioral thermoregulation and color change to maintain temperatures of around 30 to 32°C (Bennett, 2004; Andrews, 2008). The preferred temperatures of chameleons are low compared to those of most other diurnal lizards, including closely related agamid lizards (Burrage, 1973; Andrews, 2008a,b) suggesting an adaptation to lower temperatures. This, in combination with the temperature-invariant function of the ballistic tongue protraction (Anderson and Deban, 2010) may have allowed chameleons to invade high-mountain habitats rarely accessible to other lizards. Moreover, chameleons in cold environments will change skin reflectance at low temperature, allowing them to heat up more rapidly and thus reduce time spent basking (Burrage, 1973; Walton and Bennett, 1993).

3.5 SKIN PIGMENTATION, COLOR CHANGE, AND THE ROLE OF ULTRAVIOLET LIGHT

Despite the long-standing interest in color change in chameleons dating back to Aristotle (350 BC), surprisingly few studies have investigated the mechanism of color change in chameleons (Brücke, 1852b; Hogben and Mirvish, 1928; Zoond and Eyre, 1934; Canella, 1963). The current consensus is that melanophores and other chromatophores in the skin (e.g., xanthophores, erythrophores and iridiophores; see Cooper and Greenberg, 1992) are under the control of the autonomic nervous system (Berger and Burnstock, 1979; but see Canella, 1963, for a possible example of hormonal color change in *T. jacksonii*) unlike those in, for example, amphibians, which are typically under hormonal control (Hogben and Slome, 1931; Camargo et al., 1999). It is thought that the melanophores are maintained in a state of tonic contraction that is inhibited by light (Zoond and Eyre, 1934). In addition, visual stimuli may suppress the tonic contraction of the melanophores, resulting in color change and background matching (Zoond and Eyre, 1934). The light-dependent color change allows chameleons to optimize heating rates and minimize basking time (Walton and Bennett, 1993). However, basking may play a role not only in thermoregulation, but it has also been suggested to play a role in the regulation of the vital hormone vitamin D_3 in chameleons and other lizards (Ferguson et al., 2003, 2005; Karsten et al., 2009a). Vitamin D_3 has an important function in the calcium–phosphorous hormonal regulation system enhancing calcium uptake through the gut when calcium levels are low (Ferguson et al., 2003, 2005). Finally, the nervous control of the chromatophores allows rapid color change. Consequently, color change is important in social-signaling contexts and is thus likely under strong sexual selection (Berger and Burnstock, 1979; Stuart-Fox and Moussalli, 2008; Chapter 6).

3.6 DEVELOPMENTAL PHYSIOLOGY

A considerable body of work has been devoted to the effects of temperature on egg development and growth in chameleons. Chameleons are unusual because in some species embryos are in the gastrula stage at the time of oviposition and may remain at this stage for several months (Bons and Bons, 1960; Chapter 5). The duration, and end, of this so-called embryonic diapause is dependent on temperature but not on moisture (Andrews and Donoghue, 2004; Andrews et al., 2008; Adams et al., 2010). Temperature also affects speed of development, growth of the yolk sac, as well as the final hatching phenotype (Diaz-Paniagua and Cuadrado 2003; Andrews, 2007, 2008; Diaz-Paniagua, 2007). Lower temperatures result in longer developmental times and larger hatchlings (Diaz-Paniagua and Cuadrado, 2003; Diaz-Paniagua, 2007). Temperature also affects the physiology and anatomy of the amnion and the rhythmic contractions thereof during development (Nechaeva et al., 2005). Although the function of these contractions is poorly understood, it is thought that they help mix the amniotic fluid, thus preventing adhesion between the embryo and the amnion (Romanoff, 1960). Hatching success is also dependent on temperature, being higher at low temperatures (Diaz-Paniagua and Cuadrado, 2003). Interestingly, the duration of the cold period is determinant. Periods of colder temperature of intermediate lengths produced the largest offspring (Diaz-Paniagua, 2007). Moderate temperatures also provide optimal embryonic and posthatching development in desert species such as *C. calyptratus* (Andrews, 2008). The temperature to which eggs are exposed during development does not, however, affect subsequent sprint performance or selected body temperatures (Andrews, 2008). The access of females to ultraviolet light has been shown to be important and to affect the development and hatching success of the eggs (Ferguson et al., 2002). Moreover, essentials vitamins such as vitamins A and E have been detected in eggs of chameleons (Dierenfeld et al., 2002), suggesting an important contribution of maternal effects to hatchling development, growth, and survival. Finally, developmental temperature in chameleons does not affect hatchling sex ratio in *C. calyptratus* and *F. pardalis,* indicating that chameleons most likely all have genetic sex determination despite scattered reports to the contrary (Andrews, 2005).

FOUR

Function and Adaptation of Chameleons

TIMOTHY E. HIGHAM and CHRISTOPHER V. ANDERSON

Lizards have often been noted for their ability to move and capture prey in complex three-dimensional habitats (Huey and Pianka, 1981; Higham et al., 2001; Vanhooydonck et al., 2002; Mattingly and Jayne, 2004; Russell and Johnson, 2007; Montuelle et al., 2008). Given this, it is not surprising that many lizards are specialized for a particular type of locomotion and/or feeding. Chameleons, however, exhibit specialized feeding *and* locomotor behaviors. Their locomotor system has garnered substantial attention, given their purported upright limb posture and ability to perform complex maneuvers within their habitat. Their feeding apparatus is equally elaborate; they have the ability to project the tongue a considerable distance in order to latch onto prey (Zoond, 1933; Wainwright et al., 1991; Wainwright and Bennett, 1992a,b; Herrel et al., 2000; Anderson and Deban, 2010). Despite their unique and somewhat flamboyant characteristics, it is surprising that we are far from uncovering the functional mechanisms underlying their unique behaviors. However, there has been a recent surge of research that is exposing some aspects of chameleon function in relation to both ecology and morphology.

Chameleons are a diverse group of lizards found in Africa, Madagascar, southern Europe, Asia Minor, India, Sri Lanka, the Seychelles and Comoro Islands of the Indian Ocean, and via introduction, areas of North America (Ferguson et al., 2004; Tolley and Burger, 2007; Tilbury, 2010; Chapter 7). They can be fairly small (*Rhampholeon, Rieppeleon,* and *Brookesia*) or quite large (*Calumma, Furcifer,* and *Trioceros*) and may inhabit a wide range of habitats, including fynbos, forest, sandy desert, and grass (Bickel and Losos, 2002; Hofer et al., 2003; Tolley et al., 2006; Herrel et al., 2011; Chapter 5). Although most are arboreal, some species live predominantly on the ground and others frequently move on the ground between clumps of arboreal substrate. The diet of chameleons consists primarily of insects, but it can include small vertebrates such as lizards, mammals, and birds. Given the incredible diversity within chameleons and their specialized behavior, they are an attractive group for studying the functional consequences of phenotypic diversity. The goals of this chapter are: (1) to integrate the current literature that exists for chameleon locomotion and feeding, and (2) to indicate areas for future studies of biomechanics and functional morphology of chameleons.

4.1 LOCOMOTION

Lizards exhibit many types of specialization for locomotion, such as adhesive systems for climbing (Irschick et al., 1996; Russell and Higham, 2009), toe fringes for enhanced traction in sandy environments (Carothers, 1986), and claws for gripping (Zani, 2000). Chameleons, in particular, are specialized for slow arboreal locomotion (Mivart, 1870; Peterson, 1973, 1984; Abu-Ghalyun et al., 1988; Abu-Ghalyun, 1990; Mutungi, 1992; Losos et al., 1993; Bickel and Losos, 2002; Higham and Jayne, 2004b; Tolley and Burger, 2007; Boistel et al., 2010; Fischer et al., 2010; Herrel et al., 2011). Given that chameleons often live in arboreal habitats, where perches are small, gripping tightly with their feet, hands, and tail is critical for maintaining stability. Effective and stable progression is vital because of the limited and narrow base of support (Fig. 4.1) (Peterson, 1973; Cartmill, 1985; Foster and Higham, 2012). This is increasingly important for larger chameleons given that larger animals tend to experience greater toppling moments when moving on a branch (Cartmill, 1985). Chameleons have solved this problem by having prehensile (also termed "zygodactylous" by some authors) hands and feet in which the metacarpals and metatarsals are grouped into two opposing bundles (Fig. 4.1) (Gasc, 1963; Gans, 1967; Peterson, 1984; Losos et al., 1993; Russell and Bauer, 2008). In the forelimb, the first three digits form one group (mesial) and the fourth and fifth form the second group (lateral). However, the hindfoot contains a different pattern. In this case, the first and second metatarsals are bundled, and the third, fourth, and fifth metatarsal form the opposing group (Rieppel, 1993). These opposing bundles, having skin fused together between digits, are effective for grasping, as they are able to exert an adduction force on a branch with a circular cross section. This adduction force essentially squeezes the branch and generates friction. In addition to increasing stability during regular locomotion, grasping onto thin branches in this fashion is well suited for maneuvers because it increases control of pitching and rolling motions (Cartmill, 1985). It is the integration of multiple specializations that permit the effective movement of chameleons. The tail of chameleons is prehensile and can act like a fifth limb, gripping perches and aiding in stability (Zippel et al., 1999; Boistel et al., 2010; Herrel et al., 2012). This, coupled with their specialized hands and feet, extremely mobile pectoral girdle, and laterally compressed body, is ideal for an animal that moves in an arboreal habitat.

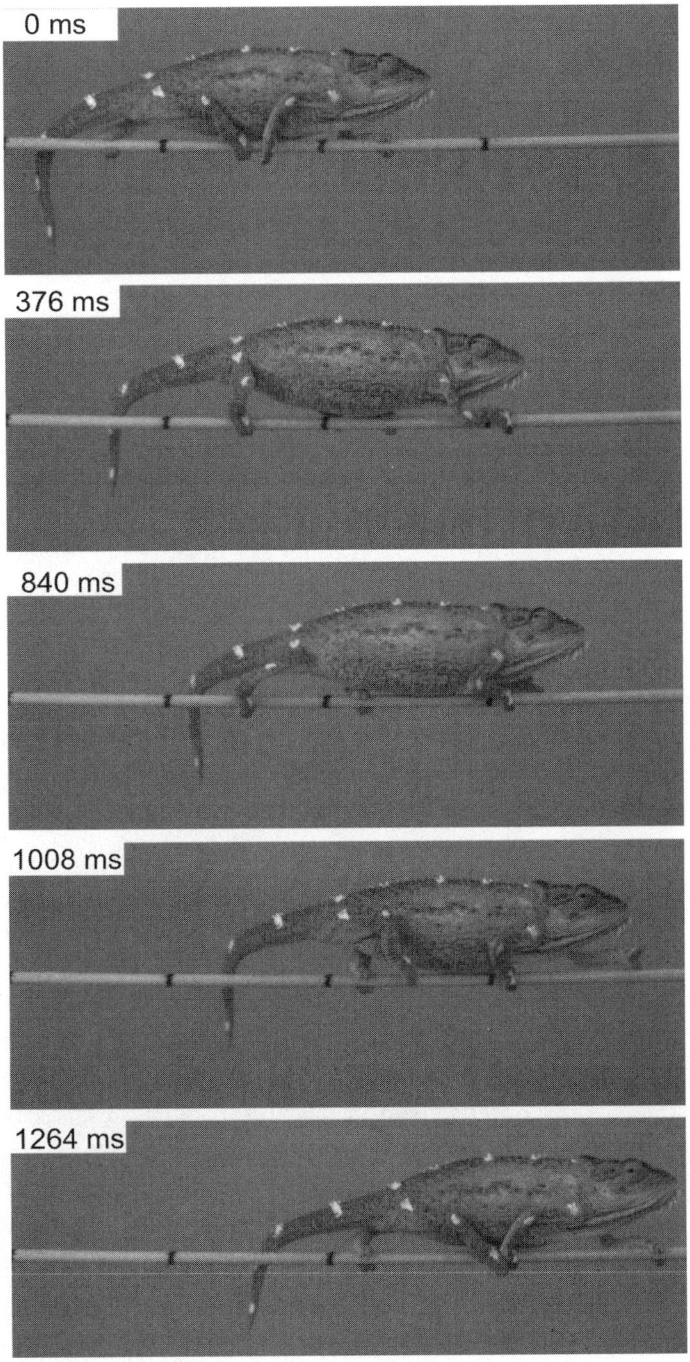

FIGURE 4.1. Lateral images of *Bradypodion occidentale* moving on a level, 3.2-mm-diameter wooden dowel at approximately 5 cm/sec. The black lines on the dowel are at 4-cm intervals. From top to bottom, the images indicate hindlimb footfall, midstance of the hindlimb, end stance of the hindlimb, midswing of the hindlimb, and the subsequent footfall of the hindlimb. Note the extreme forward reach of the hindlimb at footfall. These video stills are from unpublished data collected by Timothy Higham, John Measey, Krystal Tolley, and Anthony Herrel using a Photron APX RS camera operating at 250 Hz.

Limb Kinematics

Forelimb and hindlimb kinematics during locomotion in chameleons have been examined in several studies spanning over 30 years (Peterson, 1984; Higham and Jayne, 2004b; Fischer et al., 2010; Krause and Fischer, 2013). The techniques used to assess limb movements range from biplanar X-ray imaging, to regular video (25 frames per second), to high-speed video (250 frames per second). In all cases, the three-dimensional patterns of limb movement were determined.

Peterson (1984) conducted the first kinematic study of chameleons, although she focused solely on the forelimb. From this study and others, it is evident that *Chamaeleo* moves slowly relative to most other lizards and exhibits a slow trot-like walk. Peterson (1984) found that *Chamaeleo* (one species is now in the genus *Trioceros*) exhibited substantially less lateral undulation of the body as compared with a generalized, yet closely related, lizard species, *Agama*. Specifically, only 6% of the step was accounted for by undulation in *Chamaeleo*, as compared to 24% in *Agama*. One of the key observations in this study is the extreme pectoral girdle movement relative to the body wall. The displacement at the shoulder joint can increase the excursion arc by as much as 33 degrees to create an additional 28% of the step length (Peterson, 1984). This facilitates the long excursion arcs that are important for chameleons during arboreal acrobatic maneuvers (e.g., bridging gaps). Another benefit of substituting lateral body undulation with girdle excursion is that the center of mass displacement relative to the perch is reduced. This likely contributes to stability by minimizing toppling moments. As discussed below, the conclusions about overall chameleon locomotion in this study were based solely on the forelimb, and it turns out that the hindlimbs and pelvic girdle are not functioning like the anterior locomotor structures.

The three-dimensional movements of the hindlimbs during locomotion in chameleons was examined by Higham and Jayne (2004b). Again, this study examined *Chamaeleo,* which is currently the only genus for which kinematic data exist. In addition to the forward reach of the forelimb, which was studied by Peterson (1984), chameleons also appear to have extensive forward reach of their hindlimbs. Rather than stemming from excursion at their hip joint, hindlimb protraction appears to be increased by large amounts of femur protraction and knee extension (Figs 4.1 and 4.2a) (Higham and Jayne, 2004b). This increased reach, coupled with the knee flexion that is observed early in the stance phase (Fig. 4.2a), suggests that the hindlimb acts to pull the body forward early and then push the body forward later in stance. This is unlike the action in other terrestrial lizards, which exhibit minimal knee flexion early in stance. How are chameleons able to pull with their hindlimbs? The gripping feet secure the limb to the perch, facilitating the effective propulsion that arises from pulling the body forward. For a lizard that does not grasp (or adhere) to the substrate, knee flexion early in stance would tend to pull the foot away from the ground. Another key result from Higham and Jayne (2004b) is that pelvic rotation, unlike pectoral rotation, is rather large in chameleons as compared with terrestrial lizards. This conflicts with the results of Peterson (1984), who suggested that lateral undulation is not a major contributor to locomotion in chameleons.

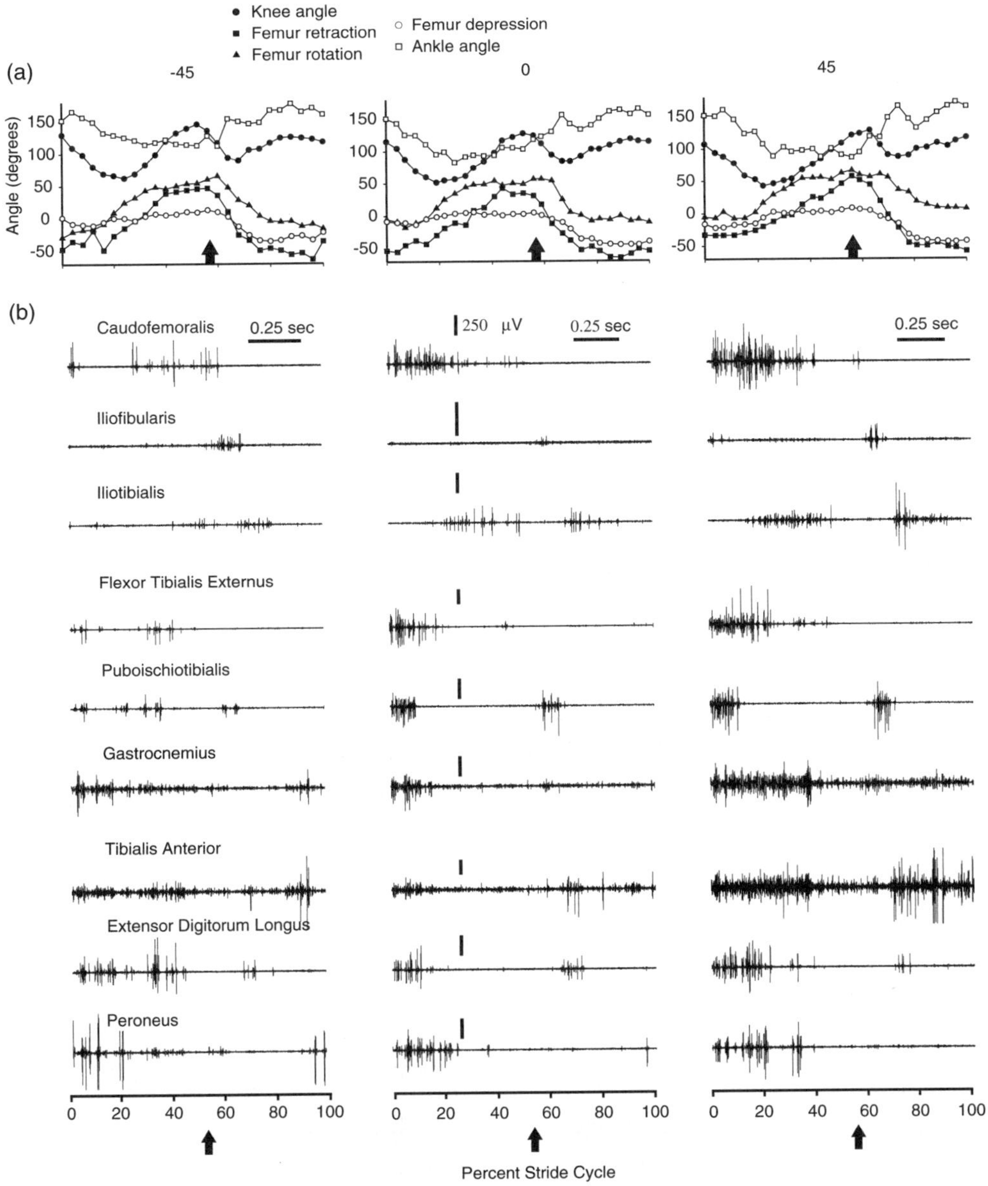

FIGURE 4.2. (a) Kinematics of the hindlimb; (b) Electromyograms versus time (as a percentage of stride cycle) for one stride from a single individual of *Chamaeleo calyptratus* on inclines of −45 degrees (left), 0 degrees (middle), and 45 degrees (right). The arrows indicate the end of stance, and footfall occurs at time 0%. The vertical bars indicate the voltage scales (250 μV), which are constant between all panels within a row. Originally published in Higham and Jayne (2004a).

Utilizing a motor-driven rope-mill (treadmill for arboreal animals) and X-ray motion analysis, Fischer and colleagues (2010) examined the kinematics of the whole body of chameleons (*Chamaeleo calyptratus*) during locomotion (see Fig. 4.3). They discovered that, unlike most lizards, chameleons exhibit a unique mode of axial undulation. Lateral bending of the vertebral column is limited to the lower spine. The anterior part of the spine remains fairly rigid, indicating that the spine does not contribute to pectoral translation but does contribute to pelvic translation (Fischer et al., 2010). This supports the conclusions of Higham and Jayne (2004b). Thus, lateral undulation is important for chameleons but is not driven by the entire spine, as is the case with other lizards. Pectoral girdle mobility compensates for the decreased anterior undulation, whereas pelvic girdle mobility is limited and posterior undulation is higher.

It is clear that understanding the three-dimensional movements of the body and limbs of chameleons is a key to linking their unique morphology and ecology with function. However, we must be circumspect when generalizing about all chameleons using the existing data. Chameleons are very diverse, yet almost all studies have focused on *Chamaeleo calyptratus* (e.g., Higham and Jayne, 2004a,b; Fischer et al., 2010). This is likely a result of their manageable size, ability to be kept in captivity, and availability in the pet trade. However, future work should explore the diversity within chameleons, and even between populations that live in diverse habitats (Hopkins and Tolley, 2011). A key group that could illuminate our understanding of chameleon locomotion is the Dwarf Chameleons from the genus *Bradypodion* (South Africa). The biogeography and phylogenetic relationships among Dwarf Chameleons have been the foci of recent studies (Tolley et al., 2004, 2006; Tolley and Burger, 2007). Thus, there is a wonderful opportunity to link morphology, biogeography, and function in a diverse group of chameleons.

Limb Posture: Are Chameleons Really Upright?

Most lizards exhibit a relatively sprawled posture, which involves limbs whose major movements are closer to a horizontally than vertically oriented plane (Rewcastle, 1981, 1983; Blob and Biewener, 1999; Russell and Bels, 2001). A hallmark of the sprawled posture is the contribution of lateral undulation to forward movement. Because undulation is in the same plane (horizontal) as the proximal segments of the limbs, lateral movement of the body will move the contact point of the limb. Given the relatively low contribution of lateral undulation to step length, as noted above, the general sense is that chameleons are relatively upright. There are a couple of other reasons that lead to the perception that chameleons employ a relatively upright posture. When chameleons are stationary on a branch, for example, they tend to hold their knees and elbows close to the body and therefore appear to be in a parasagittal plane. Second, chameleons are often observed moving on narrow perches that necessarily require their limbs to make contact more medially than those of other terrestrial lizards. In contrast, evidence suggests that chameleons do not exhibit an upright posture relative to other lizards. In a comparison of a terrestrial specialist and a terrestrial generalist, Higham and Jayne (2004b) found, by looking at locomotion on a flat surface, that

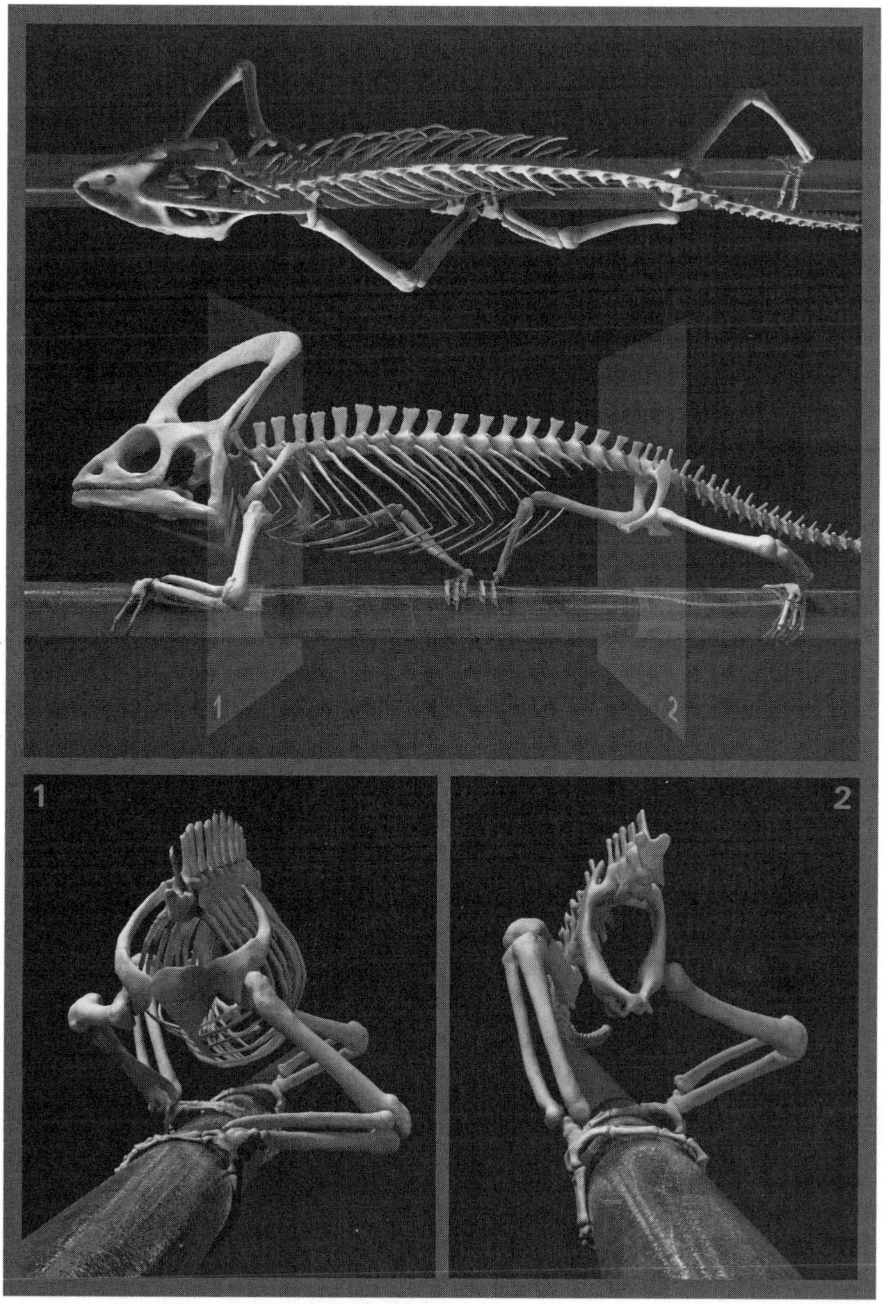

FIGURE 4.3. Dorsal (top) and lateral view (bottom) of an animated chameleon skeleton (X-ray Reconstruction of Moving Morphology, or XROMM) based on a biplanar cineradiographic record. The left forelimb and right hindlimb are in stance phase and the right forelimb and left hindlimb on the verge of lift off. The pictures at the bottom show details in transversal perspective at the height of the shoulder joint (1) and the hip joint (2). Note the high degree of abduction of the forelimb during stance phase and of the hindlimb during liftoff (left extremities). Forelimbs are in a parasagittal position during liftoff and hindlimbs during touchdown (right extremities). Prepared by J. Lauströer und A. Andikfar and reproduced (with permission from Elsevier, license 2892041208388) from Fischer et al. (2010).

maximum femur depression was greater in *Dipsosaurus dorsalis* (terrestrial) than *Chamaeleo calyptratus* (arboreal). Greater femur depression indicates a femur that is more aligned with the vertical plane, indicative of a more upright posture. Fischer et al. (2010) confirmed that maximum limb abduction is as high as in other, more terrestrial, lizards. A basic question that still lingers in comparative biomechanics is how to define posture. Vertebrates commonly move in a dynamic fashion, which means their limbs are moving into different positions throughout a stride. As noted by Fischer et al. (2010), chameleons exhibit parasagittal movements for at least part of the stride. Thus, perhaps the more upright posture that is often noted for chameleons is a result of transient parasagittalization (Fig. 4.3). Future work should aim to solidify a broadly applicable method for defining posture in lizards and use this to more extensively test whether chameleons are indeed more upright.

Limb Mechanics

A single study has examined three-dimensional substrate reaction forces during chameleon (*Chamaeleo calyptratus*) locomotion (Krause and Fischer, 2013). Both the forelimbs and hindlimbs were measured using a force-sensitive element mounted to a force plate. This, coupled, with two high-speed video cameras (500 Hz) facilitated a cohesive assessment of chameleon locomotor mechanics. On a level pole, with a diameter of 2 cm, the forelimbs and hindlimbs exhibit a peak vertical force at 74% and 27% of stance, respectively. Medially directed forces on a level perch are quite small and the magnitudes are comparable between the limbs. When moving uphill at 30 degrees and 60 degrees, the propulsive impulses from both the forelimbs and hindlimbs increase and the braking impulses almost disappear. The vertical impulse increase on the 30- and 60-degree inclines and on the −30-degree decline, but do not continue to decrease as the decline increases to −60 degrees (Krause and Fischer, 2013). Overall, it is clear that future work detailing the mechanical function of the forelimbs and hindlimbs will continue to reveal interesting functional ramifications of morphological and ecological diversity among chameleons.

Muscle Function

Given the strong differences in muscle anatomy and physiology in chameleons relative to other lizards (Abu-Ghalyun et al., 1988; Mutungi, 1992; Russell and Bauer, 2008; Chapter 2), it is commonly assumed that the function of the muscles during locomotion might also differ in relation to other, more generalized, terrestrial lizards. A single study has examined and muscle activation patterns of limb muscles during locomotion in chameleons (Fig. 4.2b) (Higham and Jayne, 2004a). In this study, the timing and magnitudes of electrical activity in nine hindlimb muscles were assessed using electromyographic (EMG) electrodes implanted directly into each muscle. Following surgery, adult veiled chameleons (*Chamaeleo calyptratus*) moved on perches of varying diameter and incline. Overall, changes in the characteristics of the EMG signals with habitat structure are driven primarily by amplitude differences rather than by changes in timing. For example, most propulsive muscles exhibit an increase in amplitude with an increase in incline, indicating an increase in recruitment and possibly force. This is important for moving up an incline given that muscle work (muscle force times muscle strain) must increase in order to increase the height of the animal. In addition, it appears that the function of the proximal muscles, as compared with the distal muscles, is altered to a greater extent when chameleons deal with changes in habitat structure.

As described above, a key aspect of chameleon locomotion that sets it apart from that of other lizards is the ability to pull the body forward with the hindlimbs during the first portion of the stance phase of the stride (Higham and Jayne, 2004a,b). Thus, it is possible that muscles that flex the knee will increase in activity when greater propulsion is required and might be enlarged relative to those of other lizards. Indeed, the amplitude of activity of the M. flexor tibialis externus and the M. puboischiotibialis (see Chapter 2) increases substantially while moving uphill as compared with move on level ground, and the peak amplitude occurs immediately after footfall at the beginning of stance (Fig. 4.2b) (Higham and Jayne, 2004a). Elucidating the underlying physiological mechanisms of chameleon locomotion is critical for understanding the evolution of this unique mode of movement. However, available data are currently limited to the hindlimb in a single species. Future work should examine forelimb muscle activation patterns and mechanics in relation to kinematics. Based on the morphological differences in the forelimb musculature of chameleons as compared with other lizards (Peterson, 1973), it is likely that muscle activity patterns also differ considerably. In addition, muscle activation patterns and mechanics in relation to morphological and ecological diversity within chameleons would reveal how chameleons move in different ways.

Tail Use

Prehensile tails are those that are involved in grasping or clinging, and they can often support the weight of the animal. Prehensile tails are found in a wide variety of vertebrates, including mammals (especially New World monkeys), reptiles (lizards and snakes), amphibians (some salamanders and caecilians) and fishes (syngnathids) (Hurle et al., 1987; Hale, 1996; Bergeson, 1998; Meldrum, 1998; Garber and Rehg, 1999; Zippel et al., 1999; Spickler et al., 2006). Despite the common occurrence of prehensile tails in lizards, there is a paucity of studies examining tail function in relation to locomotion. Tail morphology can be significantly different between chameleons, and even between populations of a single species (Herrel et al., 2011; Hopkins and Tolley, 2011).

A study by Hopkins and Tolley (2011) examined the morphological differences between populations of the Cape Dwarf Chameleon (*Bradypodion pumilum*) that occupy habitats that differ considerably in structure. One morph lives in relatively closed (forested) vegetation, whereas the other morph lives in more open (fynbos) vegetation in South Africa. The former is larger and more colorful than the latter. One of the main differences between the two morphs was tail length, which was longer in the closed-habitat morph. Longer tails could enhance locomotion through the canopy of a closed habitat, whereas longer limbs (not tail) might enhance locomotion in an open habitat.

Tails are often shorter in more terrestrial species of chameleons, including Malagasy Dwarf Chameleons from the genus *Brookesia* (Boistel et al., 2010). Although these species have prehensile feet for grasping, they often ambulate over broad surfaces. Under these circumstances, the distal portion of the tail is used to stabilize the body by making contact with the substrate. Morphologically, *Brookesia* appear to have fewer vertebrae and the mobile regions are restricted to the distal portion (Boistel et al., 2010). Thus, even chameleons that are not arboreal can use their tails for enhancing stability, although stability in a terrestrial habitat is achieved in different ways.

Ecology and Locomotion

Although the impact of morphology on performance is often examined in lizards (Bonine and Garland, 1999), habitat structure can dictate both performance and morphology in diverse groups (Garland and Losos, 1994; Irschick and Losos, 1998; Melville and Swain, 2000; Goodman et al., 2008; Higham and Russell, 2010). However, linking habitat use with a predictable suite of morphological characteristics can sometimes be challenging or may not yield differences (Vanhooydonck and Van Damme, 1999). For groups of lizards that contain both terrestrial and arboreal species, there can be a trade-off between climbing performance (on relatively narrow perches) and sprinting speed (on a level surface) because of the conflicting demands of these behaviors (Losos and Sinervo, 1989; Sinervo and Losos, 1991). However, a key question is whether there is a trade-off between climbing and locomotor speed in chameleons. For two species of chameleons from Kenya (*Chamaeleo dilepis* and *Trioceros jacksonii*), this seems to be the case (Losos et al., 1993). In this study, clinging ability, measured by pulling each chameleon along a dowel attached to a spring scale, decreased for both species as the diameter was increased from 2 mm to 11 mm. Conversely, sprinting performance steadily increased as the diameter was increased from 1.5 mm to a flat surface (Losos et al., 1993). Although clinging ability was used as a proxy for climbing ability, it is still unclear whether there is a direct relationship between these two. However, it is assumed that the ability to grasp a perch will likely be positively correlated with climbing ability given that locomotor stability will increase with increasing grasping ability (Cartmill, 1985).

4.2 FEEDING

Historical Perspectives

As one of the more extraordinary features of chameleon biology, the way chameleons feed has been the subject of scientific attention for centuries. Hypothesized mechanisms of tongue projection have included pneumatic action, tumescence, and muscle action. Despite the long history of scientific attention, however, we are still gaining insight into this highly specialized system today.

Some of the first hypothesized mechanisms of tongue projection in chameleons involve pneumatic extension of the tongue. In 1676, Claude Perrault proposed that an inflow of air from the lungs into the hollows of the tongue caused the extension of the tongue, while its retraction was caused by withdrawal of that air (Houston, 1828; Gnanamuthu, 1930). This view was similarly held by A.M.C. Duméril in 1836 (Gnanamuthu, 1930; Altevogt and Altevogt, 1954; Bell, 1989), while Antonio Vallisneri in 1715 and Mauro Rusconi in 1844 proposed that a cecum, or bladder of the trachea would inflate and thrust the tongue forward (Gnanamuthu, 1930). Subsequent research, however, revealed a lack of connection between the lungs and the tongue (Duvernoy, 1836) and suggests that a tracheal cecum would not be able to project the tongue the observed distances (Gnanamuthu, 1930).

A second proposed mechanism of tongue projection involves tumescence and vascular erection of the tongue. In addition to pneumatic action, Perrault suggested that inflow of blood into the hollows of the tongue, and the resultant tumescence, could serve as an alternative mechanism for tongue projection (Gnanamuthu, 1930). Houston (1828) suggested a similar mechanism, whereby blood flows into a vast vascular network within the tongue, causing tumescence and projection in a manner similar to erectile organs. It has been pointed out, however, that such a mechanism would cause thickening rather than thinning of the tongue as it extends and that it occurs too slowly to explain the rapid projection (Mayer, 1835; Gnanamuthu, 1930).

As early as 1805, however, Cuvier proposed that muscle action was a possible mechanism for tongue projection, but the nature of the suggested muscle action has varied considerably. Among some of these theories, Mayer (1835) proposed a muscular hydrostat model in which contraction of circular fibers in the tongue would cause the tongue to become thinner and longer, thus projecting the tongue out of the mouth. Duvernoy (1836), on the other hand, proposed that rapid protrusion of the hyolingual apparatus from the mouth would cause the tongue to be launched as the tongue skeleton suddenly stopped. A similar theory was also proposed by Dewevre (1895), who indicated that the entoglossal process would act like a billiard cue on the tongue as it was rapidly protruded forward. Finally, work by Brücke (1852a) suggested that the radial fibers of the M. accelerator linguae contracting around the tapered portion of the entoglossal process would push the tongue forward.

With these and subsequent studies, our knowledge of the morphology and mechanism of tongue projection has improved greatly. Despite all this interest in chameleon feeding, however, there are still unanswered questions about the exact mechanism of tongue projection and retraction in chameleons.

Functional Specializations

Given their proclivity to life in an arboreal setting, and on an often precarious substrate, it seems likely that ballistic tongue projection in chameleons may have evolved as a means to minimize lunge and chase during prey capture (Schwenk, 2000). Such a specialized feeding method, however, clearly necessitates a number of functional and anatomical adaptations. While the anatomical conditions of chameleons have been reviewed (see Chapter 2), their functional consequences are important to the mechanics and behavior of chameleons' feeding.

The tongue in nonchameleon iguanians, for instance, is unable to protrude far out of the mouth because the base is tightly adhered to the basihyal (Schwenk, 2000). In chameleons, however, the base of the tongue is free from the basihyal and overall is attached to the hyobranchial apparatus only at the origin of the M. hyoglossus (Schwenk, 2000; Herrel et al., 2001b). This frees the tongue to protrude greater distances.

In order to protrude the tongue greater distances, however, the length of the tongue itself must increase as well. In chameleons, the M. hyoglossus has become greatly elongated to cover the span of observed projection distances (Schwenk, 2000). In order to pack this lengthy muscle within the buccal cavity and onto the base of the hyobranchial apparatus when the tongue is retracted, the M. hyoglossus is pleated at rest and is comprised of supercontracting muscle fibers with perforated Z discs (Rice, 1973; Schwenk, 2000; Herrel et al., 2001a,b, 2002). These supercontracting muscle fibers allow the tongue to lengthen up to 600% of its resting length and exert high force during contraction across a broad range of sarcomere lengths (Herrel et al., 2001a, 2002).

The development of a rapid projectile mechanism itself is also important to reduce the opportunity for prey to evade capture while the tongue is traveling increased distances from the mouth to the prey. This rapid mechanism is accomplished by the combined specialization of aspects of the tongue skeleton, tongue muscles, and collagenous elements. The M. accelerator linguae of chameleons is formed by the M. verticalis linguae, which has hypertrophied and become modified to encircle the lingual process of the tongue skeleton (Schwenk and Bell, 1988; Wainwright and Bennett, 1992b; Meyers and Nishikawa, 2000; Schwenk, 2000). Rather than being tapered along its entire length, the lingual process of the tongue skeleton, called the "entoglossal process" in chameleons, has become more robust and parallel sided along most of its length, except for a tapered tip (Wainwright and Bennett, 1992b; Schwenk, 2000; Herrel et al., 2009). Finally, the space between the entoglossal process, and the M. accelerator linguae and M. hyoglossus contains a layer of dense connective tissue (Gnanamuthu, 1930; Zoond, 1933; Gans, 1967; Bell, 1989; Herrel et al., 2001b; de Groot and van Leeuwen, 2004) and a viscous lubricating fluid similar histochemically to synovial fluid (Bell, 1989; Schwenk, 2000). This connective-tissue layer is comprised of a nested series of collagenous intralingual sheaths with the anterior portion of the sheaths containing helical collagen fibers, which can store and release energy (de Groot and van Leeuwen, 2004). These specializations allow the tongue muscle to load elastic elements of the tongue without projecting out of the mouth before the elastic elements are completely loaded.

Feeding Performance

Chameleons are able to project their tongues relatively long distances at extremely high performance. Published estimates of the maximum length chameleons are able to project their tongues vary but have tended to indicate projection distances from one to two times the snout–vent length of the animal (Zoond, 1933; Gans, 1967; Altevogt, 1977; Schwenk and Bell, 1988; Bell, 1990; So et al., 1992; Wainwright and Bennett, 1992a; Schwenk, 2000;

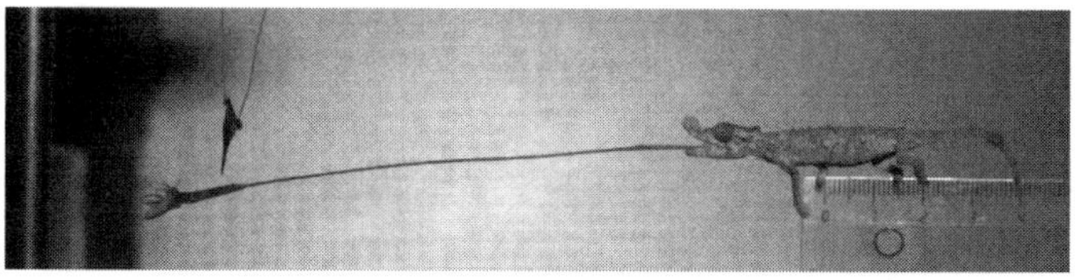

FIGURE 4.4. Lateral image of maximal tongue projection length during a feeding in *Rhampholeon spinosus*. The length of the snout vent in this specimen is 4.7 cm and it projects its tongue 10.4 cm from the mouth, for a projectile distance of 2.21 times the body length. From unpublished data collected by Christopher V. Anderson using a Photron Fastcam 1024 PCI camera operating at 3000 Hz.

Herrel et al., 2001b, 2002, 2009; de Groot and van Leeuwen, 2004; Müller and Kranenbarg, 2004). Recent research, however, has indicated that interspecifically, the length of the M. hyoglossus, and thus projectile distance, may scale with negative allometry relative to snout–vent length (Anderson et al., 2012). This would result in proportionately longer maximum tongue projection distances in smaller chameleon species than in larger species (Anderson et al., 2012). Because many studies have used larger chameleon species (Bell, 1990; Wainwright et al., 1991; So et al., 1992; Wainwright and Bennett, 1992a,b; de Groot and van Leeuwen, 2004; Herrel et al., 2009), the potential projectile distance of some smaller species may not yet have been recognized. For instance, *Rhampholeon spinosus,* a species smaller than those used in most studies, has been observed projecting its tongue 2.21 times its snout–vent length (Fig. 4.4).

The performance of the tongue as it is projected from the mouth is similarly impressive. While muscle alone is known to produce peak mass-specific power outputs of only 1121 W kg^{-1} (Askew and Marsh, 2001), chameleons project their tongues with peak mass-specific power outputs that more than double this value, thus indicating that tongue projection is not powered by muscle activity alone, but also by elastic recoil (de Groot and van Leeuwen, 2004). Incorporation of an elastic recoil mechanism in tongue projection imparts spectacular performance with mass-specific power outputs in chameleons recorded up to 3168 W kg^{-1} (de Groot and van Leeuwen, 2004), peak accelerations of up to 486 m sec^{-2} (Wainwright et al., 1991), or 50 g (50 times the acceleration due to gravity), and peak velocities of up to 5.8 m sec^{-1} (Wainwright et al., 1991). These performance parameters tend to be positively correlated with the projection distance within an individual (Bell, 1990; Wainwright et al., 1991; Anderson and Deban, 2010), indicating that as an individual projects its tongue longer distances, the performance tends to increase. Further, while velocity is expected to be size-independent (Hill, 1950), acceleration and power output are negatively correlated with size so smaller individuals would be expected to produce higher accelerations and power outputs than larger individuals (Hill, 1950; Herrel et al., 2009; Anderson et al., 2012). As discussed previously, since most studies use larger species, potential peak acceleration and peak mass-specific power output values attained by chameleons may be underestimated in the literature. Kinematic analysis of *Rieppeleon brevicaudatus,* for instance, has shown peak acceleration values of up to 1642 m sec^{-2}, or 170 g, and peak mass-specific power output values up to 11,392 W kg^{-1} (C.V. Anderson, unpublished results).

Prey Capture Kinematics

Previous studies on the kinematics of chameleon feeding have identified five phases of prey capture: (1) fixation, (2) tongue protrusion, (3) tongue projection, (4) tongue retraction, and (5) hyobranchial retraction (Fig. 4.5) (Altevogt and Altevogt, 1954; Schuster, 1984; Bell, 1990; Wainwright et al., 1991; Schwenk, 2000). These phases, however, do not correspond with standard gape-cycle phases (Fig. 4.5) (Bramble and Wake, 1985) and are not all distinct (Wainwright and Bennett, 1992a; Schwenk, 2000). That said, despite having a highly specialized hyolingual apparatus, with the addition of the projectile phase in the feeding sequence, and a prolonged retraction phase, the kinematics of feeding in chameleons is otherwise relatively similar to that of generalized iguanians (Bramble and Wake, 1985; Schwenk and Bell, 1988; Schwenk and Throckmorton, 1989; Bell, 1990; Wainwright et al., 1991; So et al., 1992; Schwenk, 2000).

Prior to the onset of mouth opening or movement of the hyobranchial apparatus, the chameleon must orient and prepare for feeding movements during the fixation phase of prey capture (Fig. 4.5). During this phase, the chameleon turns toward the prey, reduces the distance between them and rotates both eye turrets forward, focusing on the prey (Bell, 1990; Schwenk, 2000). Once oriented toward the prey, the chameleon braces the head and body to be able to withstand the forces imparted during aiming, tongue projection, and tongue retraction (Schwenk, 2000).

Tongue protrusion occurs during the slow open I (SO I) and slow open II (SO II) phases of the gape cycle (Fig. 4.5) (Bramble and Wake, 1985; Bell, 1990; Wainwright et al., 1991; Schwenk, 2000). During the SO I phase, the gape of the jaws slowly increases and the anterior portion of the tongue is slowly protruded through the forward margin of the mouth via hyoid protraction (Bell, 1990; Schwenk, 2000). Hyoid protraction and tongue protrusion continues into SO II and may cease in a plateau of movement where the tongue is held in a protracted position prior to projection (Bell, 1990; Schwenk, 2000). Also during the SO II phase, the folded membrana glandulosa of the tongue pad is evaginated, exposing the glandular inner surface on the tongue (Altevogt and Altevogt, 1954; Altevogt, 1977; Bell, 1990; Herrel et al., 2000; Schwenk, 2000).

Toward the end of SO II, tongue projection and tongue–prey contact occurs (Fig. 4.5). At the onset of tongue projection, the hyoid experiences a short yet rapid forward thrust (Wainwright et al., 1991; Wainwright and Bennett, 1992a; Meyers and Nishikawa, 2000; Herrel et al., 2001b; de Groot and van Leeuwen, 2004). The tongue is then projected rapidly from the mouth, and a slight decrease in gape is observed (Bell, 1990; Wainwright et al., 1991; Wainwright and Bennett, 1992a; Schwenk, 2000). This decrease in gape may be the result of the inertial reaction forces from the tongue projection or of changes in the muscle activity pattern (Schwenk, 2000), but it is unique to chameleons (Wainwright et al., 1991; Schwenk,

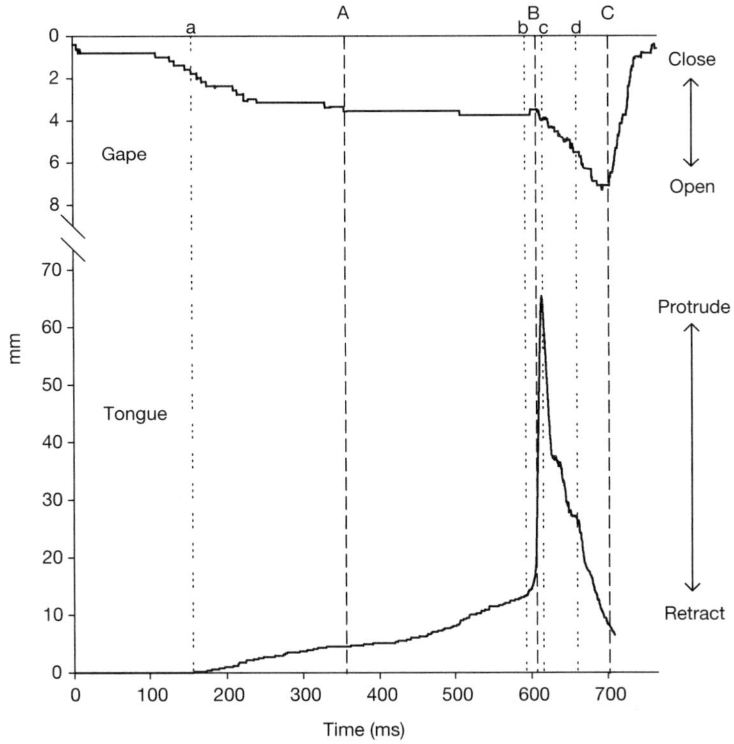

FIGURE 4.5. Kinematic profile of a feeding sequence in *Rieppeleon kerstenii*. Gape width is shown above on inverted y-axis with time correlated tongue protrusion distance below. Gape width and tongue protrusion distance are relative to the tip of the snout. Fine dashed lines and associated lower-case letters correspond with start points of chameleon feeding phases (*sensu* Altevogt and Altevogt, 1954; Schuster, 1984; Bell, 1990; Wainwright et al., 1991; Schwenk, 2000). The fixation phase occurs prior to the onset of the tongue protrusion phase (a), which continues until the onset of the tongue projection phase (b). The tongue retraction phase (c) begins at maximal tongue projection and continues beyond the onset of the hyobranchial retraction phase (d). Broad dashed lines and associated upper-case letters correspond with start points of standard gape cycle phases (*sensu* Bramble and Wake, 1985). The transition between slow open I and slow open II is somewhat ambiguous in some feedings and corresponds approximately with (A). The onset of the fast open phase begins following a sudden decrease in gape after the onset of tongue projection and corresponds with (B). The onset of the fast close phase begins following maximal gape and corresponds with (C). From unpublished data collected by Christopher V. Anderson using a Photron Fastcam 1024 PCI camera operating at 3000 Hz.

2000). The final portion of the SO II phase is the onset of prey prehension (Bell, 1990; Schwenk, 2000) and the beginning of tongue retraction.

After tongue retraction begins, the gape cycle progresses into the fast open (FO) phase with a rapid increase in the gape (Fig. 4.5) (Bell, 1990; Schwenk, 2000). Tongue retraction occurs throughout FO (Bell, 1990; Schwenk, 2000) but likely extends into the beginning of the fast close (FC) phase as well. Maximum gape and the transition between FO and FC coincides with hyobranchial retraction, which thus overlaps with tongue retraction and begins at the end of FO (Wainwright et al., 1991; So et al., 1992; Wainwright and Bennett, 1992a; Schwenk, 2000). Most of the hyobranchial retraction occurs during FC, however, when feeding on large prey, during which initial jaw closure results in a bite and a slow close-power stroke (SC-PS) phase is present, hyobranchial retraction can extend into SC-PS (Schwenk, 2000).

Mechanism of Hyobranchial Protrusion and Retraction

At rest, the tongue skeleton is positioned in the throat with the ceratobranchials and ceratohyals positioned nearly perpendicular to the entoglossal process (Herrel et al., 2001b). The tongue skeleton is protruded from the mouth by muscle action that pulls the hyoid horns backward and downward and brings the entoglossal process upward and forward as the gape of the mouth opens (Gnanamuthu, 1930; Bell, 1990; Herrel et al., 2009). As the lower jaw is depressed, the M. sternohyoideus profundus and M. mandibulohyoideus show a burst of activity corresponding with tongue protrusion (Wainwright and Bennett, 1992a). Activity of the M. sternohyoideus profundus draws the distal end of the ceratobranchials ventrocaudally toward the sternum, causing the ceratobranchials to rotate with respect to the basihyal until positioned nearly parallel with the entoglossal process (Gnanamuthu, 1930; Wainwright and Bennett, 1992a). Meanwhile, the activity of the M. mandibulohyoideus draws the distal end of the tongue apparatus forward through the opening gape of the mouth (Gnanamuthu, 1930; Zoond, 1933; Wainwright et al., 1991; Meyers and Nishikawa, 2000). This burst of activity from these muscles continues through the slow protraction phase and ends prior to tongue projection during the stationary phase of SO II (Wainwright and Bennett, 1992a).

During tongue retraction and at the end of FO, the M. sternohyoideus superficialis shows a burst of activity corresponding with hyobranchial retraction (Wainwright and Bennett, 1992a). Shortly after the onset of M. sternohyoideus superficialis activity, jaw adductor muscles become active as the FC begins (Wainwright and Bennett, 1992a). Activity of the M. sternohyoideus superficialis draws the basihyoid posteriorly toward the sternum, drawing the entoglossal process and tongue into the mouth (Gnanamuthu, 1930; Zoond, 1933; Wainwright and Bennett, 1992a) and presumably in the process, allowing the ceratobranchials to articulate back into their resting position.

Mechanism of Tongue Projection and Retraction

Following tongue protrusion, the posterior, circular portion of the M. accelerator linguae becomes active and remains active until immediately before tongue projection (Wainwright and Bennett, 1992a; Herrel et al., 2000; Anderson and Deban, 2012). This results in the posterior, circular portion of the M. accelerator linguae contracting around the parallel-sided posterior portion of the entoglossal process (Wainwright et al., 1992b; de Groot and van Leeuwen, 2004). During this contraction, the diameter of the M. accelerator linguae is reduced, causing elongation of the muscle along the length of the entoglossal process because that muscle must maintain a constant volume (Herrel et al., 2000; de Groot and van Leeuwen, 2004). Because the tongue is unable to elongate posteriorly because of the presence of the M. hyoglossus, this elongation occurs in the forward direction toward the tapered tip of the entoglossal process (de Groot and van Leeuwen, 2004). In the process of this elongation, it is hypothesized that the intralingual sheaths between the entoglossal process and the M. accelerator linguae are loaded with elastic potential energy by compression and shear stress from the contraction and elongation of the M. accelerator linguae (de Groot and van Leeuwen, 2004). This elongation stretches the intralingual sheaths and causes the angle of their helically wound collagen fibers to change, resulting in energy storage in a longitudinal plane (de Groot and van Leeuwen, 2004). Once tension has been built up in the muscle and intralingual sheaths, activity of the M. accelerator linguae ceases (Wainwright and Bennett, 1992a,b). Tension in the muscle and intralingual sheaths is maintained, however, for a considerable period of time, allowing the onset of projection to occur after the muscle activity has ceased but without a loss of force (Wainwright and Bennett, 1992b; Anderson and Deban, 2012).

The onset of tongue projection is triggered by a second, sudden contraction of the M. mandibulohyoideus, which produces a forward thrust of the entoglossal process (Wainwright et al., 1991; Wainwright and Bennett, 1992a; Meyers and Nishikawa, 2000; de Groot and van Leeuwen, 2004). This thrust is responsible for only a small portion of the total momentum of the tongue (de Groot and van Leeuwen, 2004), but activity of the M. mandibulohyoideus braces the hyobranchial apparatus and resists the posterior reaction force exerted by the tongue as it is projected off the entoglossal process (Meyers and Nishikawa, 2000; de Groot and van Leeuwen, 2004).

The forward thrust of the entoglossal process presumably causes the M. accelerator linguae and intralingual sheaths, still under high tension from the activity of the muscle itself, to slide forward over the tapered tip of the entoglossal process (Wainwright and Bennett, 1992b; de Groot and van Leeuwen, 2004). As the radial forces exerted by the M. accelerator linguae around the parallel sides of the entoglossal process are transferred into longitudinal forces at the tapered tip of the entoglossal process, the M. accelerator linguae rapidly pushes itself off the entoglossal process (Wainwright and Bennett, 1992a; de Groot and van Leeuwen, 2004). In the process, the elastic energy stored in the intralingual sheaths is also released to help power projection as the helically wound fibers are able to recoil to their resting fiber angles (de Groot and van Leeuwen, 2004). The sudden contraction of the M. mandibulohyoideus in conjunction with the largely parallel-sided entoglossal process with a tapered tip thus act as a passive trigger for the onset of tongue projection rather than having an antagonistic muscle that releases the tongue (Wainwright and Bennett, 1992a,b). Once the M. accelerator linguae has lost contact with the entoglossal process, the tongue travels forward toward the prey under its own momentum.

Immediately following the onset of tongue projection, the anterior, noncircular portion of the M. accelerator linguae becomes active and the posterior, circular portion of the M. accelerator linguae shows a second burst of activity (Wainwright and Bennett, 1992a). The activity of the M. accelerator linguae after tongue projection is presumably to provide a rigid structure for the M. "retractor pouch" of the tongue pad to act against during prey prehension (Wainwright and Bennett, 1992a). In addition, the M. hyoglossus becomes active between 10 ms before and 20 ms after the onset of tongue projection (Wainwright and Bennett, 1992a). The two portions of the M. accelerator linguae and the M. hyoglossus remain active throughout tongue and hyobranchial retraction (Wainwright and Bennett, 1992a). The early activity of the M. hyoglossus is presumably to decelerate the tongue as it approaches the prey and prevent overshoot (Bell, 1990; Wainwright and Bennett, 1992a). The M. hyoglossus remains active following prey contact and is responsible for retracting the tongue back onto the entoglossal process (Wainwright and Bennett, 1992a,b; Herrel et al., 2009; Anderson and Deban, 2012).

Mechanism of Prey Prehension

During rest the membrana glandulosa of the tongue pad is folded inward to create a lingual pocket or dimple (Altevogt and Altevogt, 1954; Altevogt, 1977; Schwenk, 1983; Bell, 1989; Bell, 1990; Herrel et al., 2000; Schwenk, 2000). During tongue protrusion, the membrane glandulosa is evaginated so that the center of the tongue pad, which is highly glandular and possesses numerous papillae, is at the apex of the tongue and becomes the contact zone for prey during feeding (Altevogt and Altevogt, 1954; Altevogt, 1977; Schwenk, 1983, 2000; Bell, 1990; Herrel et al., 2000). The tongue pad remains evaginated during tongue projection until just before contact with the prey, at which point the contact zone begins to invaginate, once again forming an upper and a lower lobe (Bell, 1990; Herrel et al., 2000; Schwenk, 2000). As the tongue contacts the prey, the center of the tongue pad is retracted at a higher rate than the upper or lower lobes, thus engulfing the prey item (Herrel et al., 2000). This invagination is caused by activity of the M. retractor pouch, which is active from just prior to prey contact through when the tongue is fully retracted into the mouth (Herrel et al., 2000, 2001b).

A combination of interactions is responsible for grasping the prey during tongue retraction. The first are surface phenomena, such as wet adhesion and interlocking, between the prey and the tongue, which are provided by serous and mucous secretions and a high concentration of filamentous papillae and plumose cells on the tongue pad (Schwenk, 1983, 2000; Bell, 1989; Herrel et al., 2000). The force of these adhesive properties is increased by the shape change of the tongue pad as it engulfs the prey, thus increasing the contact area with the prey and possibly reorienting the tongue's papillae (Herrel et al., 2000). Finally, more than two thirds of the total force grasping the prey is generated by suction resulting from negative pressure produced as the center of the tongue pad is retracted by the M. retractor pouch (Herrel et al., 2000).

Thermal Effects on Feeding Performance

Chameleons inhabit a broad range of environments (see Chapter 5), from deserts, where body temperature can reach over 39°C (Burrage, 1973), to alpine zones, where temperatures can drop below freezing (Reilly, 1982). Within this range of environments, some chameleon species are known to feed at remarkably low body temperatures (Burrage, 1973; Hebrard et al., 1982; Reilly, 1982; Bennett, 2004; Andrews, 2008), including as low as 3.5°C (Burrage, 1973) and at body temperatures below which sympatric lizard species are active (Hebrard et al., 1982).

Temperature, however, exhibits strong effects on a wide range of physiological processes, some of which can have strong impacts on whole-organism performance. Thermal effects on muscle physiology, for instance, can have a profound impact on the contractile rates, and thus locomotor capabilities, of organisms (Huey and Stevenson, 1979; Bennett, 1985; Huey and Bennett, 1987; Rome, 1990; Lutz and Rome, 1996; Herrel et al., 2007a). Because their body temperature is dictated by environmental conditions, ectotherms, like chameleons, are particularly vulnerable to the effect of low environmental temperatures.

While sprint speed in lizards, jump distance in frogs, and swimming speed in fish declines at least 33% with a 10°C decline in body temperature (Huey and Bennett, 1987; Rome, 1990), tongue-projection performance in chameleons declines significantly less (Anderson and Deban, 2010). In fact, as body temperature declines from 25°C to 15°C, peak velocity of tongue projection declines less than 11%, while peak power output declines only 19% (Fig. 4.6) (Anderson and Deban, 2010). Tongue retraction, on the other hand, exhibits a decline in peak velocity of more than 42% across the same temperature range, and a decline in peak power output of more than 64% (Fig. 4.6) (Anderson and Deban, 2010).

While tongue retraction is powered by muscle contraction alone, tongue projection is powered largely by elastic recoil of collagenous tissue that was preloaded by muscle contraction (de Groot and van Leeuwen, 2004). The observed difference in temperature effects is because muscle contractile rate properties are strongly affected by temperature (Bennett, 1985), while static contractile properties, like peak tension, are weakly affected by temperature (Bennett, 1985; Rome, 1990; Lutz and Rome, 1997), and elastic properties exhibit almost no effect of temperature (Rigby et al., 1959). As a result, the M. accelerator linguae is able to exert close to the same level of force during contraction to load the intralingual sheaths prior to the onset of tongue projection (Anderson and Deban, 2010). This loading occurs at a slower rate, but because the mechanism is preloaded, it is able to contract fully prior to the onset of tongue projection and the elastic elements then recoil at close to the same rate as they would at warmer temperatures once projection occurs (Anderson and Deban, 2010). Tongue retraction, on the other hand, is strongly affected by temperature as it relies directly on muscle contractile velocity (Anderson and Deban, 2010).

With an effective prey-prehension mechanism able to capture and hold large prey (Herrel et al., 2000) and a weak effect of temperature on the tension produced by muscle (Bennett, 1985; Rome, 1990; Lutz and Rome, 1997; Anderson and Deban, 2012), chameleons are at a

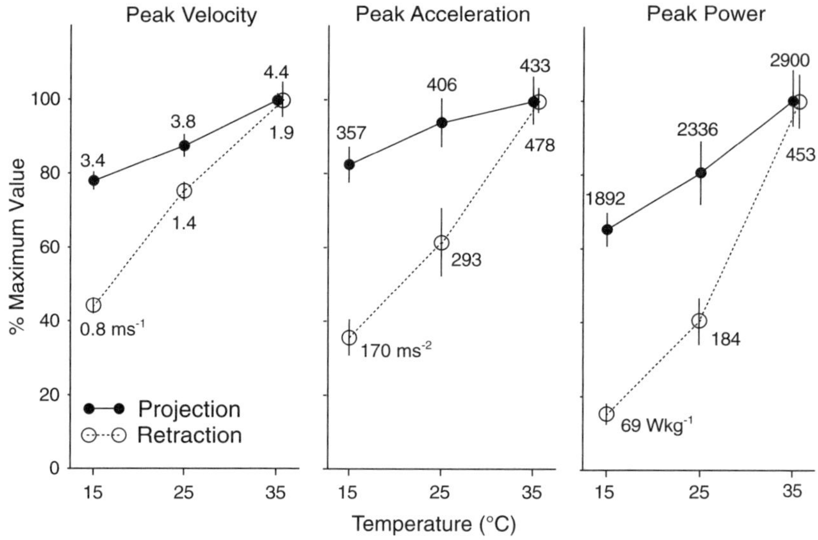

FIGURE 4.6. Performance parameters (mean ±SE) as a percent of maximum for tongue projection (open/dashed) and retraction (solid) in *Chamaeleo calyptratus* showing low thermal dependence of projection as compared with retraction. Absolute values of means are shown in native units. Modified from Anderson and Deban (2010).

limited risk of losing prey once it has been grasped by the tongue. Thus, low thermal dependence of tongue projection serves to increase the thermal breadth of their feeding mechanism (Anderson and Deban, 2010). This increased thermal breadth likely grants chameleons a temporally expanded thermal niche (Anderson and Deban, 2010) and allows them to feed early in the morning before behavioral thermoregulation can elevate their body temperature (Reilly, 1982) and sympatric lizard species are rendered inactive (Hebrard et al., 1982).

Future Directions for Chameleon Feeding Research

While it is now clear how each muscle is involved during the process of tongue projection and retraction, our understanding of the elastic energy storage component is extremely limited. Clearly an elastic power amplifier is involved in powering tongue projection, as is evident by the power output values exceeding that known to be produced by muscle contraction alone (de Groot and van Leeuwen, 2004), but the proposed mechanism has never been shown definitively.

Rectifying how the nested series of collagenous intralingual sheaths interact with each other is one issue that needs continued research. Because of their size, it is difficult to even determine accurately the exact number of sheaths present (de Groot and van Leeuwen, 2004). Further, because the more superficial sheaths, which are attached at their proximal end to the inner fascia of the M. accelerator linguae, are connected to each other by collagenous trabeculae, while the deeper sheaths, which are attached at their proximal end to the fascia of the M. hyoglossus, have no structural connection at their distal end (de Groot and van Leeuwen, 2004), it is unclear how different sheaths interact with each other, particularly those that are not connected by trabeculae.

Of particular interest is how energy stored in the intralingual sheaths is imparted on the tongue to power forward projection. Because the intralingual sheaths do not appear to be connected at their distal end to the entoglossal process, it is unclear how the recoil of these sheaths acts against the tongue skeleton to project the tongue forward. This is particularly confusing because the sheaths must be well enough lubricated so as to slide along each other and the entoglossal process without losing energy to friction, but at the same time, they must exert a force in the posterior direction in order to impart a forward force. Further, direct attachment to the entoglossal process would limit projectile distance because of the short length of each sheath.

Finally, direct measurement of the mechanical properties of the intralingual sheaths still needs to be done. Based on the weight of the sheaths, the energy needed to be stored in them in order to produce the observed energy output during tongue projection is calculated to be significantly lower than the observed capacity of other tendons under maximal stress (de Groot and van Leeuwen, 2004). This indicates that the intralingual sheaths should have the capacity to power tongue projection; however, direct measurement of their mechanical properties is needed to confirm this.

ACKNOWLEDGEMENTS

We thank Anthony Herrel, John Measey, and Lance McBrayer for extremely helpful comments on earlier drafts of this chapter.

FIGURE 1.1. *Trioceros goetzei* shows the dexterity of the chameleons' gripping hands and feet.

FIGURE 1.2. *Trioceros johnstoni* from Rwanda demonstrates the independently moving eyes typical of all chameleons.

FIGURE 1.3. The visual system of chameleons is highly tuned for judging distance and movement, as shown by the focused gaze of *Chamaeleo dilepis*.

FIGURE 1.4. *Rhampholeon viridis* from Tanzania is typical of chameleons, which are primarily terrestrial.

FIGURE 1.5. Most chameleons are cryptic and can blend in with the background. In an extreme example, *Trioceros laterispinis* resembles lichen on a branch.

FIGURE 1.6. A male *Bradypodion transvaalense* from South Africa shows a colorful display and distended ornamentation.

FIGURE 1.7. A male *Kinyongia tavetana* from Tanzania displays aggressive behavior.

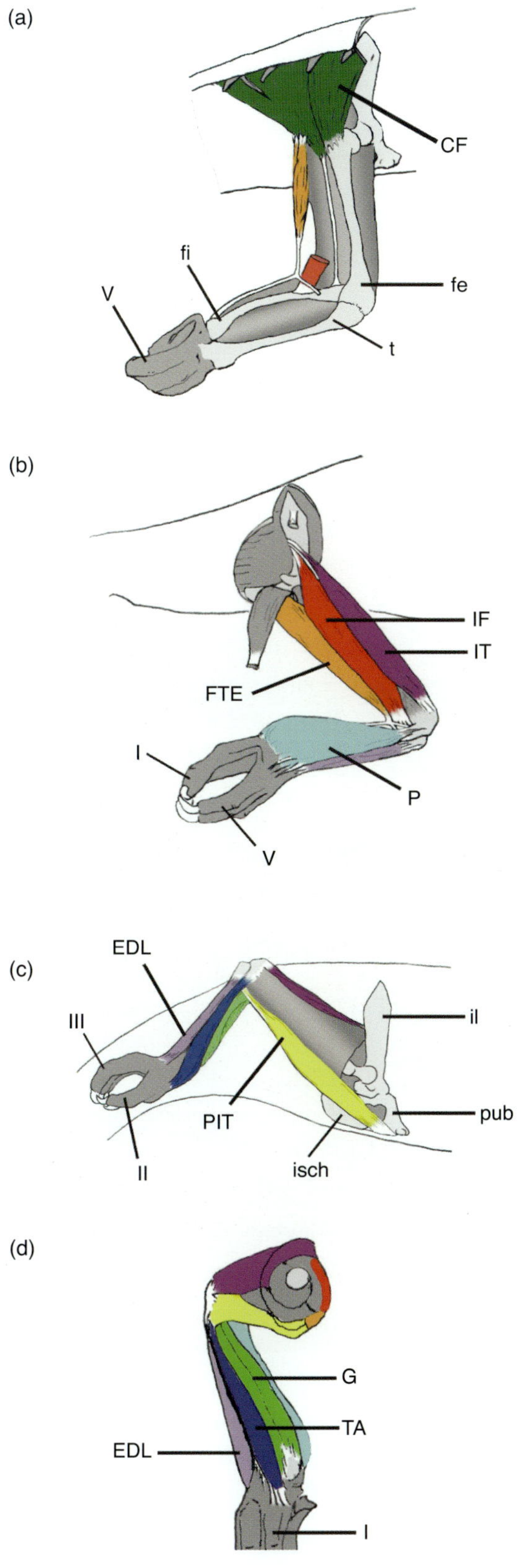

FIGURE 2.4. Right hindlimb muscles of *C. calyptratus* represented by a lateral view of the deeper musculature (a), a lateral view of a protracted and depressed limb (b), a fully retracted and abducted limb (c), and an anterior view of a fully retracted and disarticulated limb (d). Bones are in light gray and connective tissue is in white. Originally published in Higham and Jayne (2004a).

LABELS: I, II, III, V = digit numbers; CF = caudofemoralis; EDL = extensor digitorum longus; fe = femur; fi = fibula; FTE = flexor tibialis externus; G = gastrocnemius; IF = iliofibularis; il = ilium; IT = iliotibialis; isch = ischium; P = peroneus; PIT = puboischiotibialis; pub = pubis; t = tibia; TA = tibialis anterior.

FIGURE 5.1. Diverse body forms and habitats of chameleons from forest, heathland, and desert habitats. A typical forested habitat (a), inhabited by *Calumma amber,* in northern Madagascar (b). The leaf-litter and associated low vegetation of forests (c), provides the habitat for leaf chameleons such as *Rhampholeon acuminatus* in East Africa (d). Some members of the genus *Bradypodion* inhabit fynbos, a southern African heathland habitat (e), like *B. taeniabronchum* from South Africa (f). Chameleons also inhabit desert environments (g) such as this area in western Namibia where *Chamaeleo namaquensis* lives (h).

FIGURE 6.1. Male *Bradypodion transvaalense* in typical display posture with throat extended, body laterally compressed, one arm held rigidly with elbow bent at right angles, and tail coiled.

FIGURE 6.2. Copulation in *Trioceros jacksonii*, showing sexual dimorphism in ornamentation.

FIGURE 6.3. Female aggressive coloration and posture in *Bradypodion melanocephalum*. Females reject unwanted male courtship attempts by displaying highly contrasting coloration, gaping, and rocking rapidly from side to side. Unreceptive females will also chase and bite males that approach.

FIGURE 6.4. Male–male combat in *Trioceros jacksonii* involves locking together of the three rostral horns. Males often lock horns and push each other along a branch in a contest of strength. They may also violently twist their horns in corkscrew fashion, and the male with longer horns can inflict damage by piercing the tissue of his rival with the tips of his horns.

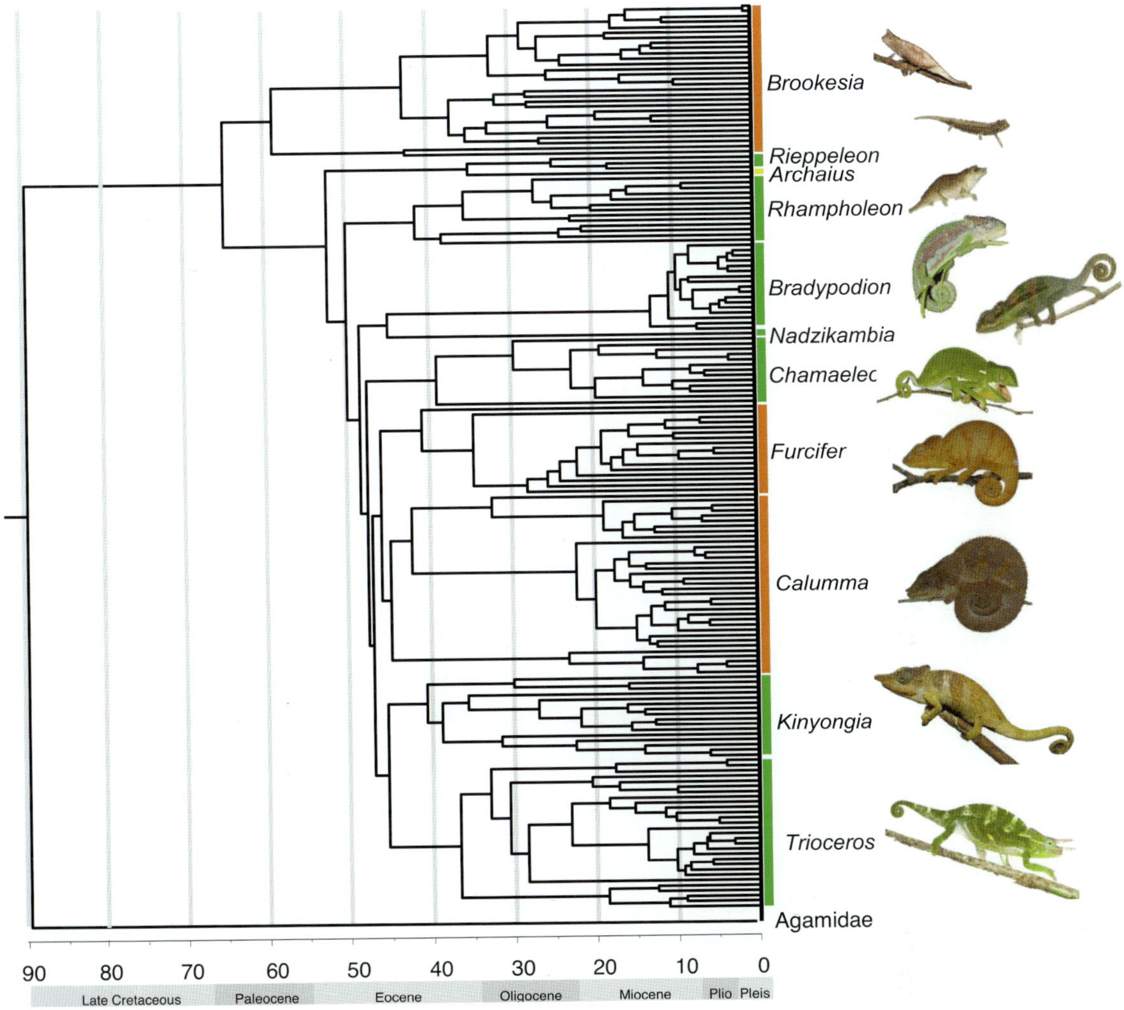

FIGURE 7.1. Schematic of the phylogenetic tree of chameleons (based on Tolley et al., 2013) shown as a chronogram. Each genus is color-coded according to its main distribution (Africa, green; Madagascar, orange), and a time scale (millions of years) is included.

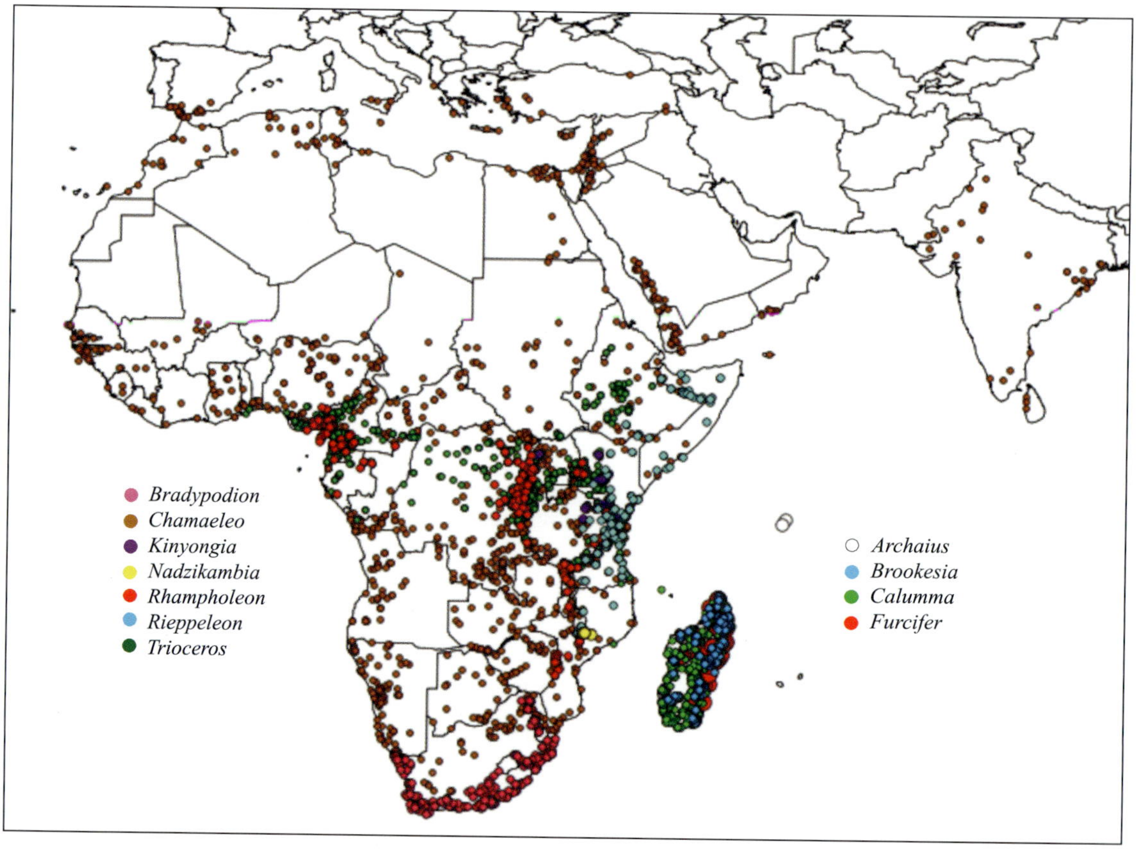

FIGURE 7.2. Point localities for all available georeferenced records of chameleons by genus, throughout Africa, Madagascar, Europe, and Asia. Records from introduced localities are not included.

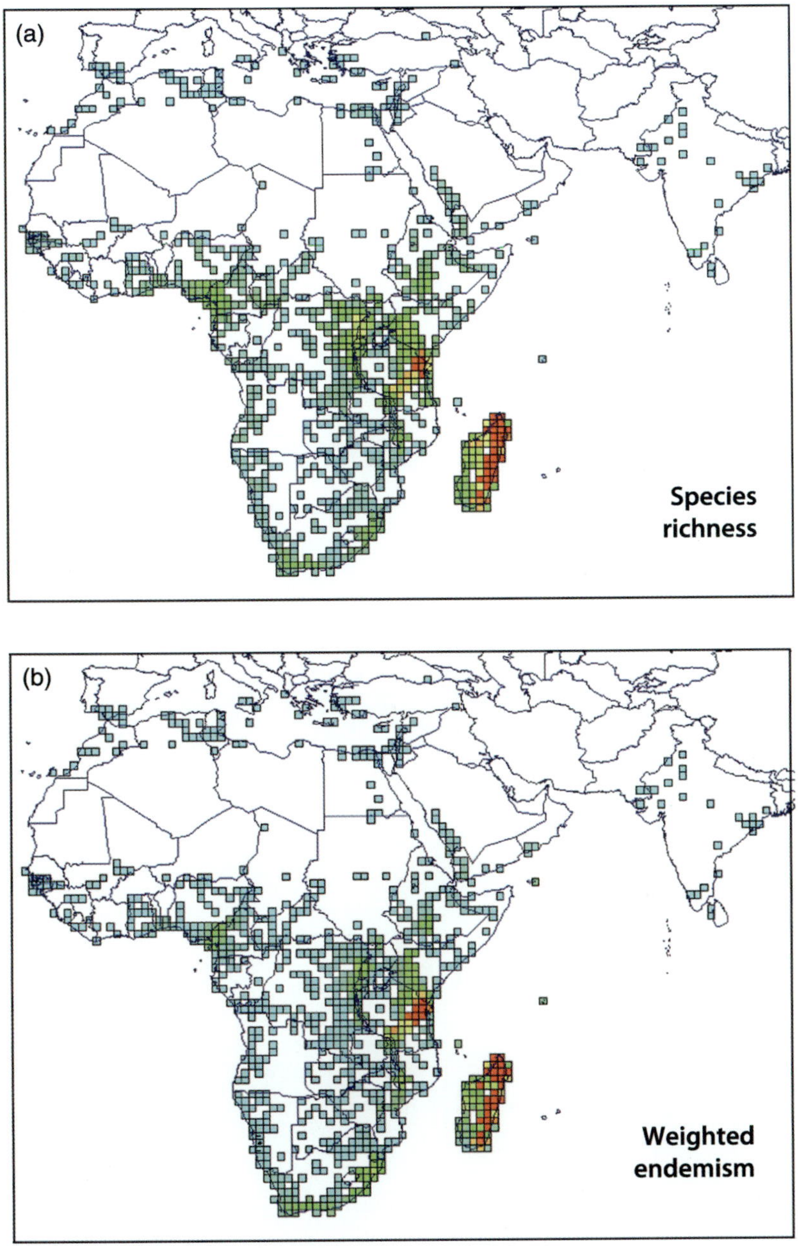

FIGURE 7.3. Patterns of (a) species richness and (b) weighted endemism for chameleons at a 1-degree resolution (Box 7.1) across the entire distribution of this family. Warmer-colored grid cells (reds) have the highest values, whereas cooler-colored cells (blues) have the lowest values.

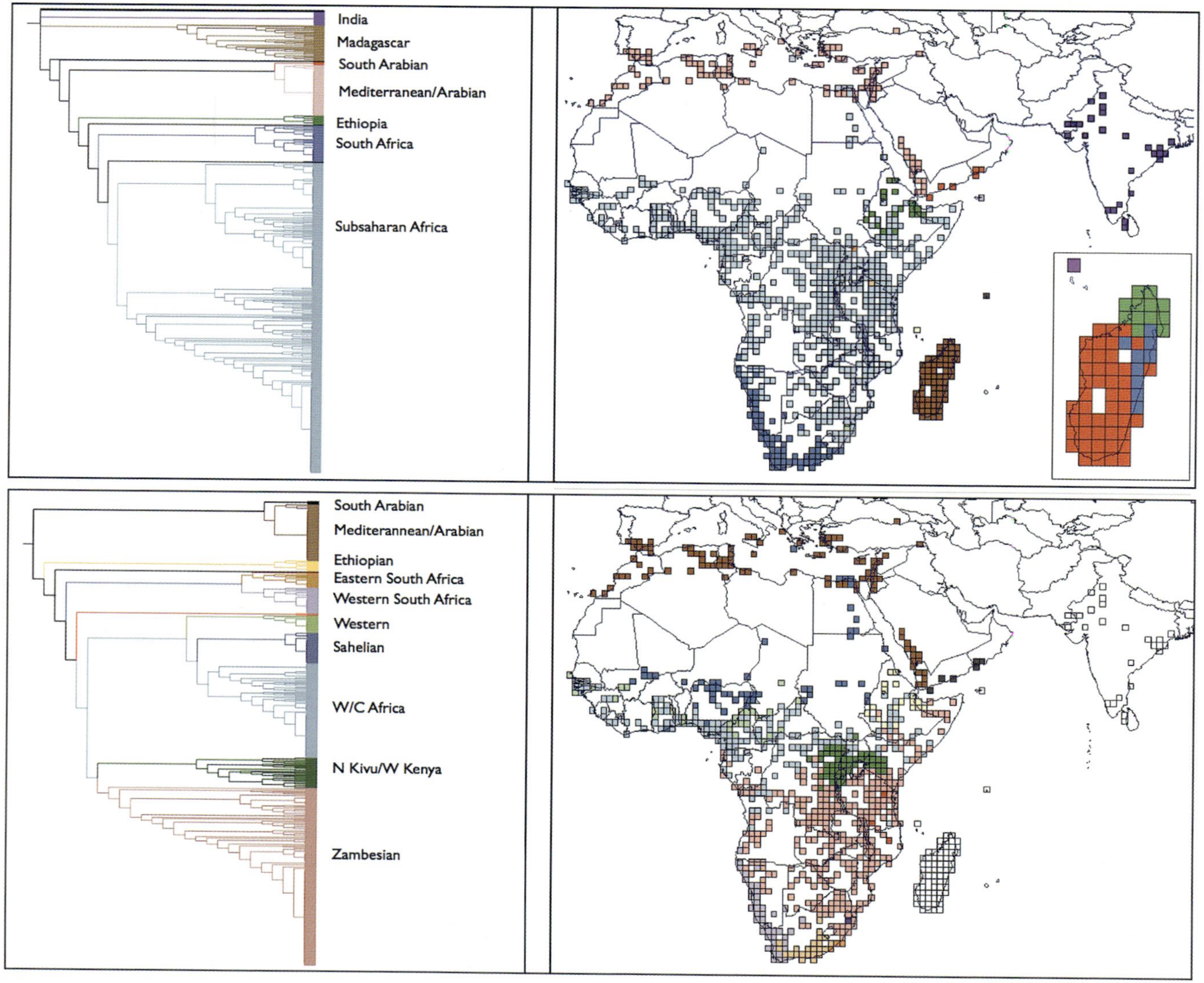

FIGURE 7.4. Clustering phenogram (Box 7.1) based on all available distribution records of chameleons: (top) including Madagascar and India (inset: analysis for Madagascar separately); (bottom) excluding Madagascar and India. The phenograms (left) show the main clusters, and these clusters are color-coded to match the biogeographic areas on the maps (right).

FIGURE 8.1. *Archaius tigris.*

FIGURE 8.2. *Bradypodion damaranum.*

FIGURE 8.3. *Brookesia superciliaris.*

FIGURE 8.4. *Calumma gastrotaenia.*

FIGURE 8.5. *Chamaeleo dilepis.*

FIGURE 8.6. *Furcifer lateralis.*

FIGURE 8.7. *Kinyongia fischeri.*

FIGURE 8.8. *Nadzikambia baylissi.*

FIGURE 8.9. *Rhampholeon platyceps.*

FIGURE 8.10. *Rieppeleon brevicaudatus.*

FIGURE 8.11. *Trioceros melleri.*

FIVE

Ecology and Life History of Chameleons

G. JOHN MEASEY, ACHILLE RASELIMANANA, and ANTHONY HERREL

Chameleons have been relatively neglected in terms of their ecology, perhaps in part, because of their cryptic nature. As detailed in this review, the majority of studies on chameleons *in situ* have been conducted relatively recently (during the past 15 years), and most of these center on the extremely diverse island of Madagascar. Although there are some data on chameleon ecology in southern Africa and Europe, mainland African chameleons represent a relatively overlooked group of lizards.

Chameleons display a unique set of morphological characteristics that set them apart from all other lizards, including gripping feet, independently moving eyes, a ballistic tongue and prehensile tail (Chapter 2), and it may be expected that these unique traits would also set them apart ecologically. While ecological data on some chameleons suggests that they adhere to many lizard generalities, the exceptions show that there is far more to learn about the ecology of chameleons.

Although some arboreal lizards are arranged into many specialist guilds (e.g., *Anolis* lizards), only two morphologically and taxonomically distinct guilds are recognized for chameleons: arboreal and ground-dwelling forms. Their cryptic behavior and camouflage continues to complicate their study, but new information on life-history traits is opening up insights into niche partitioning. Chameleon life-history traits exhibit many of their most unusual features. Like other lizards, chameleons lay eggs with late stage embryos and have viviparity in extremes of altitude and latitude. Unlike any other lizards, some chameleons lay eggs with early gastrula, which develop slowly and can undergo diapause in order for hatchlings to emerge during optimal seasons. Unlike any other tetrapod, one chameleon species (*Furcifer labordi*) is known to have an extreme annual life cycle, in which the population exists as developing eggs for 8 to 9 months of the year. However, like other lizards, many chameleons survive less favorable seasons by seeking out dry and stable conditions in which to aestivate.

Chameleons are lizards that do not conform to either sit-and-wait or active foraging strategies. Instead, it has been proposed that they have an unusual intermediary behavior, termed "cruise foraging." Most chameleons studied appear to be generalist opportunists, increasing their range of food sizes as they get larger, so that the largest include both vertebrate and invertebrate food items in their diet. Like other lizards, chameleons (especially xeric-adapted species) regularly ingest plant matter, presumably in order to supplement their water intake.

5.1 HABITAT

Existing data suggest that chameleon assemblages are divided according to habitat (Fig. 5.1 in the color insert), with the major division falling between open-canopy habitats (savanna, heathland, grassland, and woodland and closed-canopy habitats (forest). Species that frequent open habitats appear to be most tolerant of disturbance, while forest species are most often reported as being habitat specialists and thus restricted. Forest-dwelling chameleons appear to make up the majority of taxa. Around 132 species (ca. 67%) of chameleons are forest dwelling (Tilbury [2010] estimates that 70% of mainland African taxa are restricted to forest habitats). It appears most likely that the ancestor of all chameleons was a forest leaf-litter specialist from mainland Africa (Tolley et al., 2013; Chapter 7). Some chameleon taxa appear to be typically forest dwelling: *Brookesia, Kinyongia, Rhampholeon,* and *Calumma.* One genus is typically open habitat: *Chamaeleo,* with a last group that appears to have members occupying both forest and open habitats: *Bradypodion, Trioceros, Furcifer,* and *Rieppeleon.* A recent phylogenetic analysis of habitat use in chameleons concludes that nonforest taxa are ecologically derived (Tolley et al., 2013).

Forest Chameleons

Chameleon communities appear to reach peaks of species diversity within forested habitats (Fig. 5.1A,C). For example, up to eight species have been found at a single site in northern Madagascar (Brady and Griffiths, 2003); four species from Mount Manengouba and Mbulu Hills, Cameroon (Gonwouo et al., 2006), and eight species in the East Usambaras, Tanzania (Patrick et al., 2011). Note that these numbers for species that occur in sympatry are much lower than those reported from relatively small areas such as 20 species in a 1-degree square in northern Madagascar (Chapter 7). Diverse sympatric assemblages have given rise to the hypothesis that chameleons are arranged into the sorts of specialist guilds observed in *Anolis* lizards. Caribbean *Anolis* lizards have radiated repeatedly into different arboreal niches (twig, trunk, crown) and show associated morphological specializations. While the convergence of these ecomorphs in the different islands of the Greater Antilles has become a textbook case for adaptive radiations, we are still largely ignorant of how forest chameleons may partition their niches beyond the division (both taxonomic and functional) of leaf-litter and arboreal guilds (but see Townsend et al., 2011 b). In addition, much of the data collected is based on sleeping (roosting) animals, as chameleons are cryptic and difficult to observe during the day. Consequently, detailed investigations exploring the relationship between daytime and nighttime substrate use are required in order to properly interpret these data.

Density

Chameleon densities have been measured in a number of studies and often involve a comparison of sympatric species within and between habitats based on roosting data. Densities of some species get particularly high, exceeding 100 chameleons ha^{-1}, but other, sympatric species (recorded in the same surveys and therefore with presumably the same accuracy) can have particularly low densities, with <1 chameleon ha^{-1}. For example, Karsten and colleagues (2009b) found *Furcifer verrucosus* at high densities (97.7 ha^{-1}; 95% confidence interval [CI], 60.2 to 158.6) in the arid southwest of Madagascar, but *F. antimena* in the same surveys reached only a fifth of this density (17.0 ha^{-1}; 95% CI, 9.3 to 30.9). In the central high plateau of Madagascar, Randrianantoandro et al. (2009) estimated a density of 39.7 ha^{-1} for *Calumma crypticum*, 27.3 ha^{-1} for *F. lateralis*, and 16.4 ha^{-1} for *F. minor* at the same site. Differences in density within a community of *Brookesia* have been observed in the western region of Madagascar, where ubiquitous species most tolerant to habitat disturbance are the most abundant (Randrianantoandro et al., 2007b). Raxworthy (1991) searched forests at mid and low altitudes in northwest Madagascar during the day, finding *Brookesia stumpffi* to be more than 10 times more abundant than three other *Brookesia* species. In dry deciduous forests of northern Madagascar, *B. stumpffi* can reach exceptionally high densities of nearly 150 ha^{-1}, more than twice as high as sympatric *F. petteri* and *F. oustaleti* (Lowin, 2012). In summer surveys in southern Madagascar, *B. nasus* was found at 37 ha^{-1} but *C. malthe* in the same area reached only 10 ha^{-1} while *C. oshaughnessyi* was at <1 ha^{-1} (Brady and Griffiths, 2003). By winter, the same authors had found clear changes in the densities of many species, and this change appeared to relate to body size groups. It would, however, be interesting to test this hypothesis at other sites and with other species assemblages.

Some changes in densities between seasons are directly attributable to the life history of the species concerned, especially where seasonal influences are strong. In such cases, chameleons may aestivate or even die (see below and Box 5.1). However, the degree to which densities vary between seasons, as well as between years, in the same habitat (e.g. Brady and Griffiths, 1999, 2003; and see below) is likely to depend on a host of biotic (e.g. predation) and abiotic (e.g. precipitation, fire) factors, none of which have been investigated using sufficiently extensive time scales.

Vertical Distribution

Most information on vertical distributions of chameleons in forests consists of data from roosting sites when chameleons are found at night using torchlight searches. Roosting sites may be selected for their size (e.g., branch diameter) and proximity to other chameleons (e.g., Randrianantoandro et al., 2007a). Anecdotal observations suggest that chameleons

BOX 5.1 Life History of *Furcifer Labordi*: An Annual Chameleon

Chameleons that inhabit highly seasonal climates are likely to synchronize their life histories with the most productive seasons. One extreme example of this is *Furcifer labordi,* which inhabits the arid southwest of Madagascar; a region that has a distinct biphasic annual climate of low temperatures and low rainfall (April to October) and high temperatures and high rainfall (November to March). During their study in the cold dry season, Karsten et al. (2008) did not find chameleons (*F. labordi, F. verrucosus*), although adult *F. labordi* have been seen in the field from October until early April (A. Raselimanana, personal observation). At the onset of the warm, rainy season, there is synchronous hatching of eggs of both species. During this relatively short (5 months) active season one of these species (*F. labordi*) undergoes juvenile growth, maturation, courtship, and death, leaving only eggs to continue through the next generation (Karsten et al., 2008). However, a captive male overwintered without taking food from June until October (A. Raselimanana, personal observation), and it may be that future studies will reveal survival of a small number of adults.

In their study of *F. labordi*, Karsten et al. (2008) documented synchronous hatching and rapid daily growth rates of juveniles wof nearly 2% snout–vent length (mass: 4% for males, 2% for females) for fewer than 60 days (November to January). At this time, growth ceased, and snout–vent length even reduced in some individuals after this time. During the next 30 days, most females in the population became gravid and laid eggs (February). Once egg-laying ceased, adults quickly disappeared from the area in as little as 2 weeks such that by early March, no adults could be found. Over the next 3 months (April to July) the eggs were thought to be in torpor, as temperatures in this region plummet, with embryonic development occurring only from August through November. The onset of rains and the concurrent rise in temperature signal the synchronous hatching of the next cohort, with few to no adults remaining in the environment. While no other chameleon is currently known to have such an extreme life history as that of *F. labordi,* very few data exist on the majority of species.

change roost sites, as well as the position on a perch, depending on weather conditions; for example, rain, high winds, and cold temperatures result in roosting sites that involve inner branches or areas under leaf cover (Raselimanana and Rakotomalala, 2003; Raholdina, 2012). Some authors suggest that roost sites are a limited resource and as such are vigorously defended (cf. Burrage, 1973). Concordantly, most chameleons are found roosting alone, although in some leaf species, males and females have been found to roost in pairs during the mating season (Wild, 1994; A. Raselimanana, personal observation). Some leaf chameleons roost vertically on stems with their head orientated upward, and some stay in mate-guarding position overnight during the breeding season (*Brookesia exarmata, B. minima, B. nasus,* and *B. ramanantsoai;* Glaw and Vences, 2007; A. Raselimanana, personal observation). Mate guarding, using roosts in close proximity, appears to be relatively common in chameleons (Toxopeus et al., 1988; Cuadrado, 2001; Chapter 6). Most authors agree that roost sites are selected to minimize nocturnal predation, especially from snakes (see below). Newly hatched or newborn chameleons also roost close to each other at the extremity of leaves, vines, or small branches. Perch diameters are not random with respect to available perches (e.g. Razafimahatratra et al., 2008), and their selection is likely related to hand and foot size because of the relationship between the latter and gripping performance (cf. Herrel et al., 2011; Chapter 4). This implies that maintaining a good grip on a perch during the night is an important roosting requirement. Perch sites for arboreal species are typically on isolated branches or leaves (often at the distal tip), presumably as the smallest branches are less likely to support the weight of many predators, and/or so that potential predators that do approach provide vibrations that give the chameleon advance warning. While there are no empirical observations to back up these inferences, some chameleons readily drop from perches if disturbed (see below). The importance of the selection of nocturnal roosting sites may have played a fundamental role in the evolution of arboreality in chameleons. Roosting most likely evolved in the chameleon ancestor to reduce predation by nocturnal ground-dwelling predators.

We do know that roosting site is disassociated from foraging habitat in leaf chameleons, which hunt in the leaf litter during the day and perch in low vegetation at night. Moreover, arboreal species move from their nighttime perches during the day, but there are few data on daytime foraging areas. One study in which arboreal forest chameleons (*Trioceros oweni, Calumma gracilis,* and *T. cristatus*) were followed moving from their roost to foraging areas suggested that all species moved up from roost sites toward the highest branches by midday and returned to lower roosts in the evening (Akani et al., 2001). In this study, the majority of feeding took place midmorning (from 9 to 11 AM) and in the evening (from 3 to 5 PM) for all species (Akani et al., 2001). These authors attribute the inactive period during the middle of the day to a time when chameleons escape from the heat (although this contradicts data on forest temperatures, see Gehring et al., 2008) and avoid predation by hiding behind large leaves. In a study of captive chameleons in a large tropical house, Gehring et al. (2008) found that radio-tracked *Furcifer pardalis* spent most of their time in the top third and on the periphery of available trees (see below), descending only to make longer lateral movements. Although it has been asserted that roost sites are good indicators of foraging locations (Carpenter and Robson, 2005), there appear to be no data on the majority of species to back up this claim.

Leaf chameleons (e.g., Fig. 5.1D) are known to forage on the forest floor, but also catch insects from low vegetation (e.g., Raxworthy, 1991; see below). Moreover, these chameleons spend the night roosting in low vegetation. In the East Usambara Mountains, leaf chameleons (*Rhampholeon temporalis*) roost close to the ground (mean $\pm$ SD, 0.60 $\pm$ 0.45 m), while larger (typically arboreal) species roosted from 2.0 to 4.5 m high (Patrick et al., 2011). In Cameroon, most *R. spectrum* were found roosting below 1 m, but some individuals were found as high as 2 m (Wild, 1994). Broadley and Blake (1979) report roosting sites up to 4.5 m for *R. marshalli,* and they suggested that this may be indicative of a difference in predators in the eastern highlands of Zimbabwe. Sympatric species may differ in the precise choice of roost site, both by vegetation type, substrate diameter, and roost height (Carpenter and Robson, 2005; Randrianantoandro et al., 2007a; Herrel et al., 2011; Patrick et al., 2011). In Madagascar, sympatric leaf chameleons show different roosting heights, with hatchlings and juveniles being found lower than adults (see Razafimahatratra et al., 2008). However, even species such as *Brookesia stumpffi,* which has relatively high roosts for a leaf chameleon (mean height, 0.43 m), still roost 2 m lower than sympatric *Furcifer angeli* (Carpenter and Robson, 2005). In two studies of leaf chameleons, no significant differences between sexes were found for roost height (Randrianantoandro et al., 2007a; Razafimahatratra et al., 2008). However, male *F. pardalis* were found to roost higher than females (Andreone et al., 2005). As we now know that gripping strength is related to perch choice (Herrel et al., 2011), future studies will need to disentangle observed perch differences in ontogeny, sex, and species from morphological constraints.

Only one member of the genus *Brookesia* (*B. ebenaui*) is always found in trees during the day, roosting between 2 to 3 m high (Glaw and Vences, 2007). Conversely, one of the large forest species from the arboreal genus *Trioceros* (*T. cristatus*) appears to inhabit the leaf litter on the forest floor or low-lying shrubs (Akani et al., 2001). Data on diet suggest that this species may rely on large prey not available to smaller sympatric leaf chameleons (*Rhampholeon spectrum*), and others have commented that stomach contents contained mainly terrestrial insects (Klaver and Böhme, 1992). *Trioceros cristatus* also has a relatively short tail, which is typically associated with ground dwelling (Klaver and Böhme, 1992; Boistel et al., 2010).

Few searches have been made specifically in forest canopies, and suggestions of partitioned use of this habitat type are based mostly on anecdotal observations and thus remain speculative. For example, large species have been suggested to be canopy specialists (Raxworthy, 1988; Nečas, 2004), yet quantitative data are lacking. Large species were rarely observed in the forest canopy of Nosy Be, Madagascar, and only a single *Furcifer pardalis* was found in a 10-day canopy search, but otherwise it appeared to be absent from the forest (Andreone et al., 2005). Parcher (1974) reported that *F. willsii* inhabits only the upper canopy, as few animals were found roosting with sympatric species but five adults were found in the upper canopy. Similarly, in the Anjozorobe Forest in the central high plateau of Madagascar, *F. willsii* was the only species recorded from the upper canopy area (A. Raselimanana, personal observation).

The available data are not sufficient to confirm or reject the existence of more than a single partition of forest chameleons into arboreal and leaf-litter guilds. Whether or not more complex partitioning exists, and existence of specializations toward the use of the forest canopy or understory niches, remains a challenge for future studies of these lizards.

Horizontal Distribution

Within forests, chameleons have been said to favor tree falls or ecotones between forest and adjacent habitat (e.g., Metcalf et al., 2005; Reisinger et al., 2006). While it is certainly true that forest chameleons can be found in gaps and on edges, this may be because the forest canopy is mostly unavailable to human observation. Therefore, the important caveat for observations made from ground surveys for chameleons (which make up the majority of studies) is that chameleons may remain unobserved in the canopy and understories.

Several authors have investigated edge effects on the abundance of chameleons in forests. The periphery of the forest appears to be an environment where both forests and non-forest species co-occur (e.g., Patrick et al., 2011). These (often unnatural) ecotones might be attractive to chameleons for a number of reasons, including increases in prey diversity and abundance, basking opportunities, visibility for intraspecific communication, and vegetation structure. Natural edge effects, such as those produced by tree falls or along streams and rivers, are present in many forests, and there is evidence to suggest that these areas also have an increased abundance of chameleons. Gaps from tree falls seem to be important, with more *Bradypodion caeruleogula* roosting in them than on forest edges or in the forest interior (Reisinger et al., 2006). *Furcifer pardalis,* which normally does not enter forest, has been found within forest along river transects (Andreone et al., 2005; Raselimanana, 2008), and *Calumma* were more abundant in riparian vegetation while for *Brookesia,* the converse was true (Andreone et al., 2005). However, riparian habitat had equal or higher densities of all of chameleon species (including *Brookesia* species) in other surveys (Jenkins et al., 2003; Rabearivony, 2012). Yet, it is not clear that it is the river or the canopy opening around which chameleons cluster. For example, *Rhampholeon spectrum* was particularly abundant in riparian forest vegetation, independent of whether streams were running or dry (Wild, 1994). Metcalf et al. (2005) investigated the edge effect from forest paths, which represent a relatively small canopy opening. They found a significant decrease in abundance away from paths for two species of *Furcifer* (*F. oustaleti* and *F. rhinoceratus*). In addition, there is a suggestion that some chameleons may migrate to riparian vegetation during the dry season (Brady and Griffiths, 1999; Rabearivony et al., 2007).

Because many forest species are restricted to forest patches, discontinuation of habitat does not allow migration between populations. This appears to be the most common cause of disjunct distributions in East Africa (e.g., Measey and Tolley, 2011), and it may also contribute to speciation (Tolley et al., 2011; Chapter 7). Exceptions also occur; for example species like *Furcifer lateralis* and *F. oustaleti* are regularly found walking across open ground from one forest patch to another (A. Raselimanana, personal observation). Where habitat is continuous, there appears to be a distinct altitudinal turnover of some species, while others inhabit a wide range of altitudes. In West Africa, *Rhampholeon spectrum* is found from 500 to 1900 m, passing through discontinuous distributions of 10 species of *Chamaeleo* and *Trioceros* (Wild, 1994; Akani et al., 2001), while in East Africa, altitudinal partitioning is reported between *R. temporalis* and *R. brevicaudatus* (Emmett, 2004). Luiselli (2006) conducted simulations based on survey data in West Africa, which suggested that chameleons in lowland forests are distributed according to food niche resources rather than spatial niche resources, while the opposite was true for montane species. In Madagascar, *Calumma* and *Furcifer* exhibit a distribution structure along altitudinal gradients (Nussbaum et al., 1999; Raselimanana et al., 2000), but this seems particularly pronounced in species of the genus *Brookesia* (Raxworthy and Nussbaum, 1995; see Andrews, 2008). Species assemblages in general can change in structure even within contiguous blocks of undisturbed forest (Brady et al., 1996; Brady and Griffiths, 1999; Jenkins et al., 1999; Rabearivony, 1999). Landscape features (e.g., rivers) are often suggested to be barriers (see Chapter 7), but as Raselimanana and Rakotomalala (2003) point out, chameleons are likely to be able to traverse even large rivers when these are bridged by fallen trees. Further research is required to help define the ecological niches of most chameleons, especially in relation to species turnover in continuous habitat.

Disturbance

Disturbed forests lack certain chameleon species while other species appear more tolerant or apparently even thrive after disturbance (e.g., Rabearivony et al., 2007; Irwin et al., 2010). Regardless, leaf chameleons generally seem to be negatively affected by habitat disturbance, which results in a reduced abundance of their leaf-litter habitat. *Brookesia minima,* for example, was absent from highly disturbed habitats (Jenkins et al., 2003). In another study on *Brookesia* distributions, abundance was generally higher at undisturbed sites (*B. thieli,* 58 ha^{-1}; *B. minima,* 7 ha^{-1}) as compared with burnt sites (*B. thieli,* 20 ha^{-1}; *B. minima,* 0.0 ha^{-1}) (Rabearivony, 1999). Small patches of forest appear to be able to maintain individuals (e.g., Wild, 1994), although it is not known whether such populations are sustainable in the long term. Nevertheless, it is also worth noting that some species, like *B. stumpffi,* reach high densities in plantations (e.g., of coffee and cacao; F. Glaw, personal communication).

A significant interaction between habitat type (high-disturbance, low-disturbance, and riparian) and the presence of *Calumma* spp. was found in eastern Madagascar (Jenkins et al., 2003). High-disturbance habitats were found to have a negative effect on all species in that study. However, at least some species, such as *C. brevicorne* and *C. nasutum,* appear to increase in density whenever the habitat is disturbed (Brady and Griffiths, 1999, 2003). Densities of *Furcifer pardalis* are much higher along roads or transformed habitats, and this species is absent from pristine or closed forest (Andreone et al., 2005; Rabearivony et al., 2007). Brady and Griffiths (2003) found low densities of *C. brevicorne* in undisturbed forest at several sites, but high densities along the forest edge. At a high-altitude site in central Madagascar, *Furcifer* species dominated in open heathland and agricultural lands, while *Calumma* dominated in humid forest (Randrianantoandro et al., 2010). While exceptions do occur, these mostly consist of individuals of *Furcifer* species being found at the forest periphery. It has been speculated that highly disturbed habitats represent a sink for some chameleon populations, being maintained by dispersing individuals or newly hatched juveniles from adjoining forest (Jenkins et al., 2003).

One consistent feature of disturbed forest habitats is that species with a preference for open-canopy habitats are quick to move in. A difference in distribution of seven species was found during transect walks in the East Usambara Mountains (Patrick et al., 2011). Interestingly, the sample contained the typical savanna species *Calumma dilepis,* which was occasionally found on the forest edge, but absent from within the forest. Hebrard and Madsen (1984) also report the presence of *C. dilepis,* becoming sympatric with forest species in deforested areas. Other species such as *Kinyongia matschiei* and *Rhampholeon temporalis* were found only within the forest (Patrick et al., 2011). In West Africa, *C. gracilis* moved into disturbed forest patches that were previously inhabited by *Trioceros* spp. (Akani et al., 2001).

Nonforest Chameleons

While the majority of chameleon species occur in forested habitats, a smaller group has radiated into habitats that have been broadly classified as: (i) bushes and heathland scrub, grassland (Fig. 5.1E), (ii) open canopy dry forest and savanna, and (iii) desert (Fig. 5.1F). The movement out of forests to open-canopy habitats occurred in multiple lineages of *Chamaeleo* and *Furcifer* in the early Miocene, while the transition into grassland and heathlands in *Bradypodion* and *Trioceros* occurred much later, at the start of the Pliocene (Tolley et al., 2013). The unifying feature of all of these nonforested or open habitats is that the chameleons have increased exposure, as this environment is less buffered from solar radiation, with a notable decrease in humidity. These chameleons (together with those from the dry forests of southwest Madagascar) have had to undergo physiological adaptations to deal with water stress as well as behavioral and morphological adaptations to cope with new substrate types. Living in a more open habitat, these chameleons may be more visible to potential predators (Herrel et al., 2013; Chapter 3); consequently, the most brightly colored and highly ornamented species are typically forest dwellers (e.g., Fig. 5.1B) and are frequently missing from open habitats, although there are exceptions. Unlike forest chameleons, open-habitat species are typically allopatric, with occasional range overlaps and sympatry of arboreal and leaf forms (e.g., *Rieppeleon* and *Chamaeleo* in East African savannas). The vegetation types they inhabit are also vulnerable to fire, which is able to impose dramatic reductions in chameleon populations, but also results in a changed landscape for any individuals that survive the fire. A total of 47 chameleon species (26% of all chameleons, not including morphs or species for which habitat is not known) are recognized as inhabiting such open habitats.

Like chameleons in forests, open-habitat species are often bounded by the vegetation types that they inhabit. This implies that in continental Africa, where open habitats are now extensive, these species are characteristically wide ranging. However, some species inhabit a restricted range of vegetation; well-documented examples of this occur in southern Africa, with ecomorphs adapted to open habitats occurring in the genus *Bradypodion* (Measey et al., 2009; Herrel et al., 2011; Hopkins and Tolley, 2011). *Bradypodion* ecomorphs adapted to open habitats are generally smaller, have less bright coloration, and have smaller ornaments than their sister taxa living in forest or woodland (Tolley and Burger, 2007; Stuart-Fox and Moussalli, 2008) (Fig. 5.1F). The history of the radiation of certain lineages of this genus of dwarf chameleons out of forest can be found in Box 5.2.

Mountaintops often emerge out of the forest zone and are frequently dominated by heathland scrub. Some chameleons that enter this habitat appear to have speciated there. The genus *Calumma* is mostly forest dwelling, but three species *C. vatosoa*, *C. peyrierasi*, and *C. jejy* all occur in shrubland adjacent to forest. *C. peyrierasi* also seems to be in the high-altitude heathland above the forest belt. Raholdina (2012) found higher densities (42.7 ha^{-1}) of *Furcifer*

BOX 5.2 *Bradypodion* Radiation out of Forests

All chameleons of the genus *Bradypodion* were once considered to be a single species (*B. pumilum*), with a bewildering array of forms (Hillenius, 1959; Mertens, 1966), although now the genus is considered to be composed of 17 species. These chameleons are allopatrically distributed from the coast to roughly 300 km inland in southern Africa, an area that passes through seven vegetation biomes (Branch, 1988; Tolley et al., 2004; Tolley and Burger, 2007). Today, arid biomes dominate the interior of the subcontinent (savanna, grassland, and Nama Karoo) following a process of aridification that occurred in two pulses of substantial uplifting of the east coast: 250 m and 900 m at 20 Mya and 5 Mya, respectively (McCarthy and Rubidge, 2005). These geological processes saw the end to much of the dominant forested vegetation in the area with aridification of the interior and a concurrent diversification of the ancestors of the chameleons.

The presence of a relatively large number of fairly recent lineages of dwarf chameleons in southern Africa, their mixture of preferred habitats in a number of biomes, and the existence of a robust phylogeny (Tolley et al., 2004, 2006, 2013), has allowed workers to investigate the evolutionary consequences of movement of these species from ancestral forested areas to biomes with open habitat types. In order to interpret these studies, it has been critical to determine the direction of this evolutionary radiation, which has been done by optimizing habitat on a phylogeny of the genus, confirming that the historical habitat for the most recent common ancestor was forest (Tolley et al., 2008).

Moreover, repeated radiations out of the forests of southern Africa into savanna (*B. thamnobates, B. transvaalense*), grassland (*B. melanocephalum*), and fynbos (ecomorphs of *B. pumilum, B. damaranum, B. gutturale*) took place. Reconstruction of ancestral vegetation for these lineages shows radiations into open habitat, corresponding with most dramatic uplifting 5 Mya (Tolley et al., 2008), a period that also corresponds to rapid diversification of the fynbos biome (Fig.5.1e).

The movement from closed to open habitat carries with it an increased exposure to predators (e.g., Stuart-Fox et al., 2006a; Stuart-Fox and Moussalli, 2008). Consequently, *Bradypodion* show a clear shift from bright to dull colors corresponding to the radiation into more open habitats (Branch, 1988; Stuart-Fox and Moussalli, 2007; Measey et al., 2009; Hopkins and Tolley, 2011). However, a study that investigated whether this change in habitat related to a change in camouflage found no evidence for increased crypsis in open habitat species; instead, *Bradypodion* inhabiting forests were found to exhibit greater changes in color associated with increasing contrast against background for social signaling (Stuart-Fox and Moussalli, 2008).

Morphologically, forest lineages are typically larger-bodied species with long tails, high casques, and long hands and feet, while those in open habitats are smaller, have lower casques, and have shorter hands and feet (Branch, 1998; Stuart-Fox and Moussalli, 2007; Measey et al., 2009, Tolley and Hopkins, 2011). There are some data (for *B. pumilum*) to suggest that these changes are adaptive, such that long tails can be used to increase grip force on wider perches (Herrel et al., 2011). However, an increased bite-force performance in open-habitat *B. pumilum* could not be attributed to a change in diet (Measey et al., 2011), but was instead postulated to be due to a reduction in ability to settle disputes by signaling.

campani in the central plateau highland of Madagascar associated with ericoid vegetation, as compared with lower densities (28.6 ha^{-1}) in other areas. In mainland Africa, *Trioceros kinetensis, T. rudis, T. hoehnelii, T. affinis, T. bitaeniatus, T. goetzei, T. harennae, T. nyirit,* and *T. schubotzi* all inhabit similar shrubby habitats at high elevations. The species of the *bitaeniatus* group of *Trioceros* are also noteworthy, as they all exhibit viviparity in addition to inhabiting regions up to and above 3000 m. *Kinyongia* are usually strongly associated with forest, although there are a few notable exceptions which suggest some Kinyongia species have flexibility. For example, *K. gyrolepis* has been found in shrubby habitats at high altitudes (Greenbaum et al., 2012), and *K. boehmei* is sometimes found in ericaceous habitat above cloud forests, in low-intensity agricultural areas, and into the alien vegetated shrubland below the forest limits in the Taita Hills (G.J. Measey, personal observation). This suggests that even forest-dwelling species may occasionally move into adjacent non-forested habitats when conditions are favourable, although there to appear to be limits. For example, despite being able to use vegetation adjacent to primary forest, *K. boehmei* are separated into distinct populations associated with forests that are separated by only a few kilometers of savanna (Measey and Tolley, 2011).

Forest ecotones appear to be strongly associated with an increased diversity of chameleons (see above), and it appears that many species utilize disturbed habitat outside of forests. Such habitat types do arise naturally; for example, hurricanes or cyclones can remove large sections of forest. Forest can also be impacted by landslides and mudslides, so perhaps it is not surprising that some species appear to be specially adapted to the fringe areas and disturbed forest (see above for examples). Broad habitat distributions are considered to result in different life histories for chameleons in Madagasar, where large sexually dimorphic species (*Furcifer*) require more open habitat for basking and intraspecific communication (Andreone et al., 2005). Many of the species that inhabit the forest edge ecotone are some of the largest, brightest species (e.g., *F. pardalis, Trioceros jacksonii*). Temperature and humidity variation should increase at the forest-edge ecotone, and many species that can tolerate this also have wide distributions, suggesting that they are also capable of migrating over larger distances. Outside the forest, different species also appear to have different vegetation height preferences. Lin and Nelson (1980) found that sympatric *T. hoehnelii* and *T. jacksonii* inhabited different vegetation within their study area: *T. hoehnelii* were found predominantly in areas of open secondary scrub, rarely above 2 m, while *T. jacksonii* were mostly in trees above 2 m.

Savanna species have the largest distribution of all chameleons, in part because the savanna now covers large areas of continental Africa (see Chapter 7). The open nature of the habitat makes them more visible, and one of the few studies of chameleon behavior during the day comes from a common sub-Saharan species, *Chamaeleo dilepis*. Unlike other chameleons *C. dilepis* is usually found alone (Toxopeus et al., 1988; see below), and a behavioral study found divergent microhabitats between the sexes. Hebrard and Madsen (1984) investigated diel perches and observed sexual differences in perch height and perch type. During the dry season, males chose higher perches (2.8 m, vs. 1.5 m for females) without leaves (but no measures of roosting sites were made). Higher male perch heights have previously been associated with the sexual displays of male lizards (Andrews, 1971). However, in the rainy season no sexual differences in habitat use were found, and males were found to move significantly further than females (males, 4 to 17 m per day; females, 1 to 3 m) (Hebrard and Madsen 1984). It is also worth noting here that male *C. chamaeleon* are thought to spend more time on the ground and that females perch lower in trees during summer (Pleguezuelos et al., 1999). Savanna species appear to be tolerant of disturbance and regularly take up residence in gardens. The majority of species of the genus *Chamaeleo* are not forest dwelling (only two species seem to inhabit wet forest: *C. necasi* in the coastal forest of the Dahomey Gap, West Africa and *C. zeylanicus* in India and Sri Lanka).

In addition to the radiation of large arboreal chameleons into the savannas, some members of one genus of leaf chameleons—*Rieppeleon*—appear to have moved out of the rainforest into adjacent savanna (Matthee et al., 2004). These small chameleons live in low bushes and grasslands and can frequently be found walking on the ground. Savannas have large areas of grassland where all these chameleons can be found, frequently sitting in isolated bushes. Madagascar has relatively small natural savanna areas, but they are inhabited with chameleons such as *Furcifer lateralis* and *F. campani*.

Another well-studied chameleon inhabits scrubland areas of southern Europe; *Chamaeleo chamaeleon* were found to select habitats with south to southwesterly facing slopes in southern Spain to take advantage of the increased radiance (Hódar et al., 2000). Moreover, these chameleons were found to favor trees with increasing density above 1 m and to avoid shrubs and dead trees.

There appear to be several evolutionary radiations of chameleons into arid habitats, and specific adaptations are associated with these. The first radiation was into seasonal forests that undergo distinct arid periods during which some chameleons are active, while others aestivate (see below). Adaptations for survival in these areas of climatic extremes include reproductive diapause (see Box 5.3), aestivation and for at least one species, and a complete change of the life cycle (see Box 5.1). Movement into extremely seasonal forests appears to be gradual. Only three species from the genus *Furcifer* (*F. major, F. oustaleti,* and *F. verrucosus*) inhabit the most arid deciduous forests in the southwestern regions of Madagascar, where mean rainfall is around 420 mm, typically confined to only 3 months, from December to February. Others appear in transitional and deciduous forest in the southwest and western region of Madagascar; including *F. antimena, F. labordi, Brookesia bonsi, B. decaryi, B. brygooi, B. perarmata,* and *B. stumpffi* (Raxworthy and Nussbaum, 1995; A. Raselimanana, personal observation). During the dry season, adult *Brookesia* aestivate under debris, or dig into loose soil (Brady and Griffiths, 2003).

A second radiation into arid habitat involves several species from the genus *Chamaeleo* that have moved into arid regions and even true deserts. For example, Swakopmund, which is within the range of *C. namaquensis,* has annual rainfall of around 14 mm (although horizontal precipitation in the form of coastal fog may be much higher) (Fig. 5.1g,h). These chameleons experience extreme diel temperatures: from 8 to 56°C (Burrage, 1973). They appear to cope with these stressors by means of behavioral and physiological adaptations, including what Burrage (1973) referred to as "ploughing." This behavior involves digging groves into the substrate where the chameleon makes contact with warmer or cooler sand. Together, body compression, color (brightness) change, and ploughing allow daily behavioral thermoregulation when ambient temperatures are

BOX 5.3 Embryonic Diapause

Embryonic diapause in chameleons is a unique form of developmental arrest that is unknown in other squamates (Andrews and Karsten, 2010). Embryonic diapause occurs when eggs are in gastrulae at the time of laying, and gastrulation occurs so slowly that development is effectively arrested for periods of several months (Bons and Bons, 1960; Andrews and Donoghue, 2004; Ferguson et al., 2004). In addition to embryonic diapause, embryos of some chameleon species also undergo cold torpor. In *Chamaeleo chamaeleon,* eggs with embryos in diapause are laid prior to the onset of winter, with the onset of winter conditions causing a second suspension of embryonic development. Despite *C. chamaeleon* egg clutches being laid over the course of several weeks, hatching is synchronous over a number of days more than 10 months after being laid. Ferguson et al. (2004) suggested that high temperatures during diapause prolongs developmental resting, whereas low temperatures during diapause, followed by an elevation in temperature, ends the diapause and accelerates development to term.

In experiments and field-measured nests, Andrews et al. (2008) manipulated the temperature of nests of *C. chamaeleon* to determine how nest temperatures and embryonic development were synchronized. By maintaining egg clutches at prewinter temperatures, they were able to prevent embryos from entering the period of cold torpor and to show that development continues to hatching. In both the field and laboratory clutches held at field temperatures, Andrews and colleagues showed that embryonic development began as soon as temperatures began to rise after winter. This general warming of all nests synchronizes the development of embryos, giving rise to synchronous hatching and emergence of juveniles in late summer, which may be important in diluting predation pressure and optimizing hatching at the time of maximum food availability.

Synchronous hatching of juveniles is known in many chameleons that inhabit highly seasonal environments. The dry deciduous forests of southwestern Madagascar are a good example, and researchers there believed that that embryo development was inhibited during the winter (dry colder) season so that hatchlings from different clutches emerge synchronously (Brady and Griffiths, 1999). Documentation of this is provided by two species that inhabit this environment (see Box 5.1).

not favorable. Burrage also notes that these chameleons are known to burrow completely into the substrate. Similarly, at night *C. namaquensis* were reported to make use of burrows, presumably to escape the cold temperature typical of the desert. This is the only species of chameleon that does not necessarily roost at night; instead, animals lie on the ground, possibly to maximize the ventral area in contact with the substrate (G.J. Measey, personal observation).

Chamaeleo calcaricarens inhabits the dry savanna and semidesert of Ethiopia, but there appear to be far fewer ecological data on this species. Spawls (2000) notes that, like *C. namaquensis, C. calcaricarens* is capable of moving rapidly and that it may aestivate throughout the dry season.

5.2 LIFE-HISTORY TRAITS

Chameleons exhibit a diverse array of life-history traits, from annual species, that spend most of their year as an egg, to large and long-lived species. The body size of extant lizards in this family covers four orders of magnitude and includes candidates for the world's smallest vertebrate (*Brookesia tristis,* 0.2 g, and the even smaller *B. micra*) (Glaw et al., 2012) to one of the largest arboreal lizards (*Calumma parsonii*, 700 g) (Abate, 1998). Live young are known to have evolved 108 times in squamates (Blackburn, 1999, 2006), and at least twice in chameleons (Tinkle and Gibbons, 1977; Andrews and Karsten, 2010; Schulte and Moreno-Roark, 2010; see also Tolley et al. 2013). Egg retention is considered to be the first step to viviparity, but most squamates either lay eggs with embryos about one-third developed (stage 30) or retain eggs until development is complete (i.e., are viviparous) (Shine and Thompson, 2006). However, chameleons lay eggs that can have one of three distinct strategies: eggs with early gastrula, eggs with well-developed embryos, and completely formed embryos (viviparous). Laying eggs with an early-stage gastrula allows chameleons to remain in the egg for unfavorable periods, and it has emerged that in some species a period of diapause allows eggs to remain unhatched for up to a year (see Box 5.3). The presence of the early embryologic stage in chameleons is thought to have evolved from an ancestor with (conventional) late-stage oviposition (Shine and Thompson, 2006). In their phylogeny, Andrews and Karsten (2010) suggest exactly the opposite: from late-stage oviposition (*Brookesia* and *Rhampholeon*) to viviparity (*Bradypodion*) to late arrested development and late early-stage oviposition (*Chamaeleo* and *Furcifer*). A model wherein the majority of squamates are prevented from this early-stage oviparity by nest sites that are not sufficiently hydrically stable (Shine and Thompson, 2006) suggests that female chameleons would have highly selective nest choice to maintain stable soil moisture levels.

Oviparity and Viviparity

Viviparity is thought to have its origins in cold environments: high latitudes or altitudes (Shine, 1985), and this hypothesis appears to holds true for chameleons (Andrews and Karsten, 2010; Schulte and Moreno-Roark, 2010). Viviparity has evolved in southern African *Bradypodion*, which occur at relatively high latitudes (up to 34.7°S), and *Trioceros*, which inhabit the highest montane areas in central Africa (up to 4 000 m asl). It is worth noting that *Bradypodion* are not at the most extreme latitude, which is taken up by an oviparous species *Chamaeleo chamaeleon* (up to 37°N), which has an embryonic diapause and goes into a state of torpor to avoid the coldest period (see Box 5.3). However, our current climate is at an interglacial stage, and some chameleon distributions are likely to have shifted during glacial periods. There are also oviparous chameleons at high altitudes, such as the Malagasy species: *Calumma hilleniusi* and *Furcifer campani*. Both are adapted to high elevations and cold montane habitats (temperatures ≤0°C during winter), but incubation periods vary, about 90 days for the former and 140 to 265 days for the latter (Glaw and Vences, 2007).

Variation in reproductive traits is thought to be an adaptation to unpredictable variation in the environment, notably rainfall (Shine and Brown, 2008). Although the majority of chameleon species live in forests, where the hydric environment is relatively stable, many species have moved into totally arid environments and yet others have moved into environments where the rainy season is particularly short. A good example of this is the dry deciduous forests of southwestern Madagascar, where the chameleons exhibit some of the most extreme life-history traits. The population of *Furcifer labordi* spends the 9-month dry season as eggs, synchronously hatching at the onset of rains (Karsten et al., 2008) (Box 5.1). The sympatric

F. verrucosus can be found throughout the year, with juveniles reaching adult size during the rains and aestivating during the long, cold, dry season (Karsten et al., 2008).

Several species of chameleons have been documented to store sperm, and it has been suggested that storage in this family of lizards may be considerably longer than in other more mobile species (Birkhead and Møller, 1993). However, this suggestion was based on a study with a low sample size (four species) and there are few new data to indicate whether duration of sperm storage is related to low densities or slow movement. Detailed studies of the phenology of viviparous and oviparous chameleon species exist, and there are examples of both species with continuous reproduction as well as species with distinct annual cycles. The entire genus *Bradypodion,* which inhabits the subtropical and Mediterranean climatic zones of southern Africa (see Box 5.3) is bear live young (Branch, 1998; Tolley and Burger, 2007). Reproduction in *B. pumilum* females is aseasonal; they can give birth year round and are capable of bearing multiple generations within a year (Jackson, 2007). Males had distinct biannual testicular peaks before and after the dry summer. This species also seems capable of sperm storage (Atsatt, 1953), and it is possible that other *Bradypodion* spp. share these life-history traits (Jackson, 2007). Similarly, *Trioceros bitaeniatus* gives birth to live young throughout the year in Kenya and *T. montium* lays eggs in both wet and dry season in Cameroon (Bustard, 1966; Herrmann and Herrmann, 2005). Lin (1980) made a detailed study of the reproductive traits of live-bearing and oviparous species of *Trioceros: T. jacksonii* and *T. hoehnelii,* respectively. *T. jacksonii* exhibited a distinctly annual reproductive cycle. Males were found to have an annual testicular cycle, peaking just before the onset of the rains in March and prior to mating observed in May. Females began the production of yolked follicles at this time, with a peak of ovulation in August. Gestation began in August and continued until February, when females were observed giving birth until March. Toxopeus et al. (1988) found that *T. jacksonii* could regularly be found in male–female pairs throughout pregnancy, suggesting some form of mate guarding. Following parturition, females entered a period of postreproductive quiescence, as fat bodies accumulated to reach their peaks in May (Edgar, 1979).

In his examination of the sympatric oviparous, *Trioceros hoehnelii,* Lin (1980) noted that this species was mostly aseasonal, with high testicular activity year round and mating observed in nearly every month. Females had yolked follicles throughout the year, with eggs with advanced embryos being laid year round. Females were also able to store sperm. Toxopeus et al. (1988) also studied *T. hoehnelii,* finding that while animals were often found in male–female pairs, these partners were frequently changed, although duration in pairs increased when females were notably gravid. Lin (1980) speculated that *T. hoehnelii* and *T. jacksonii* had evolved in allopatry, although they are now sympatric in the anthropogenically altered habitat of the central Kenyan highlands. Based on the presence of juveniles throughout the year, Wild (1994) suggested that *Rhampholeon spectrum* probably breeds all year round. Dominancy of juvenile *Furcifer campani* (91% vs. 3% subadults and 9% adults) has been reported at the beginning of the warm and rainy season in Madagascar, while adults were abundant during the wet season (80% vs. 20% subadults and 0% juvenile) and in winter (85.4% vs. 4.1% juveniles and 10.5% subadults; Raholdina, 2012).

In their study of another oviparous species, *Chamaeleo chamaeleon* in southwestern Spain, Andrews et al. (2008) again found a distinctly annual reproductive cycle. Males in this species were also found to practice mate guarding for a short period (around 2 weeks) during a distinct mating season when females were receptive to mating (Cuadrado, 2001) and leave once females show specific body coloration and behavior to indicate that they were gravid (Cuadrado, 2000). Chameleons are capable of continuous reproduction or annual cycles irrespective of whether the species is oviparous or viviparous. However, any species with diapause in eggs would be expected to exhibit an annual cycle.

Clutch size and hatchling size have both been shown to be strongly correlated with female body size. This has been shown intraspecifically for *Trioceros hoehnelii* and *T. jacksonii* (Lin, 1980), *T. montium* (Herrmann and Herrmann, 2005), *Chamaeleo chamaeleon* (Diaz-Paniagua et al., 2002) and over 33 other species (Andrews and Karsten, 2010). The largest recorded clutch sizes are from some of the largest egg-laying species, which may lay nearly 100 eggs (*C. calyptratus* and *T. melleri*), while those giving birth to live young reach approximately half this number (*T. jacksonii*). The smallest chameleons have clutches of 2 eggs (e.g., *Brookesia tristis*). Residuals of body size and hatching size indicated the existence of a trade-off between these traits as has been observed for most lizards: larger clutches result in smaller hatchlings and smaller clutches in larger hatchlings (Andrews and Karsten, 2010). Further, the strength of this relationship was found to increase when independent phylogenetic contrasts were included. Hence, like many other lizards, chameleons exhibit a continuum of reproductive strategies between large clutches with small hatchlings (<10 mm) and small clutches with large hatchlings (around 100 mm).

Egg-laying chameleons bury their eggs meticulously in the ground. Some ground-dwelling species deposit eggs singly within depressions under large leaves, while others dig holes of varying depths. Eggs of *Brookesia stumpffi* can be found in depressions under dead leaves on the forest floor (Raxworthy, 1991). Many people have observed female chameleons laying eggs, noting the vulnerability of the female during this long process. Hódar et al. (2000) and Brain (1961) both remarked on the search by females of *C. chamaeleon* and *C. dilepis* (respectively) for suitable ground conditions in which to dig holes. As many species may select different habitats for their offspring (see below), females move over considerable distances in order to oviposit. It is also the only time when chameleons have been seen to be active at night (*C. dilepis*, Brain, 1961). Egg chambers are dug obliquely into the soil, and eggs in small clutches are normally deposited singly, with soil separating each one, while large clutches are deposited together. Optimal nesting sites may be in high demand as precisely the same nesting site has been seen to be used multiple times by different individuals of *Kinyongia boehmei* (Measey, 2008).

Temperature changes with soil depth, decreasing and becoming more stable as depth increases. For example, temperature in areas inhabited by *Kinyongia boehmei* was constant by a depth of 30 cm (irrespective of whether the soil was in forest or cultivated areas). Yet, temperature changed by nearly 4°C over 700 m of altitude (Measey et al., 2009). *Chamaeleo calyptratus* was found to have consistent developmental rates between 28 and 30°C.

Development was significantly slower, yet hatchlings were significantly heavier at 25°C (Andrews, 2008). In southwestern Spain, nests of *C. chamaeleon* are laid up to 45 cm deep, where they undergo a seasonal change in temperature of nearly 20°C (Andrews et al., 2008). In contrast, nest temperature for *K. boehmei* eggs buried at 18 cm in a Kenyan forest underwent a change of only 4°C over a year (Measey, 2008). No temperature-dependent sex determination takes place in chameleons, as suggested by experiments with *C. calyptratus* (Andrews, 2005). *Archaius tigris* is notable in that females deposit their eggs in the leaf funnels of the introduced wild pineapple plants that are common in the Seychelles (Van Heygen and VanHeygen, 2004). Whether other plants (such as palms) would have been used before the arrival of the bromeliads is not known.

Once laid, chameleon eggs are susceptible to any number of vertebrate and invertebrate predators (see below). Likewise, hatchlings are likely to have high mortality between emergence from the nest and their first movement into surrounding vegetation. Chameleons are known to exhibit a diverse period of development, ranging from a few weeks to as long as a year. Andrews and Karsten (2010) proposed three groups with respect to other squamates for these highly divergent developmental rates: normal (50 to 70 days), slow (70 to 175 days), and arrested (175 to 365 days; see Box 5.3).

Ontogenetic Habitat Shift

Ontogenetic habitat shifts are hypothesized to occur widely within chameleons as they do in other arboreal lizards, as this prevents interactions between adults and juveniles (e.g., Irschick et al., 2000; Vanhooydonck et al., 2005). Ontogenetic habitat shifts may avoid cannibalism (see below) and competition and also aid in dispersal. Moreover, juveniles may inhabit areas with large numbers of small prey, which may not be suitable for adults (see below). In both experimental studies and observations on free-ranging *Chamaeleo chamaeleon*, juveniles were found to forage and sleep at lower heights than adults, which generally climbed to the top of available vegetation (Keren-Rotem et al., 2006). Juveniles were found to avoid adults (but not other juveniles), and when provided with an opportunity, most adults attempted to feed on juveniles.

Ontogenetic shifts in perch size may be facultative if, as observed in other arboreal reptiles, branch selection is dependent on animal size (Irschick and Losos, 1998). Keren-Rotem et al. (2006) found that adult *Chamaeleo chamaeleon* foraged and slept on thicker branches than juveniles. Specifically, most juveniles were found sleeping on grasses while adults slept on woody vegetation. Stratification of roosting level is well illustrated by *Brookesia decaryi*, whose hatchlings roost almost among the leaf-litter followed by juveniles, then by adults at the top (Razafimahatratra et al., 2008). Herrel et al. (2011) studied the perch selection of *Bradypodion pumilum* ecomorphs in heathland and wooded environments, concluding that perch and hand/foot size were well correlated. Moreover, larger hands were found to be stronger on larger perches (see Chapter 4). These authors excluded juveniles and subadults from their study, but observations suggest that juveniles of *B. pumilum* regularly occupy microhabitats distinctly different from those of adults, such as grasses at the periphery of wooded areas (G.J. Measey and A. Herrel, personal observation). Similarly, gravid *B. thamnobates* have been observed to deposit their young in grassland before returning to woodland type gardens (J. Herd, Howick, South Africa, personal communication). When eggs or juveniles are placed in a different environment by females, their migration becomes obligatory as they age.

Seasonality

Many of the studies that have been conducted on chameleons over time have suggested some degree of seasonality in density, reproduction, and other life-history traits. Lin and Nelson (1980) studied two Kenyan species, finding that *Trioceros hoehnelii* was aseasonal, while *T. jacksonii* showed distinct seasonal patterns of reproduction in both males and females. *T. jacksonii* gave birth in January, when insect abundance peaked. Despite the comparative aseasonality, the reproductive output of *T. hoehnelii* was found to be significantly reduced in the dry season (Lin and Nelson 1980), suggesting that reproduction is linked to climatic parameters.

Chameleon densities have been found to be lower in winter than in summer at one eastern Malagasy site for some species (*Brookesia nasus, B. superciliaris, Calumma brevicorne,* and *C. oshaughnessyi*) (Brady and Griffiths, 2003), while for others there was no change (*C. gastrotaenia*), or even an increase in winter densities (*C. nasutum* 51.7 ha^{-1} in winter as compared with 16.7 ha^{-1} in summer). Density fluctuations between seasons are suggested to correlate with body size (Brady and Griffiths, 2003). Similarly, at Ranomafana National Park, density estimates were greater in the summer (*B. superciliaris,* 39 ha^{-1}; *B. nasus,* 41 ha^{-1}; *B. thieli,* <0.1 ha^{-1}) as compared with the winter (*B. superciliaris,* 14 ha^{-1}; *B. nasus,* 6 ha^{-1}; *B. thieli,* <0.1 ha^{-1}) (Rabearivony, 1999). Smaller-bodied chameleon species at Andranomay (*C. glawi* and *C. nasutum*) displayed increases in population density between the summer and winter seasons, while the densities of larger-bodied species (*C. brevicorne* and *C. oshaughnessyi*) tended to decline over the same period (Brady and Griffiths, 1999).

Despite the considerable seasonal variation in densities of chameleons, there is little explanation of why some individuals are active and others inactive during unfavorable seasons. An unpublished study on *Bradypodion pumilum* suggests that reduced winter densities are not attributable to a reduction in population size, but rather to a reduction in the number of animals seen on exposed perches during winter. Different animals were found roosting on different nights, with the total number of observations related to weather conditions (K. Dicks, personal communication). Conversely, some species really do have a dramatic change in their seasonal abundance (Box 5.1), and it remains to be seen how widespread this phenomenon is.

Individual chameleons may disappear during winter in very cold or high-elevation areas, when they are thought to hibernate. Anecdotal reports suggest *C. calcaricarens* may aestivate during winter in Jijiga, Ethiopia (Spawls, 2000). *Bradypodion thamnobates* in the Drakensberg have been seen to move inside houses or under boxes in garages, remaining inactive for many months. *Rhampholeon marshalli* has diapause for around 6 months in the

Vumba Mountains of Zimbabwe (1730 m asl) (Broadley and Blake, 1979). Aestivation may also occur in extremely dry seasons (see Box 5.1). Like other lizards, chameleons appear to seek out dry and stable conditions in which to aestivate.

Growth and Longevity

Most reports of chameleon growth appear to come from animals bred in captivity, with relatively few studies conducted in natural populations. One notable exception compared the growth of caged juveniles (but not adults) with those from recaptures of *Trioceros hoehnelii* and *T. jacksonii,* finding that caged individuals had a significantly lower average growth rate (~0.05 mm · d^{-1}) than recaptured animals (~0.1 mm · d^{-1}) (Lin and Nelson, 1980). Burrage (1973) found a mean growth rate of 0.17 mm · d^{-1} for *Bradypodion pumilum* between birth and maturity, which he considered to be at 50 mm snout–vent length (SVL) (but see Jackson et al., in review, and below). He also found that growth rates changed during the year, reaching up to 0.29 mm · d^{-1} for animals born in November at the start of the austral summer. For *Chamaeleo namaquensis,* Burrage (1973) found sexually different growth rates with 0.25 mm · d^{-1} for males and 0.38 mm · d^{-1} for females. Burrage (1973) commented on a number of other studies (e.g., Brain, 1961) of growth rates of captive animals, noting that none compared with animals in the wild. It is likely that an advance in husbandry techniques has resulted in many captive chameleons able to grow at rates equivalent to or even faster than those in the wild (C. Anderson, personal communication). Yet, the information on natural growth rates is generally lacking (but see Box 5.1).

An important life-history variable is the time to maturity, which in lizards is generally longer with increasing body size. Chameleons appear to conform to this rule, with one of the largest species, *Chamaeleo parsonii,* taking 3 to 5 years to reach maturity (Brady and Griffiths, 1999). Lin and Nelson (1980) calculated growth rate and size of smallest mature female or male based on birth size. Their results suggested that *Trioceros hoehnelii* males and females reached maturity within a year, while *T. jacksonii* took just under 2 years. In both cases, females matured up to 20% more quickly than males. Burrage (1973) calculated that male *C. namaquensis* matured in 210 days, while females took only 150 days to achieve a slightly larger size at maturity. Jackson (2007) found the opposite for *B. pumilum,* in which males mature at a smaller size (41 mm) than females (53 mm); yet, both were mature in 18 months. Wager (1958; in Schaefer, 1971) states that *C. dilepis* can mature within a year. However, many Malagasy chameleons appear to reach maturity at large sizes within a single season, such as *Furcifer labordi* (see Box 5.1) and *F. campani,* which reach maturity in 3 months (Raselimanana and Rakotomalala, 2003). In captivity, other species are similarly reported to reach maturity relatively quickly: *F. willsii,* 4 months; *F. minor,* 5 months; *C. brevicorne,* 8 months (Le Berre, 1995).

The life span of most tetrapods has generally been found to correlate with body size (Blanco and Sherman, 2005), and there are some data to suggest that the largest chameleons are long-lived: *Trioceros melleri* may live as long as 12 years in captivity (Le Berre, 1995), while *T. montium* often lives beyond 9 years (Klaver and Böhme, 1992).

Longevity of wild chameleons is very poorly documented, with the majority of studies coming from animals bred in captivity. As chameleons are notoriously hard to keep in captivity, it is hard to estimate the bias these kind of data may introduce.

Longevity has been calculated from maximum sizes observed in the field, and growth rate of mature individuals (Lin and Nelson, 1980). Male and female *Trioceros hoehnelii* were found to have a similar longevity of around 4.5 years, while for *T. jacksonii*, smaller males have a shorter life span by as much as a third of the 6.6-year expected life span of females. *Furcifer pardalis* was studied using skeletochronology and showed that although adults were large, most individuals had a single line of arrested growth (LAG). Andreone et al. (2005) interpreted this as evidence that animals were around 1 year old (the largest individual in their sample was the only animal with 2 LAGs). There is some suggestion, that chameleons that grow fast may also die young (see Box 5.1), although exactly how widespread this phenomenon is within chameleons remains unknown.

Survival rates for smaller *B. pumilum* were found to be lower (from 0.49 for 45 mm SVL) than larger animals (0.98 for 80 mm SVL) over a single season (9 weeks) within the Cape Town metropolitan area (Tolley et al., 2010). While there have been no other formal studies of survival, Burrage (1973) estimated that 40% of his marked *B. pumilum* survived the 3-year duration of his study, and Bourgat (1968) recorded survival of 43% of *F. pardalis* after a single year.

5.3 FORAGING AND DIET

Diet has been poorly documented for chameleons, but they are known to eat, at least occasionally, relatively large prey (e.g., Broadley, 1973; Luiselli and Rugiero, 1996; Herrel et al., 2000; Keren-Rotem et al., 2006). Yet, most studies of fecal remains suggest they are opportunistic predators of invertebrates (Burrage, 1973; Pleguezuelos et al., 1999; Akani et al., 2001; Hofer et al., 2003). Chameleons have excellent visual acuity, which allows the assessment of prey from a distance (Ott and Schaeffel, 1995; Chapter 2). They have large heads and exhibit strong tongue retractors with supercontractile properties that can relay large items into the mouth (Herrel et al., 2001b; Chapter 4). Chameleons are also known to have a relatively high bite force (Vanhooydonck et al., 2007), which may be related to a diet of hard or oversized prey items, and/or to intrasexual and intersexual combat involving fighting and the biting of opponents (Bustard, 1967; Stuart-Fox and Whiting, 2005; Tolley and Burger, 2007; Measey et al., 2009; Chapter 6). Like other lizards, many chameleons are known to drink free-standing water by lingual protrusion.

Cruise Foraging

Chameleons are lizards that do not conform to either sit-and-wait or active foraging strategies. Instead, it has been proposed that they have an unusual intermediary behavior, termed "cruise foraging" (Butler, 2005; see also, Williams and McBrayer, 2011 for an alternative explanation). A cruise forager examines its environment, moves a short distance, and then conducts more scans (Regal, 1978). As a cruise forager makes these short movements within its environment, it would be expected to encounter sedentary prey in addition to active prey. Increases in movement would logically result in more encounters with sedentary prey and a commensurate reduction of active prey in the diet. Based on behavioral observations, the South African Cape Dwarf Chameleon, *Bradypodion pumilum*, conformed to this intermediary foraging mode (Butler, 2005), and more recently the same behavior was found in an invasive population of *Trioceros jacksonii* in Hawaii (Hagey et al., 2010). Cruise foraging suggests a similar proportion of active and passive prey types as compared with those in the environment, and Measey et al. (2011) found exactly this for *B. pumilum* from heathland habitats. However, ecomorphs from wooded habitats were found to have more active prey, suggesting that the degree of cruising may change in proportion to the availability of active and passive prey types. There is also a suggestion of ontogenic adjustment as Keren-Rotem et al. (2006) found that prey of adults were more sedentary than those of juvenile *C. chamaeleon*.

Although lingual capture appears to be the dominant method of chameleon feeding, there are also anecdotal reports that chameleons pursue and capture prey in their jaws, although these need to be confirmed (see Takashi, 2008, and below). *Chamaeleo namaquensis* is notable in that it inhabits an area with very low abundance of prey and appears to have moderated its behavior. Burrage (1973) described *C. namaquensis* running parallel with prey and then taking the prey in their jaws; however, other observations record these chameleons running after prey and simultaneously using lingual capture (M. Burger, personal communication). It was also noted that *C. namaquensis* are capable of searching for or ambushing prey that hides during pursuit (Burrage, 1973). Neither of these behaviors appears to fit into the cruise-foraging mode described by Butler (2005), and it may be that the Chamaeleonidae exhibit a range of foraging strategies.

Chameleon Diet

In their study of the diet of three sympatric chameleons in a Nigerian forest, Akani et al. (2001) suggested that some were more generalist (*Chamaeleo gracilis*) than others (*Trioceros cristatus*), based on relative niche overlap estimates (see Pianka, 1986). However, this method presumes good sample sizes for all taxa compared, and it is notable that the most specialized species had the smallest sample size (only 15 fecal pellets). Only two studies have attempted to compare potential prey to those ingested by chameleons. In the first, three montane chameleons from Cameroon (*Trioceros montium, T. pfefferi,* and *T. quadricornis*) had a niche breadth almost equal to the resources available (Hofer et al., 2003). Measey et al. (2011) analyzed prey in terms of hardness and evasiveness for two ectomorphs of *B. pumilum*, finding that the ecomorph in open habitat was neutral with respect to both measures, while those from woodland appeared to select more soft items (avoiding hard ones) and consume less sedentary prey. They interpreted these differences to differing availability of prey abundance and suggested that this may change in different seasons.

Seasonal changes in the diet of *Bradypodion pumilum* (woodland ecomorph) were most prominent in the change in the proportion of dipterans, which peaked at 80% in autumn and winter and sunk to 13% in spring (figures calculated from Burrage, 1973). Similarly, a peak in ground-dwelling carabids (up to 15%) also occurred in the winter months (Burrage, 1973). Marked seasonal changes in diet have also been observed in a population of *Chamaeleo chamaeleon* in southeastern Spain (Pleguezuelos et al., 1999), where hard hymenopteran taxa (mostly bees and wasps) made up the greatest part of *C. chamaeleon* diet in spring. By summer through to autumn, the major dietary component was orthopterans, the largest of prey in their study. Pleguezuelos et al. (1999) suggest that this shift represents a change in orthopteran availability and the more terrestrial nature of animals in summer, when males spend more time on the ground and females are found lower in the trees (see above). The other surprising finding in their study is the small size of some prey in relation to that of the chameleon, to which they conclude that chameleons typically rely on many small items (from 1.1 mm or 0.7% of SVL), rather than taking larger more infrequent meals. A study of invasive *Trioceros jacksonii* in Hawaii similarly remarked on the large number of small prey eaten by even the largest animals (Kraus et al., 2012). Similarly, small-volume prey were found in all sizes of *C. dilepis* dissected from museum specimens (Reaney et al., 2012). Bringsøe (2007) observed a subadult *Archaius tigris* predating on worker ants (*Technomyrmex* cf. *albipes*, 2 to 2.5 mm) at the start of the dry season on Praslin in the Seychelles Islands, and Keren-Rotem et al. (2006) observed adult *C. chamaeleon* eating a fruit fly. Although not analyzed by Hofer et al. (2003), their data suggest neutral selection on prey hardness across all species but with some selection toward prey evasiveness for *Trioceros montium*. Although there is little evidence for specialization in the chameleon diet, future studies may find that in optimal conditions chameleons may select particular prey types.

Dietary differences are expected where chameleons inhabit different habitats, partition microhabitats, and/or vary greatly in body size. The most obvious dietary divergence might be expected between small, ground-dwelling genera (*Rhampholeon* and *Brookesia*), and large arboreal species within the same forest (i.e., *Trioceros, Kinyongia, Furcifer,* and *Calumma*). Akani et al. (2001) found that *R. spectrum* had the least food niche overlap with three other sympatric arboreal species (*C. gracilis, T. cristatus,* and *T. owenii*). Perhaps unsurprisingly, this significant dietary difference extended to prey size, which was significantly smaller for *R. spectrum* (most items <3 mm). Indeed, leaf chameleons are able to make use of abundant social insects, such as termites (Wild, 1994). The shift to smaller foods in smaller chameleons is expected, as large lizards tend to have a wider range of food sizes available (Vitt, 2000). Hofer et al. (2003) examined fecal pellets of several different chameleon species in Cameroon, finding that prey size was significantly smaller in the smaller species studied. This indicates that most small chameleons are size-restricted in their feeding. This could be considered surprising, as chameleons are known for their high bite force (Vanhooydonck et al., 2007), which may help to reduce large prey items to an ingestible size (Measey et al., 2009). However, a more recent study of diet and bite force in two ecomorphs of *Bradypodion pumilum* suggested that diet is not dependent on the use of high bite forces (Measey et al., 2011).

There is some evidence that chameleons may position themselves in places where prey is abundant. Animals are often found in tree clearings and forest-edge ecotones, where invertebrate activity is also increased. Forest streams and ponds may be frequented for the same reason (Bringsøe, 2007; Jenkins et al., 2003; G.J. Measey, personal observation—see above), although there may be other reasons that these areas are selected (Jenkins et al., 2003). In urban Madagascar, *Furcifer pardalis* is well known to use flowering plants (e.g., *Lantana camara*) or trees (e.g., Jacaranda or fruit trees). Gardeners often remark that chameleons will favor particular plants in bloom (see also, Parcher, 1974), and this has been borne out in surveys (e.g., Tolley and Measey, 2007). *Bradypodion pumilum* and *B. occidentale* have both been observed moving to the supratidal zone to feed on abundant flies and tenebrionids there (Burrage, 1973). *Chamaeleo namaquensis* has been observed moving into the intratidal zone to feed on the abundant arthropods present (Burrage, 1973). Loveridge (1953) remarked that *Rieppeleon brachyurus* was noted to gather around fruit (Mikwambi) eating the small fruit flies that are attracted to it. Similarly, *R. kerstenii* were found gathered around fresh goat feces in Kenya, eating the flies that were attracted to it (J. Measey, personal observation).

Vertebrates in the Diet

While chameleon prey can generally be described in terms of opportunistic selection of invertebrates of appropriate size, there are many examples noted in the literature of chameleons ingesting vertebrate prey. The largest chameleons, such as *Trioceros melleri*, have been kept in captivity and are widely reported to consume small birds and mice (Broadley, 1966; Nečas, 2004). Nigerian *T. cristatus* held in captivity are known to readily eat frogs and newly metamorphosed toads (Reid, 1986). However, documented examples of natural predation of vertebrates are more unusual. Capture of a bird (presumably an adult *Foudia madagascariensis*) by one of the largest chameleons (*Furcifer oustaleti*) went unseen, although the chameleon was seen with the bird in its mouth, which was then swallowed whole (García and Vences, 2002). Widespread reports of *T. melleri* eating red-billed firefinch (*Lagonosticta senegala*) and a Cordon Bleu (*Uraeginthus bengalus*) (e.g., Pitman, 1958; Broadley, 1973; Hockey et al., 2005), all stem from reports in Loveridge (1953). Although not in the original report, Pitman (1958) asserts that that these small birds were captured through lingual projection. There are assertions that other large chameleons, such as *F. parsonii*, also eat small birds and day geckos (*Phelsuma* spp.: Le Berre, 1995; Brady and Griffiths, 1999; Abate, 1998; Raselimanana and Rakotomalala, 2003). Although there are no reports of chameleons predating on nestlings, this does not seem unreasonable, and support for this hypothesis comes from reports of nesting adult birds mobbing chameleons (Paxton, 1991; Masterson, 1994, 1999). *Chamaeleo namaquensis* is documented to consume lizards, including day geckos (*Rhoptropus afer*) and lacertids (presumably, *Meroles* spp.). Moreover, small feathers and hairs have been found in fecal pellets (Burrage, 1973). Indeed Burrage (1973) reported that this chameleon has been observed to capture a Namib dwarf sand-adder (a 200-mm *Bitis peringueyi*). These records are all unsurprisingly from large chameleons (the mean snout–vent length of these species is 255 mm, as compared with 94 mm for other chameleons, or 110 mm, excluding leaf chameleon species), as large lizards often tend to include both vertebrate and invertebrate food items in their diet (Meiri, 2008).

Vertebrates may not be unusual dietary items of chameleons when they are abundant in the environment, and this is borne out by several of the relatively few dietary studies finding small vertebrate prey: frogs and lizards (Luiselli and Rugiero, 1996; Akani et al., 2001; Measey et al., 2011; Reaney et al., 2012).

Cannibalism

Chameleons are also widely reported to be cannibalistic (e.g., Broadley, 1966), although most instances of this are known from captivity, where animals are kept in confined conditions (but see Parcher, 1974). Similarly, large chameleons may eat smaller chameleons of different species, although again, the only records are from captive individuals (e.g., Ionides, 1948, in Loveridge, 1953). Keren-Rotem et al. (2006) observed an adult *Chamaeleo chamaeleon* predate a juvenile in the wild, and suggested that the risk of cannibalism in may be avoided by shifts in habitat use (see above).

Herbivory

Insectivorous lizards are also known to be, at least occasionally, herbivorous (Cooper and Vitt, 2002; Herrel, 2007). Chameleons are not widely recognized to deliberately ingest plant matter, with most instances attributed to accidental ingestion with prey (Burmeister, 1989; Schwenk, 2000; Cooper and Vitt, 2002). However, both *Chamaeleo calyptratus* and *C. chamaeleon* are regularly observed to eat plants and fruit in captivity (A. Herrel, personal observation), and reports of this can also be found in the literature (Lutzman, 2000). It may be expected that xeric-adapted chameleons may regularly ingest plant matter in order to supplement their water intake, as is generally true of other lizards in these conditions (Herrel et al., 2007b). Support for this view comes from the data of Burmeister (1989), who consistently found vegetal matter (seeds, leaves, flowers, leaf buds) in the diet of *C. chamaeleon* in arid Libya, and Keren-Rotem et al. (2006) who found fruit in the diet of most adults (but not juveniles) of *C. chamaeleon* in Israel. However, a detailed dietary study in less arid south-eastern Spain did not report any plant matter (Pleguezuelos et al., 1999). Interestingly, *C. namaquensis* caught in the coastal areas of the Namib ate a higher proportion of plant matter than those caught inland (Burrage, 1973). Moreover, Burrage (1973) noted that of the plant matter ingested, principle items were the fleshy leaves of the dollar bush (*Zygophyllum stapffi*). It is possible that, like other lizards, chameleons in xeric habitats (e.g., *Chamaeleo calcaricarens*) will be found to regularly utilize plant matter in their diet.

Takahashi (2008) made a remarkable study of frugivory in *Furcifer oustaleti*. He observed an adult approaching a bunch of fruit (red fruit, 20 by 15 mm, of *Grangeria porosa*) and pulling them toward the mouth with a foot, whereupon half a fruit was bitten off and ingested. Burrage (1973) also notes that *C. namaquensis* often used its front feet to assist manipulation of oversized prey items. Takahashi (2008) saw a juvenile *F. oustaleti* deliberately take and ingest three small round black fruits of *Chassalia prince*. While both of these observations involved jaw prehension of the fruit, a further observation was made of an adult using lingual projection on a red fruit of *Malleastrum gracile*. One attempt was successful, but when the tongue failed to loosen another fruit, the chameleon walked up and used jaw prehension to take it. In an experiment with the same species, Takahashi (2008) found that jaw prehension was the norm for fruit, while tongue projection was generally used for flies, but he found exceptions to each. Not only is the observation that tongue projection was used on a fruit of interest, but the whole study took place during the wet season, when food was relatively abundant, suggesting that this large chameleon from an arid region of Madagascar regularly ingests fruit.

5.4 PREDATORS

A comprehensive review of predation on chameleons has not been undertaken. Here, the principle types of predators that consume chameleons are considered, with more attention to those that specialize in this prey type. Accounts of birds predating on chameleons may well be disproportionate in the literature because of observational bias. Similarly, there are many reports of the gut contents of snakes, and these may inflate their importance as predators as compared with some other groups, especially the invertebrates.

Chameleons are not likely to be able to flee from predators (although several authors remark at how fast chameleons are able to move: Spawls, 2000; Cuadrado et al., 2001; cf. Herrel et al., 2011) and instead need to rely on crypsis or active defense involving threatening behavior. As many are arboreal lizards, their chief predators are considered to be climbing mammals, birds, and snakes (Branch, 1998; Spawls, 2000). In addition, many authors comment on the increased vulnerability of female chameleons to predators while in the process of laying eggs (see above). In their review of predation of chameleons in Madagascar, Jenkins et al. (2009) found that birds outnumbered all other taxa recorded as chameleon predators. Yet, as noted earlier, this may reflect an investigatory/reporting bias. In addition to 19 species of birds, they found records for 5 snakes, 2 frogs, 1 primate, and 1 carnivore.

Invertebrates

Of all chameleon predators, the invertebrates probably exert the highest of all predatory pressures. Chameleon eggs are very vulnerable to many subterranean invertebrate predators, especially army ants (*Dorylus* spp.)m which predate upon most of what they encounter both above and below the ground. Ants are easily capable of overwhelming juveniles and possibly even adults if they do not flee their swarming raids (Lin, 1980). Viviparous chameleons are likely to avoid a large portion of predation risk from ants, but even their offspring are vulnerable, especially immediately after parturition (e.g., Bustard, 1966). Juvenile chameleons and small adults are also known to fall prey to large spiders and mantises (Parcher, 1974).

Amphibians

Medium-sized and large anurans occasionally predate upon small chameleons (Wild, 1994, Jenkins et al., 2009). *Amietophrynus camerunensis* is known to predate upon *Rhampholeon spectrum,* and there is an inverse relationship in the abundance of these toads and the chameleons (Wild, 1994), although Wild did not suggest predation as the cause, but rather competition. In Madagascar, one chameleon hatchling (genus *Furcifer* or *Calumma,* measuring 40 mm in total length) was found in a stomach of a *Mantidactylus femoralis* (42 mm SVL: Vences et al., 1999) and a *Ptychadena mascareniensis* preyed upon a juvenile *Furcifer lateralis* (D'Cruze and Sabel, 2005).

Snakes

Snakes probably have the largest predatory impact on adult chameleons, given that they do not solely rely on visual stimuli to find prey, thus allowing them to find cryptic, stationary prey, including chameleons. Many colubrid snakes are arboreal hunters that may rely on chameleon prey, although only few instances have been reported (Jenkins et al., 2009). In addition, terrestrial snakes (like the Malagasy *Pseudoxyrhopus ambreensis*) are known to prey upon chameleon eggs (Knoll et al., 2009).

In sub-Saharan Africa, the arboreal boomslang (*Dispholidus typus*) is a venomous colubrid snake that is known to take considerable numbers of chameleons (Haagner and Branch, 1993; Branch, 1998) and that in some situations may rely solely on chameleon prey (Loveridge, 1953). Vine snakes (*Thelotornis kirtlandii, T. capensis, T. mossambicanus,* and *T. usambaricus*) are all known to predate on chameleons, including terrestrial species (Loveridge, 1923; Menegon et al., 2009). Around half of all prey of *T. capensis* were found to be arboreal, including chameleons and day geckos (Shine et al., 1996), and Broadley (1983) suggested that these snakes are lizard specialists. Arboreal snakes from the genus *Philothamnus* (including *P. irregularis, P. semivariegatus,* and *P. angolensis*) are also widely reported to predate on chameleons (Lin and Nelson, 1980; Broadley, 1983). *Rhamnophis aethiopissa* and *Hapsidophrys lineatus* were both reported to be predators of *Rhampholeon* in West Africa (cf. Luiselli et al., 2000, 2001; Akani et al., 2001). Juveniles of large species, such as the green mamba (*Dendroaspis angusticeps*), also rely on small prey such as birds, their eggs, chameleons, and geckos (Broadley, 1983; Lloyd, 1974). These and other arboreal snakes are likely to be predators of most chameleons; for example, adult *Furcifer oustaleti* have been ingested by both terrestrial and arboreal snakes (*Madagascarophis colubrinus* and *Ithycyphus oursi*) (A. Raselimanana, personal observation; Crottini et al., 2010). A nighttime observation of predation of a *Brookesia superciliaris* by a *Parastenophis betsileanus* (Kaloloha et al., 2011) shows that roosting behavior is not always successful to escape from predation by snakes.

Chameleons often respond to the presence of snakes by dropping off their perch, both during the day and at night. The meticulous selection of perch sites, on isolated or distal branches or leaves, may be to avoid snakes by receiving advance warning of their presence, and/or because the perch cannot support the weight of the snake (or other predators). Stuart-Fox et al. (2006a) also found that chameleons become brighter when threatened by model snake predators (see Chapter 6). Lloyd (1974) made some interesting observations of two different arboreal snakes presented with *Chamaeleo dilepis*. Snakes were presented with a lateral view, heightened compression, extended gular region, and straightened legs, giving the impression of larger size (see also Stuart-Fox et al., 2006a). Moreover, the chameleon swayed and became very pale. Close inspection by the snake resulted in rapid jerking movements by the chameleon. Although this behavior allowed the chameleon to avoid predation by a green mamba, a boomslang quickly approached and ate the chameleon (Lloyd, 1974). These observations suggest that antipredator responses may be effective only against nonspecialist predators.

In addition to arboreal snakes, most snakes that eat lizards or frogs and that come across a chameleon of appropriate size are likely to eat it. For example, *Bradypodion ventrale* was found in the gut of *Crotophopeltis hotamboeia* and *B. dracomontanum* was regurgitated by *Psammophis crucifer* (Haagner and Branch 1993). The Namaqua dwarf adder (*Bitis schneideri*) was found predating on *B. occidentale* (Wessels and Maritz, 2009). In Israel, chameleons are predated by several snake species (e.g., *Malpolon monspessulanus, Hemorrhois ravergieri;* Keren-Rotem et al., 2006).

Mammals

Many small mammals may opportunistically take chameleons. For example, in Namibia *Chamaeleo namaquensis* is predated by the jackal *Canis mesomelas* (Burrage, 1973). Small arboreal carnivores, such as *Martes foina,* take *Chamaeleo chamaeleon* in Israel (Keren-Rotem et al., 2006). In addition, arboreal carnivorous mammals, such as civets and genets are likely to predate chameleons. Andriatsimietry et al. (2009) reported the presence of chameleon in the feces of the mongoose, *Galidictis grandidieri*, in southwestern Madagascar. *Calumma brevicorne* eggs have been reported to be vulnerable to predation by invasive *Rattus rattus* (Parcher, 1974). Broadley and Blake (1979) report that likely predators of *Rhampholeon marshalli* include civets and genets. Jenkins et al. (2009) report predation on chameleons by the Malagasy civet (*Cryptoprocta ferox*) and a lemur (*Lemur catta*). It is possible that many other primates would eat chameleons, but there are few reports of this in the literature.

A Malagasy mongoose *Galidia elegans* was observed attacking a large male *Calumma ambreense* in Montagne d'Ambre (A. Raselimanana, personal observation). The chameleon stayed in a vertical position along a small tree at 1.30 m from the forest floor. This terrestrial carnivore attempted to catch the chameleon by jumping several times before climbing onto a fallen log to get close. The mongoose was observed to attack the eyes first, and then tried to remove the hands of his victim from the support.

Birds

Although there are a large number of bird species that occasionally eat chameleons (e.g., *Larus hartlaubii;* see Hockey et al., 2005), there are considerably fewer that regularly have chameleons as part of their diet. Birds that regularly take chameleon prey include various raptor species, shrikes (Laniidae), helmetshrikes (Prionopidae), puffback shrikes, bush shrikes, tchagras and boubous (Malaconotidae), cuckoo-shrikes (Campephagidae), cuckoos (Cuculidae), rollers (Brachypteraciidae and Coraciidae), hornbills (Bucerotidae), and barbets (Lybiidae). Jenkins et al. (2009) reported a taxonomic division between avian predators and chameleon prey corresponding to the main division within terrestrial and arboreal chameleons. In the canopy, both specialists (e.g., *Eutriorchis astur* and *Falco zoniventris*) and generalists (e.g., *Buteo brachypterus*) predate on arboreal chameleons (*Furcifer* and *Calumma*). While on the ground, generalists (e.g., *Brachypteracias leptosomus*) forage in the leaf litter consuming terrestrial species (*Brookesia*). Interestingly, nocturnal birds were rarely found to predate on Malagasy chameleons.

Shrikes are predators of southern African *Bradypodion* chameleons, and are famed for hanging their prey on thorns (e.g. Wager, 1986; Branch, 1998; Tolley and Burger, 2007). The common fiscal (*Lanius collaris*) was used as a model predator by Stuart-Fox et al. (2006a), who found that *Bradypodion transvaalense* readily avoided these predators (see Chapter 6), and these authors considered that common fiscals are likely to exert a strong selection on chameleon antipredator responses. However, the substantial grip of some individual chameleons (see Herrel et al. 2011; Chapter 4) may help them to escape predation from at least some shrikes. In Cape Town, an adult common fiscal was seen trying to remove an adult *Bradypodion pumilum* from its perch on a reed by using its weight to hang from a single limb (G.A. Millar, personal communication). Despite this predation attempt going on for over 10 minutes, the bird finally flew away leaving the chameleon apparently unscathed. Most shrikes inhabit open habitats and are therefore likely to predate on savanna, grassland, and heathland chameleons (see above), but there are a few species that also inhabit forests. Other predatory birds inhabit dense forests, where chameleons may also be vulnerable to these predators. Broadley and Blake (1979) suggested that shrikes were important predators of *Rhampholeon marshalli*, and Hockey et al. (2005) record chameleons as prey items for the grey-headed bushshrike (*Malaconotus blanchoti*). Similarly, hornbills mostly inhabit more open habitats such as woodland and savanna, where they are regularly seen taking chameleons as prey (see Tolley and Burger, 2007). Hockey et al. (2005) report five species of hornbills to have chameleons in their diets (*Tockus damarensis, T. alboterminatus, T. leucomelas, T. monteiri,* and *Bucorvus leadbeateri*). Other generalists likely to be significant predators of chameleons are the barbets and boubous, of which Hockey et al. (2005) report chameleon prey for *Centropus burchellii* and *Laniarius aethiopicus*.

Jenkins et al. (2009) list several raptors as the principle avian predators of chameleons on Madagascar. In mainland Africa, three raptors stand out as likely to exert substantial predation pressure on chameleons. The African cuckoo-hawk (*Aviceda cuculoides*) has been found to specialize on chameleons in southern Africa. *Chamaeleo* and *Bradypodion* made up 6 of 25 prey items in South Africa, *C. dilepis* 20 of 39 prey items in Zimbabwe and 32 of 51 prey items in Kenya (Hockey et al., 2005; W. Tarburton, personal communication). Lin and Nelson (1980) suggest that the main chameleon predators in Kenyan highlands include lizard buzzards (*Kaupifalco monogrammicus*), which have a large distribution over much of sub-Saharan Africa. Lastly, the forest buzzard (*Buteo trizonatus*) is probably a substantial predator of chameleons. As chameleons range into more specialized terrain, the predatory avifauna will also change. Indeed, Burrage (1973) stated that the arid zone *Chamaeleo namaquensis* is predated upon by several opportunistic raptors (*Falco rupicolis, F. terinunculus,* and *Melierax musicus*).

ACKNOWLEDGMENTS

Many people helped with compiling this chapter by sharing their knowledge of chameleons and their predators, drawing our attention to and finding literature, and commenting on the text. We are indebted to Chris Anderson, Marius Burger, Kara Dicks, Frank Glaw, Justin Herd, Ian Little, Graham Millar, Philip Shirk, Rob Simmons, Warwick Tarburton, Krystal Tolley and James Vonesh. Krystal Tolley, Marius Burger and Tania Fouche generously provided images for Figure 5.1.

SIX

Chameleon Behavior and Color Change

DEVI STUART-FOX

Although in many respects chameleon behavior resembles that of other lizards, there are also unusual characteristics associated with their unique morphology (Chapter 2). Chameleons are perhaps most famous for their ability to change color, a characteristic that has seen the term *chameleon* adopted in popular language to mean someone changeable or who unconsciously mimics others. Color and color change are integral to all aspects of chameleon behavior, including social, thermoregulatory, antipredator and foraging behavior. Consequently, this chapter focuses on both coloration and behavior in chameleons and will outline the modes of communication in chameleons, highlighting the importance of vision to all aspects of chameleon behavior. Chameleon color change, including the mechanism, triggers, and general function and evolution of color change are covered, as well as a detailed review of coloration and social behavior associated with reproduction. Finally, this chapter synthesizes knowledge of antipredator behavior in chameleons, including camouflage and escape behaviors.

One aspect of social behavior that is of central importance for many taxa is parental care and parent–offspring interactions more generally; however, there is no evidence of direct parental care in chameleons. In terms of indirect parental care, the placement of offspring at parturition could conceivably influence their subsequent survival in live-bearing species. Another form of parent–offspring interaction is cannibalism, which is relatively common in *Chamaeleo chamaeleon*, and may account for differences in habitat preferences of juveniles and adults (Keren-Rotem et al., 2006). However, many other factors, such as body size, diet, thermal requirements, and predation risk could also account for age-specific habitat differences, which appear to be quite common in chameleons. Cannibalism may be relatively common, particularly among larger species. Apart from potential cannibalism, parent–offspring interactions are minimal or entirely absent, particularly for egg-laying species.

Although there are other general reviews of the biology of chameleons, which cover many aspects of behavior described here (Martin, 1992; Nečas, 2004; Pianka and Vitt, 2003; Tilbury, 2010), the goal of this chapter is to provide a relatively detailed synthesis of coloration and behavior. This is placed within a broader ecological and evolutionary context where possible, highlighting important gaps in our understanding. Hopefully, this review will stimulate additional research on coloration and behavior in these remarkable animals.

6.1 SENSORY SYSTEMS AND MODES OF COMMUNICATION

Given the central importance of color and color change to chameleons, it is not surprising that vision is their most important sense (see Chapter 2) and that communication is primarily via the use of visual signals. The optical system of chameleons is covered in greater depth in Chapters 2 and 4; however, it is worth emphasizing several features that give chameleons exceptional eyesight. First, chameleons have the highest density of visual cells (cones) in the retina recorded (756,000 per square millimeter) (Wall, 1942), giving them extreme visual sensitivity. Second, their independently moving, protruding eyes have at least 180-degree movement in all directions, providing an exceptional field of view (Mates, 1978). Third, chameleons have an apparently exclusive reliance on accommodation cues to judge distance (Harkness, 1977). Accommodation cues can be used by each eye independently, allowing chameleons to accurately judge distance with one eye (Ott et al., 1998) and to gain monocular parallax cues simply from eye rotation, without moving their head or body (Pettigrew et al., 1999). Lastly, the optical power of the eye is contributed exclusively by the cornea rather than the lens (Ott and Schaeffel, 1995). Because the cornea is further forward than the lens, the nodal (focal) point is separated from the axis of rotation of the eye (Pettigrew et al., 1999). This results in increased magnification of the retinal image (the largest known for vertebrates). In other words, chameleons have long-focus or telescopic vision (Pettigrew et al., 1999). The highly developed vision of chameleons is likely to influence the evolution of social communication using visual signals and also poses a challenge to understanding these signals from the point of view of chameleons.

In contrast to the highly developed visual system, the olfactory and auditory systems in chameleons are relatively insensitive. The Jacobson organ used for olfaction in squamate reptiles is vestigial or lacking (see Chapter 2). Although some species possess temporal glands at the corners of their mouths that produce strong-smelling secretions, these may be more likely to be used as a predator deterrent or prey attractant than for intraspecific communication. There are, however, reports of males following the path taken by a female through dense vegetation, despite not being able to see her (Tilbury, 2010), which suggests the possible use of olfactory signals. The auditory system in chameleons is similarly underdeveloped, being less sensitive than that of humans (Wever, 1968; Wever, 1969b). Vocalization is sometimes incorporated into threat displays, with some species exhibiting a hissing exhalation. According to Tilbury (2010, p. 83), *Trioceros goetzei* produces a "high-pitched exhalation squeaky wheeze" when first handled. However, there is no evidence that these auditory signals are used for intraspecific communication.

Chameleons may use tactile signals in the form of substrate (plant-borne) vibration, which is commonly used by insects but rare in vertebrates. Some chameleons vibrate in short bursts at a high frequency. This behavior has been documented in an antipredator context in several species of *Brookesia* (Raxworthy, 1991) and *Chamaeleo calyptratus* when disturbed by humans and also during courtship (Barnett et al., 1999). Barnett et al. (1999) quantified the vibrational signals in *C. calyptratus*, showing that there are two types with initial dominant frequency of around 150 Hz and 20 Hz. These are transmitted to the receiver either directly or via vegetation. Vibration has also been anecdotally reported in *Rhampholeon*, *Rieppeleon*, and several species of *Trioceros* and *Chamaeleo* (Lutzmann, 2004; Tilbury, 2010). This has led some authors to suggest that vibrational signaling is likely to be widespread among the Chamaeleonidae (Barnett et al., 1999; Tilbury, 2010).

Because of their overriding importance, the remainder of this chapter will focus on visual signals.

6.2 COLOR CHANGE

Mechanism

Coloration and color change in chameleons is a function of specialized cells called "chromatophores." Chromatophores are cells in the dermis of ectotherms that are responsible for generating skin and eye color (Bagnara and Hadley, 1973; Fox, 1976). There are four main types of chromatophore, each containing different types of pigment. Xanthophores and erythrophores are very similar, distinguished primarily by their color (yellow-orange and red, respectively). They are located nearest the surface and contain carotenoid or pteridine pigments (generating reds, oranges, and yellows). Beneath the xanthophores are iridophores (also called "guanophores"), which contain colorless crystals of guanine pigment, arranged as stacks of platelets separated by cytoplasm. Their appearance is a function of the structural arrangement of platelets. For instance, depending on their spacing, the stacked guanine platelets may preferentially scatter shorter wavelengths of light and transmit longer wavelengths (Tyndall scattering), resulting in a blue appearance. Melanophores comprise the deepest layer and, as their name suggests, contain melanin pigment, generating black and brown coloration. The melanophores are large, stellate (star-like) cells with long dendrites ("arms") that extend between the iridophores and overlay the xanthophores.

Chameleon coloration is a function of the type, density, and arrangement of chromatophores. For example, green coloration may be generated by the combination of yellow xanthophores underlain by blue iridophores. The extent of color change and range of colors exhibited is species-specific, as it depends on the types and distribution of chromatophores characteristic of that species.

Color change occurs because of the movement of pigment-containing organelles within chromatophores. For example, the darkening of the skin is a result of the concentration or dispersions of motile vesicles ("packets") containing melanin pigment (the melanosomes) within the melanophores (Bagnara and Hadley, 1973). When the melanosomes are aggregated within the center of the cell, the skin appears very pale, whereas when they are dispersed throughout the dendrites to the skin's surface, the skin appears dark. Varying the degree of dispersion of the melanosomes blocks reflectance of the iridophores but not the xanthophores or erythrophores, so the skin appears yellow-red. Color change may also occur because of changes in the spacing of platelets within the iridophores, a phenomenon that has been demonstrated in fish (Clothier and Lythgoe, 1987), amphibians (Bagnara and Hadley, 1973), and the ornate tree lizard, *Urosaurus ornatus* (Morrison et al., 1996). However, it is currently not known whether movement of platelets in iridophores or pigments in xanthophores and erythrophores is involved in color change in chameleons specifically.

Color change may occur because of multiple "triggers" that fall into four main categories: (1) in response to temperature; (2) as a reflexive response to light (via photoreceptors in skin); (3) as a function of physiological state (e.g., receptivity); and (4) as a response to sensory input (surroundings, presence of prey, predators, or conspecifics). Most discussions of color change in chameleons state that it reflects their "mood"; however, the precise meaning of this phrase is unclear. Temperature-dependent color change is discussed briefly below, and the latter three triggers of color change are elaborated on in subsequent sections.

Color change in response to temperature is well known in iguanian lizards, including chameleons. However, temperature-associated color change in chameleons appears to be dependent on habitat. Specifically, among Kenyan populations of three species, *Trioceros jacksonii* and *T. ellioti,* which occur in low-latitude, midelevation evergreen forests, show limited or no change in reflectance with temperature, whereas *Chamaeleo dilepis,* which occupies a greater range of seasonally variable, subtropical savannah environments shows significant changes (Walton and Bennett, 1993). Chameleon species from higher latitudes or altitudes have been observed to perch on the top of a bush first thing in the morning, laterally compress their body and orient one flank toward the sun, and adopt a dark-brown to black coloration, often only on the flank oriented toward the sun (Burrage, 1973; Walton and Bennett, 1993). Walton and Bennett (1993) estimated that this behavior increased the rate of radiant heat gain in *C. dilepis* by 7%. This is likely to have advantages in terms of increasing activity time and locomotor performance. Chameleons may also become pale to increase reflectance during exposure to high levels of solar radiation, such as observed for *C. dilepis* on roads and in open areas at midday (Walton and Bennett, 1993).

Color change is controlled by neural and hormonal mechanisms in chameleons. Color change has been shown to occur not only in response to direct nervous stimulation (electrical stimulation of spinal nerves) in *Bradypodion pumilum* (Hogben and Mirvish, 1928) but also to hormones of pituitary origin in *Trioceros jacksonii* (Bagnara and Hadley, 1973). As in other reptiles, the primary hormone affecting aggregation or dispersion of melanosomes is likely to be melanophore-stimulating hormone (MSH) but adrenocorticotropic hormone (ACTH) has also been shown to induce darkening in *T. jacksonii* (Bagnara and Hadley, 1973). Control of color change in chameleons is poorly understood. It may be that a range of neurotransmitters and hormones of neural, adrenal, gonadal or thyroid origin (e.g., melatonin from the pineal gland or the catecholamines epinephrine and norepinephrine), and complex interactions between them, influence color change as in other reptiles (reviewed in Bagnara and Hadley, 1973; Cooper and Greenberg, 1992)

Function and Evolution

As for many other animals, coloration is strongly related to behavior and affects social communication, camouflage from predators and prey, and thermoregulation. Color change can be seen as a compromise between opposing selection resulting from these multiple, often conflicting functions. A widespread solution to the problem of needing to be conspicuous and cryptic at different times is to have bright display colors that can be revealed when needed but remain concealed at other times (Cooper and Greenberg, 1992). Display colors may be located on body regions that can be flashed (e.g., dewlap) or revealed through display postures (e.g., ventral or ventrolateral coloration). Color change is a special case of this strategy, allowing chameleons to exhibit conspicuous color patterns during social communication while remaining concealed from predators and prey at other times (Stuart-Fox and Moussalli, 2011).

Within the Chamaeleonidae, the ability to change color varies markedly—from limited changes in brightness (e.g., shades of brown) to remarkable changes in both color and pattern. Why is the ability to change color so much more developed in some species than in others? The answer to this question cannot be inferred from the function of coloration, as color has multiple purposes in chameleons. Although color change in chameleons is often associated with camouflage, it also functions in social communication and thermoregulation. Thus, the evolution of color change may be driven by selection for camouflage against different backgrounds, selection for signaling functions, or thermoregulatory requirements. These selective forces generate different, testable predictions, which are detailed elsewhere (Stuart-Fox and Moussalli, 2009, 2011).

The camouflage and social signaling hypotheses have been tested in dwarf chameleons (*Bradypodion* spp.). Stuart-Fox and Moussalli (2008) tested whether the capacity for color change is associated with particular habitat features or with higher variance in background color (relative to the animal's movement) as predicted by the camouflage hypothesis. They also tested a prediction of the social-signaling hypothesis that species showing the greatest color change should use sexual signals that are more conspicuous to conspecific receivers. To measure color change, color was measured in response to different stimuli, including another male or female and model predators (bird and snake). As the greatest color change occurred during male–male contests, color change was quantified as the difference between an individual's dominant coloration (submissive opponent) and his submissive coloration (dominant opponent). Color change was estimated for 21 lineages of dwarf chameleons (*Bradypodion* spp.). The phylogenetic comparative analysis showed that those with greatest capacity for color change had social signals that were more conspicuous to the chameleon visual system but did not occupy habitats with greater variance in background color. Although color change clearly serves a camouflage function in chameleons, results of this study suggest that the remarkable ability for chromatic change in dwarf chameleons may have evolved to facilitate social signaling rather than background matching. Ideally, this should be corroborated with additional comparative evidence for an association between degree of color change and other indicators of the strength of sexual selection, within and between species. Such indicators of strong sexual selection may include highly skewed reproductive success or mating systems that promote skewed reproductive success, and greater sexual dimorphism (Stuart-Fox and Moussalli, 2011).

6.3 SOCIAL AND REPRODUCTIVE BEHAVIOR

Mating System and Territoriality

Chameleons are polygamous. Males may mate with more than one female and females may mate with different males during the same or different ovarian cycles. Females are known to mate repeatedly with the same male during the relatively brief period of receptivity within an ovarian cycle (Tilbury, 2010). Whether a female will mate with more than one male within an ovarian cycle is likely to depend on male density (and therefore encounter rate) and the intensity of mate guarding or territoriality. If given the opportunity, however, females will mate with more than one male when receptive (Cuadrado and Loman, 1997).

The mating system has been studied in the most detail in the common chameleon, *Chamaeleo chamaeleon* (Cuadrado, 2001). In this species, males guard females, with the period of guarding ranging from 0 to 46 days (mean, 13.2 days). Males may sequentially guard up to eight females, although not all guarding episodes result in successful copulation or reproduction. Males cease guarding shortly after mating, when the female shows clear signs that she is no longer receptive. Spatial organization in this species is complex and varies depending on the nature of female home ranges. Some males defend stable nonoverlapping home ranges (i.e., territories) that incorporate the home range of one or more females when females have small, stable home ranges. Other males simply follow and defend an area around a guarded female if her movements are more erratic (Cuadrado, 2001).

Mate guarding appears to be common in chameleons. For example, male *Trioceros jacksonii* and *T. hoehnelii* actively defend an area around the female and stay with the female for up to several months (Toxopeus et al., 1988). However, the extent of mate guarding appears to vary among species, even within the same genus, with *Chamaeleo dilepis* appearing to be solitary, tolerating the presence of another individual only during mating (Brain, 1961; Toxopeus et al., 1988). In species of *Brookesia* and *Rhampholeon,* in which females are substantially larger than males, mate guarding may take the form of the male riding on the female's back. Tilbury (2010) has recorded male *Rhampholeon gorongosae* being carried by the female for up to 10 days.

The prevalence of territoriality, the consistent defense of area against conspecifics, is unclear in chameleons. Unambiguous territoriality has only been documented in *Chamaeleo namaquensis,* in which both sexes vigorously defend territories (Burrage, 1973). Male

Ch. chameleon may show defense of stable territories associated with mate guarding (Cuadrado, 2001; see above). *Calumma brevicorne* and *Ca. oshaughnessyi ambreensis* maintain small, discrete home ranges, often being faithful to a single roosting site on consecutive days, although the extent to which these species actively defend territories is unclear (Kauffmann et al., 1997). By contrast, *Ca. oshaughnessyi oshaughnessyi* have relatively large and overlapping home ranges and seldom return to the same roost site (Kauffmann et al., 1997). Absence of roost-site fidelity has been documented in several other species (Hebrard, 1980). However, roost-site fidelity provides little indication of territoriality. *Bradypodion pumilum*, for example, defend roosting sites from conspecifics (specific site defense; Stamps, 1977), but do not show defense of a consistent area surrounding them and, therefore, cannot be considered territorial (Burrage, 1973). *Ch. dilepis* show no indication of territoriality, often moving considerable distances and showing no indication of having a stable home range, although they vigorously repel any conspecifics that they encounter except potential mates (Brain, 1961). Studies of territoriality in arboreal species with three-dimensional home ranges are particularly challenging. Consequently, relatively little information is available on territoriality and its relationship to mating systems in chameleons.

Courtship, Copulation, and Sperm Storage

In most species studied to date, males persistently court both receptive and nonreceptive females (see following section on "Female Reproductive Status"). Males usually begin courtship from a distance with a lateral display, involving lateral compression, legs rigidly beneath the body and tail coiled (Cuadrado and Loman, 1999; Kelso and Verrell, 2002; Stuart-Fox and Whiting, 2005) (Fig. 6.1 in the color insert). This lateral display is similar to that used in aggressive encounters, and its primary purpose seems to be to maximize the signaler's apparent size, potentially allowing size assessment by the receiver. Courtship also includes approach with head movements, the precise nature of which varies among species. For instance, male *Chamaeleo calyptratus* exhibit both a slow head roll and rapid vibration (Kelso and Verrell, 2002). Courting male *Bradypodion* shake the head rapidly from side to side (personal observation). The speed, duration, and amplitude of the side-to-side head movements in chameleons may be analogous to the species-specific stereotyped head-bobs of iguanian lizards (Carpenter, 1977). As in iguanian lizards, the head movements are often displayed during both courtship and contests; although whether they differ consistently between these two contexts is unknown.

Following initial courtship displays, the male will attempt to approach the female and position himself behind her. He may then attempt to mount straight away, or, if the female does not accept mounting, the male may repeatedly nudge or rub her. For example, male *Chamaeleo calyptratus* courting nonreceptive females approached the female from behind, gave short bursts of vibrations and repeatedly nudged (head butted) and rubbed their chins on the female. This sequence of behaviors was primarily shown toward nonreceptive females and after unsuccessful attempts to mount (Kelso and Verrell, 2002). Similar persistent courtship and nudging behavior toward behaviorally rejecting females has been recorded in *Furcifer labordi* and in one case, the female changed back to passive coloration and allowed copulation (Karsten et al., 2009c). During copulation, receptive females generally remain relatively drably colored and passive when the male approaches, allowing him to mount, while nonreceptive females exhibit characteristic conspicuous coloration and courtship rejection behaviors (described in the section below on "Female Reproductive Status").

Successful mounting and copulation involves the male grasping the female's flank or dorsal keel with his front feet, coiling his tail around hers and inserting a hemipene into her cloaca (Fig. 6.2 in the color insert). Copulation lasts from less than a minute (*Chamaeleo chamaeleon;* Cuadrado and Loman, 1997) to up to an hour (*C. calyptratus;* Kelso and Verrell, 2002). Males and females adopt drab coloration during mating, the male visibly darkening from his previous bright courtship/display coloration. During mating, females of some species, such as *C. calyptratus,* exhibit short bursts of vibration (Kelso and Verrell, 2002) and males may repeatedly stroke the female's flank with his hindfoot (Brain, 1961; Tilbury, 2010). The female appears to initiate disengagement by moving (personal observation; Brain, 1961; Tilbury 2010).

Chameleons exhibit long-term sperm storage (weeks to months and across female ovarian cycles). The occurrence of sperm storage in chameleons is reviewed by Tilbury (2010). Multiple broods in the absence of mating between broods has been recorded in *Chamaeleo africanus, C. chameleon, Furcifer lateralis, Bradypodion pumilum,* and several species of *Trioceros* (Tilbury 2010, and references therein, and sperm storage is likely to be the rule rather than the exception in chameleons. However, the consequences of female sperm storage for male and female fitness have not been studied. The most obvious explanation for female sperm storage is to ensure fertilization when encounter rates with males are very low or uncertain. Alternatively, it may allow females with asynchronous/continuous breeding to progress rapidly from one ovarian cycle to the next without incurring the potentially high costs of copulation. A third possibility is that it facilitates sperm competition and cryptic female choice, whereby females may have some control over the use of ejaculates for fertilization when there are ejaculates from two or more males within her reproductive tract (Uller et al., 2010). Thus, sperm storage is likely to exert strong selection on both male and female mating behavior and the evolution of mating strategies in chameleons.

Female Reproductive Status

Female chameleons show characteristic colors and behavior to signal reproductive status, in particular to signal that they are gravid and nonreceptive. Although male courtship toward receptive females is often more frequent and intense (Cuadrado, 1998b; Kelso and Verrell, 2002), males also persistently court nonreceptive females (although this may not be the case in all species, e.g., *F. labordi,* Karsten et al., 2009c). In species in which males court nonreceptive females, they may presumably gain reproductive benefits due to female sperm storage. Females exhibit conspicuous color patterns and specific behaviors to signal nonreceptivity, presumably to reduce costs associated with male courtship and harassment (Cooper and Greenberg, 1992). Interestingly, in *Chamaeleo chamaeleon,* female movement rate (horizontal distance covered per observation period) was significantly higher once they had developed sexually receptive coloration, together with much higher rate of copulations and rejections. Cuadrado (1998b) hypothesized that the higher movement rate was perhaps as a response to the more intense courting.

Female coloration in relation to reproductive status has been studied in depth in a few species of *Chamaeleo*. In *C. calyptratus*, receptive females have faint blue spots, while nonreceptive females have bright orange markings that appear 24 hours after mating (Kelso and Verrell, 2002). Female common chameleons (*C. chamaeleon*) develop yellow spots when receptive, and shortly after copulation they exhibit dark body coloration with bluish and yellow spots, combined with aggressive rejection displays (Cuadrado, 1998b; and see below). Females that did not develop yellow spots during the breeding season skipped reproduction despite persistent courtship by males (Cuadrado, 1998b). In *Furcifer labordi*, female coloration appears to be associated with sexual maturity. Early in the breeding season, smaller, nonreceptive females exhibit a conspicuous yellow spot on the anterior flank and smaller faint yellow lateral spots on a green background. In larger, sexually mature females the anterior flank spot is bright red, the lateral spots are violet, and the background coloration is lighter green (Karsten et al., 2009c). These sexually mature females are potentially receptive but may either accept or reject male courtship and copulation. By contrast, coloration of *F. verrucosus* females appears to be associated with sexual receptivity rather than maturity. Females showing gray and light green-yellow allow nearly all courting males to attempt copulation, whereas brick-red females reject all male courtship attempts (Karsten et al., 2009c). Characteristic coloration of gravid females has also been documented for other species, such as yellow-red patches on the lower half of the body in *C. zeylanicus* (Singh et al., 1983) and black spots and yellow stripes in *C. africanus* (Bonetti, 1998, in Tilbury, 2010).

Nonreceptive females tend to be highly aggressive toward courting males, exhibiting an open mouth, often with hissing, swaying or vigorous rocking, and a laterally flattened body. These behaviors have been observed in a range of species, including dwarf chameleons (Stuart-Fox and Whiting, 2005), *Chamaeleo zeylanicus* (Singh et al., 1983), *C. chameleon* (Cuadrado, 1998b, 2000), *C. gracilis* (Bustard, 1967), and *C. calyptratus* (Kelso and Verrell, 2002). The aggressive rejection behavior is generally accompanied by characteristic color change involving dark or highly contrasting coloration (Fig. 6.3 in the color insert). However, this is not the case for all species, even those that are relatively closely related. For example, among potentially receptive females, behaviorally rejecting *Furcifer labordi* females show black background coloration with highly contrasting orange, purple, and red spots, whereas behaviorally rejecting *F. verrucosus* females do not change color (Karsten et al., 2009c). This difference may be related to differences in predation pressure on the two species. Alternatively, it may be associated with differences in the degree of male harassment, which may be greater in *F. labordi* because of more intense competition over mates as a result of more synchronous reproduction (Karsten et al., 2009c).

If the aggressive display is not a sufficient deterrent, females may chase and bite courting males. For instance, in *Bradypodion pumilum*, courting males were bitten by nonreceptive females in 28% of laboratory behavioral trials, and males that attempted mounting the female were more likely to be bitten (Stuart-Fox and Whiting, 2005). Female aggressive rejection appears to successfully deter males, as males reduced courtship with increasing intensity of female rejection and males were more likely to approach and court smaller females (Stuart-Fox and Whiting, 2005). Male *Chamaeleo calyptratus* adjust both the length and content of their courtship displays toward receptive versus nonreceptive females (Kelso and Verrell, 2002). Males courted nonreceptive females for significantly longer periods, even though they courted and mated with receptive females much more frequently than with nonreceptive females (Kelso and Verrell, 2002).

There is little evidence of active female mate choice in chameleons, with receptive females of many species appearing to accept all male advances. However, females of some species may exhibit mate choice by rejecting or accepting males when receptive, as appears to be the case in *Furcifer labordi* (Karsten et al., 2009c). In *Chamaeleo chameleon,* the female initiates long walks when courted by a male. During the walks, the pair is exposed to interference from other males and is more exposed to predation. Cuadrado and Loman (1997) hypothesized that given the presumed cost of this behavior, it may benefit females by allowing indirect female mate choice or at least mate assessment.

Male Aggressive Behavior

Male chameleons engage in ritualized aggressive displays, which may escalate to physical combat in some species and may sometimes result in injury. The prevalence and intensity of male aggressive behavior in chameleons suggests that male–male competition is likely to be important in gaining access to receptive females and preventing other males from doing so.

Ritualized aggressive displays involve characteristic color change. For example, on encountering another male, male *Chamaeleo gracilis* from Sierra Leone initially become paler and more uniform in coloration (blotches disappearing) but with intensified spots (Bustard, 1967). Courting male *Furcifer labordi* and *F. verrucosus* increase the contrast of their coloration, and in the latter species, males developed bright green and blue over the lower half of their bodies (Karsten et al., 2009c). In many species of dwarf chameleon (*Bradypodion* spp.), coloration intensifies dramatically, resulting in high-contrast color combinations (Stuart-Fox et al., 2007). For example, *Bradypodion transvaalense* males become orange and black. Among *Bradypodion* spp., species that have greater capacity for color change have male display colors that contrast more with the background vegetation and comprise more contrasting color combinations (to the chameleon visual system) as compared with species with less capacity for color change (Stuart-Fox and Moussalli, 2008). Some of these color combinations are likely to appear more conspicuous to the chameleon visual system than to humans. The display coloration of *B. damaranum* includes both green and ultraviolet-blue-green, which contrast highly in the chameleon visual system but appear similar to humans (Stuart-Fox et al., 2007). While display coloration varies substantially among species, submissive coloration generally comprises drab grays and browns, often darker and drabber than cryptic coloration.

Ritualized aggressive behavior in chameleons invariably involves lateral displays comprising lateral compression, extension of the legs vertically beneath the body, and extension of the gular region to appear as large as possible (Fig. 6.1 in the color insert). In some species, such as *Bradypodion*, this posture is accompanied by horizontal extension and/or coiling of the tail, which may add to the appearance of large size. Many species accompany this posture with head shaking. However, in *Furcifer labordi* and *F. verrucosus*, males used head movements only in a courtship context (Karsten et al., 2009c).

Following approach and display, one individual may assume submissive coloration. Alternatively, if the contestants are closely matched, the contest may escalate to lunging and biting. In extreme cases, chameleons may lock their jaws together and wrestle, each attempting to push its opponent along the branch in a contest of strength. Male *Trioceros jacksonii* lock their three horns together and push each other along the branch in an obvious contest of strength (personal observation) (Fig. 6.4 in the color insert). They also violently twist their horns into each other in a corkscrew fashion and the male with longer horns can inflict damage by piercing the tissue of his rival with the tips of his horns (M. Whiting, personal communication). Whether rostral appendages are used in male–male combat appears to vary among species. Species with keratinized appendages may be more likely to use them in combat rather than courtship, whereas species with flat, fleshy appendages, such as *Furcifer labordi*, may use them only during courtship (Karsten et al., 2009c). Keratinized rostral appendages and bites can cause surface wounds, but more serious injury from male contests is likely to be rare in most species. Escalated fights have been observed in the field in *Bradypodion pumilum*, with males locking jaws and falling from perches to the ground in prolonged wrestling matches (Burrage, 1973). The contest may continue for extended periods (many minutes and potentially up to an hour) until one male retreats and adopts submissive coloration. Escalated physical fights may be rare or absent in some species, with contests confined to ritualized displays.

The intensity of male aggression is likely to vary within and among species. Within species, the intensity of male–male competition is likely to be more intense during the breeding season, particularly in seasonally reproducing species (Singh et al., 1983). Intrasexual selection may also vary among populations depending on density and/or habitat, with important consequences for the evolution of male ornamentation and behavior. For example, closed-habitat ecomorphs of *Bradypodion pumilum* have a central pink flank patch and rugose casques, and the size of these traits predict fighting ability (Stuart-Fox et al., 2006b). By contrast, the open-habitat ecomorph is smaller and less ornamented but has a stronger bite force for its size (Measey et al., 2009). The intensity of male–male aggression can also differ among closely related species. For example, *F. labordi* exhibits much more male–male aggression than *F. verrucosus*, which may be related to competition over mates, as reproduction in the former is much more synchronous (Karsten et al., 2009c). Variation in the intensity of intrasexual selection associated with ecology and life history is likely to be important in explaining the morphological diversity—particularly in coloration and ornamentation—apparent in chameleons.

6.4 SEXUAL DIMORPHISM: BODY SIZE AND ORNAMENTATION

Sexual dimorphism in body size (SSD) varies from female-biased to male-biased. In *Rhampholeon* and *Brookesia* for example, females are larger than males, whereas in *Chamaeleo, Calumma,* and *Furcifer,* males tend to be larger than females (Nečas, 2004). In some genera, such as *Bradypodion,* both forms of dimorphism are found, although female-biased SSD predominates (Stuart-Fox, 2009). SSD is likely to reflect a complex combination of selective forces, including sexual and natural selection on both male and female body size. Natural selection is likely to act directly on both male and female body size. For example, both male and female *Bradypodion* spp. are smaller in habitats with a higher density of perches, such as grasslands and heaths with short dense shrubs. Males, but not females, are also relatively larger in forested habitats (Stuart-Fox and Moussalli, 2007). Sexual selection on male body size appears to vary among species, with larger size not necessarily being advantageous. For example, in *Bradypodion pumilum,* larger males are not necessarily likely to win contests (Stuart-Fox et al., 2006b). In *Ch. chamaeleon,* success rate per courtship attempt was no greater for larger males, although larger males put more effort into courting and had higher overall reproductive success (Cuadrado and Loman, 1997; Cuadrado, 2001). Selection for fecundity is likely to favor larger female body size, since larger females tend to bear more offspring (Burrage, 1973; Lin and Nelson, 1981; Cuadrado, 1998a), although reproductive effort also decreases with size and age in several species (Lin and Nelson, 1981; Cuadrado, 2001). The variation in natural and sexual selection on male and female body size is likely to account for the great diversity of sexual dimorphism apparent in the Chamaeleonidae. In this regard, chameleons are similar to most other lizard families (e.g., Iguanidae, Agamidae, Lacertidae, Teiidae, Scindidae, Gekkonidae), in which sexual size dimorphism varies from female-biased to male-biased, depending on demographic and ecological factors (reviewed in Fitch, 1981).

In contrast to body size, which shows various forms of sexual dimorphism across the chameleons, males are either similar to females or more ornamented. Ornaments include rugose cranial casques, occipital lobes, a wide range of keratinized or fleshy rostral appendages, a range of gular ornamentation (lobes, flaps, and spines) and enlarged scales on the flank (detailed in Nečas, 2001; Tilbury, 2010). Even in species or populations with little ornamentation, males tend to have relatively larger heads, which is likely to result from intrasexual selection, as males engage in physical combat (described above) and head size is related to bite force (Measey et al., 2009). Male ornaments are used in male–male contests both directly (e.g., the rostral horns of *Trioceros jacksonii*) and indirectly via emphasis during lateral displays and head movements.

6.5 ANTIPREDATOR BEHAVIOR

Chameleons' primary defense against predators is camouflage. Chameleons can change both their colors and their patterns to varying extents (depending on the species) in order to match their backgrounds or perhaps disrupt the body outline. Many arboreal chameleons are able to exhibit uniform or mottled green hues to match moss or leaves as well as brown and gray hues to match twigs and vines. Chameleons that are mostly terrestrial or adapted to environments with little vegetation (e.g., *Chamaeleo namaquensis*) may have a more restricted color repertoire, mostly consisting of changes in brightness from cream, gray, brown, and black. Many chameleons also adopt a light and dark botched color pattern, associated with antipredator behaviors. For example, *Brookesia superciliaris*, adopting the freeze and roll behavior (see below), show blotched coloration on the exposed upper flank, which according to Raxworthy (1991, p. 22), "disrupts the body profile against the dead leaves." However, there have been no experimental demonstrations that blotched coloration in chameleons is a form of disruptive camouflage, which hinders recognition of the body outline by creating the appearance of false edges and boundaries (Stevens and Merilaita, 2009), rather than background matching. Some species may exhibit specific patterns that aid camouflage. For instance, grassland species often show horizontal stripes on the flank and underside, resembling the vertical stripe pattern of grass when the chameleon is holding onto a vertical grass stalk (Tilbury, 2010).

As chameleons have exceptional vision, they may often detect a threat before they are themselves detected. In addition to camouflage, many arboreal chameleons flatten themselves laterally and flip to the other side of the branch or twig to which they are clinging. During experiments in which a model snake (boomslang [*Dispholidus typus*]) and a stuffed fiscal shrike (*Lanias collaris*) were presented to chameleons (Stuart-Fox et al., 2006a, 2008), my colleagues and I were struck by their ability to accurately maneuver so that the twig remained between themselves and the line of sight of the predators. This behavior is widespread among cryptic, arboreal iguanian lizards and geckos.

The primary predators of chameleons are arboreal snakes and birds. However, chameleons are also preyed upon by mammals, particularly mongooses, larger reptiles and amphibians, and some invertebrates such as large mantids and spiders (see Nečas, 2001; Tilbury 2010, and Chapter 5 for a more detailed list of chameleon predators). These predators differ in their foraging tactics and sensory capabilities. For example, birds rely exclusively on visual cues for prey detection (Kassarov, 2003), whereas most snakes use olfactory cues in addition to visual cues (Schwenk, 1995). Birds have greater visual acuity (the spatial frequency resolvable at maximum contrast), which is due to greater cone density in the retina than that of snakes (Osorio et al., 1999). Birds also have superior color vision, as they are tetrachromats, having four types of visual pigment (Hart, 2001), rather than the three types of visual pigments of most snakes studied to date (Sillman et al., 1997, 1999, 2001).

Stuart-Fox and Moussalli (2009) showed that among 21 populations of dwarf chameleon (*Bradypodion* spp.), including all currently recognized species and several morphologically and genetically distinct lineages (Stuart-Fox et al., 2007; Stuart-Fox and Moussalli, 2008), 13 differed in the camouflage response to model bird and snake predators. Chameleons showed consistent behavioral responses to the two predators, consisting of lateral compression, flipping to the opposite side of the branch and occasionally dropping from the perch (see below). The frequency of these behaviors did not differ toward the two predators (Stuart-Fox et al., 2006a), but in all cases in which chameleons showed different color responses to the two predators, they showed greater background matching (lower contrast against the background) in response to the bird than the snake. Although they showed closer background matching in response to birds, they appear more camouflaged (i.e., less chromatically contrasting against the background) to snakes because snakes have poorer color discrimination. These results raise the possibility that chameleons fine-tune their color response to maximize camouflage in response to different predators (see Stuart-Fox et al., 2006a, 2008; Stuart-Fox and Moussalli, 2009, for discussion of this phenomenon).

Chameleons may also employ a form of movement-based camouflage, which may reduce the probability of detection by either predators or prey. Chameleons often rhythmically rock backward and forward as they walk—the characteristic "jerky walk." A common explanation is that they are imitating a swaying leaf or vegetation moving in the breeze; however, this has never been demonstrated experimentally. The behavior is widespread in highly cryptic, generally slow-moving, ambush predators, notably chameleons and some snakes (Fleishman, 1985) and mantids (stick insects). An alternative potential explanation is that chameleons rock back and forth while they walk to create parallax, enabling depth perception in both eyes simultaneously. This explanation, however, is unlikely because rocking behavior is not seen in the pygmy genera but is restricted to the arboreal genera (Tilbury, personal communication). Furthermore, chameleons use accommodation cues (changes in optical power of the eye to maintain focus with varying distance) to judge distance (Harkness, 1977). This allows them to accurately judge distance with one eye, without the use of parallax.

At night, chameleons become pale (cream to almost white) and very easily visible in flashlight. This reflects the resting state of the melanophores with melanosomes concentrated rather than dispersed. Photoreceptors in the skin of chameleons respond to light, triggering dispersion of melanosomes. Thus, chameleons darken rapidly in response to light, even when remaining with their eyes closed. Many chameleon biologists have observed that when a chameleon is captured at night, the side on which the flashlight was shone becomes dark, while the opposite side of the chameleon remains pale. *Brookesia* and *Rhampholeon* appear to be exceptions to this generalization, retaining their brown coloration to some degree at night. Raxworthy (1991) has proposed that this may have a cryptic function even when the chameleons roost among green leaves, since most nocturnal predators are color-blind and the brown and green may match in luminance rather than color. However, *Brookesia* also have much more limited capacity for color change, which may equally explain their relatively lesser degree of pallor when roosting.

Because of their pallor, chameleons stand out against the background at night and may therefore be vulnerable to predators. Perhaps as a result, chameleons perch in places that may be difficult for predators to reach, such as on the ends of vines and at the tips of thin twigs and grass stems. These perches may be just strong enough to support the weight of a chameleon but not a nocturnal snake or mammal predator. If the twig or vine moves excessively, disturbing the chameleon, it may let go of the perch and drop to the ground, where it either remains perfectly still or writhes until is it buried out of site among dead leaves (Tilbury, 2010). Many arboreal species accompany this "dropping" behavior with a "hopping" motion as they initiate their fall, perhaps to aid the probability of escape (Tilbury, personal communication).

The behavior of dropping from perches may vary depending on species and degree of threat. In *B. transvaalense,* my colleagues and I observed it in less than 10% of trials in which we presented chameleons with model predators (10 of 116 trials = 8.6%; Stuart-Fox et al., 2006a). However, falling from perches is common in *Chamaeleo chamaeleon,* having been observed in the majority (>60%) of simulated predator attacks (approach by humans), when they were perched in open bushes in which they had a higher probability of detection. When perched in denser bushes, "free falling" was observed less frequently overall and more commonly in hatchlings than in juveniles or adults (Cuadrado et al., 2001). Such dropping or death-feigning behavior may also be shown by juveniles or subadults in response to a threat from a much larger conspecific (Brain, 1961). Thus, chameleons appear to alter their antipredator behaviors based on probability of detection and age class, which may reflect vulnerability to predators or threat from a conspecific.

Chameleons sometimes also appear to show death-feigning behavior, remaining curled up and perfectly still and appearing to be dead when handled. Raxworthy (1991) described seven types of antipredator behavior, including what appears to be death-feigning, in which the chameleon would freeze in an upright stiff posture when handled. Raxworthy (1991) also described behaviors that appear to be leaf mimicry. These were freezing and rolling, in which the chameleon would fold the legs under the belly, roll to one side, and remain motionless and dorsal flattening, which gave the chameleon "a very flattened leaf-like appearance."

When chameleons have been detected by a predator or are cornered or stressed, they may exhibit a range of secondary defenses against potential predators. These include threat displays, vibration, and "spine-thrusting" in *Brookesia*. Spine-thrusting entails vigorous thrusting out of dorsolateral spines during lateral convulsions and is observed only in species with dorsolateral spines (Raxworthy, 1991). The threat displays include lateral compression, swaying, open mouth, and distended throat. Open-mouthed threat displays may be accompanied by a hissing sound (Cuadrado et al., 2001). These displays often reveal brightly colored gular interstitial skin, the color of which varies greatly among species and even within species. For instance, among the dwarf chameleons (*Bradypodion* spp.), interstitial gular skin may be white, red, orange, purple or black, and bright orange or black within *B. occidentale* (personal observation). Although the behavioral postures are similar to threat displays during male contests, males do not show the same conspicuous display colors during contests and under threat-induced stress (personal observation). The latter tend to be highly contrasting light and dark blotches rather than the bright hues often seen during male courtship displays and contests.

Chameleon coloration and behavior is diverse, with exceptions to almost every generalization that can be made. This is despite the fact that coloration and behavior in most contexts has been studied in very few species. Color lability in chameleons is associated with complex social and antipredator behaviors. Coloration, ornamentation, and associated social behaviors are likely to vary in relation to ecology and life history, which are diverse in the Chamaeleonidae. In particular, factors such as seasonality and synchrony of reproduction, spatial organization and density, and habitat-associated predation pressure are all likely to influence the intensity of both intersexual and intrasexual selection, with important consequences for morphological and behavioral adaptations.

Another factor that may influence the evolution of coloration, ornamentation, and behavior in chameleons is the number of sympatric chameleon species. Discussion of the function of coloration and behavior in this chapter has been limited to intraspecific interactions, because experiments specifically testing a species-recognition role for visual signals in chameleons are lacking. However, interspecific competition could also potentially drive the evolution of signal diversity, as appears to be case for the dewlaps of *Anolis* lizards (Ord and Stamps, 2009, and references therein). It would be interesting to test whether the degree of difference in visual signals (coloration, ornamentation, and behavior) among closely related species is positively related to the number of sympatric species (discussed in Chapter 5). Experiments testing whether individuals respond more strongly to conspecific or heterospecific signals would also be informative.

Observational field studies of chameleon behavior are particularly challenging because of the difficulty of locating these exceptionally well-camouflaged animals. Telemetry can be used for larger species but has so far not been possible for smaller species. This makes it difficult to study the ecological context of behavior, which is important for understanding its diversity and evolution. However, chameleons are well suited to laboratory and field manipulative behavioral experiments, as evidenced by many of the studies cited in this chapter and extensive laboratory research on reproduction and thermoregulation in some species (e.g., *Chamaeleo calyptratus*) (Andrews, 2007, 2008; Andrews et al., 2008; Andrews and Karsten, 2010). Consequently, there is much scope for studying the functional significance of visual signals, particularly in relation to sexual selection. Specifically, given the substantial life-history variation among chameleons (e.g., Karsten et al., 2008), they may be excellent models for studying life-history effects on the strength of sexual selection. As chameleons have the most developed capacity for color change among terrestrial vertebrates, they are also ideally suited to understanding adaptive, dynamic camouflage and the visual features important for effective camouflage (see Stuart-Fox and Moussalli, 2011).

ACKNOWLEDGMENTS

I thank Cissy Ballen, Colin Tilbury, Krystal Tolley, and Martin Whiting for helpful comments. I was supported by an Australian Research Council Australian Research Fellowship.

SEVEN

Evolution and Biogeography of Chameleons

KRYSTAL A. TOLLEY and MICHELE MENEGON

7.1 EVOLUTIONARY RELATIONSHIPS

Both the fossil record and molecular dating show that most major squamate families were well established by the mid-Cretaceous and that a few stem lineages were present as early as the Jurassic (Kumazawa, 2007; Chapter 9). For example, skinks, cordylids and lacertids were present in the early Cretaceous or the late Jurassic (Vidal and Hedges, 2005, 2009), and gecko diversification is attributed to Gondwanan fragmentation in the early to mid-Cretaceous (Gamble et al., 2008). Chameleons (family Chamaeleonidae) however, are a young clade as compared with most other squamates. Together with Agamidae and Leiolepididae (collectively known as "acrodonts"), they diverged from a common ancestor with the Iguanidae ("pleurodonts") more than 120 Mya (Townsend et al., 2011a). Within the acrodonts, divergence between chameleons and the Leiolepididae / Agamidae occurred approximately 90 Mya (Kumazawa, 2007; Townsend et al., 2011a; Tolley et al., 2013). Although there is some uncertainty regarding the monophyly of Agamidae and the placement of Leiolepididae relative to chameleons, the chameleons themselves are clearly monophyletic (Macey et al., 2000b; Townsend et al., 2011a,b). Although a Malagasy origin for the family has been proposed (Raxworthy et al., 2002), an alternative scenario is that of an ancestral chamaeleonid clade in Africa that diverged from agamids around the late Cretaceous, which dispersed to Madagascar and also radiated within Africa (Tolley et al., 2013). Given the direction of prevailing oceanic currents at the time (Ali and Huber, 2010) combined with ancestral-area reconstructions, an African origin for the family is likely (Tolley et al., 2013).

The stem lineage of chameleons split into two sister clades approximately 65 Mya (Tolley et al., 2013); the result was two deeply divergent clades, one of which contains 10 of the 11 extant genera (ca. 155 species), while the other is comprised of only the Malagasy leaf chameleons,

Brookesia (>30 species). This ancient split occurred around the Cretaceous–Paleogene boundary, coinciding with the mass extinction at the end of the Mesozoic Era. Although numerous invertebrate and vertebrate lineages went extinct during that time, all major squamate lineages survived (Macleod et al., 1997). After the end of the Mesozoic, the two chameleon lineages persisted for nearly 15 Myr through the Paleocene epoch, without any major diversification (Fig. 7.1 in the color insert). Although it is possible that lack of diversification is obscured by extinctions during that time, it is also possible that opportunities for diversification were limited.

By the end of the Paleocene (ca. 55 Mya), a rise in global temperatures due to the release of carbon into the atmosphere produced a "Greenhouse Earth" climate, which progressively began to reverse again ca. 50 Mya (Maley, 1996; Zachos et al., 2001, 2003, 2008). For chameleons, the end of this greenhouse climate corresponds to a burst of lineage diversification in the Eocene epoch (ca. 50 to 45 Mya), giving rise to most of the extant genera (Matthee et al., 2004; Townsend et al., 2009; Tolley et al., 2011, 2013). Most notably, early diversification (ca. 50 Mya) was between the terrestrial clades (*Brookesia, Rieppeleon,* and *Rhampholeon*), with arboreal clades arising later, around the end of the Eocene (ca. 45 Mya). Because the burst of diversification corresponds with the reversal of Greenhouse Earth conditions, chameleon radiation may be linked to increased ecological opportunity as a result of environmental changes (e.g., Losos and Mahler, 2011) brought about during the thermal maximum. While vacant niches certainly would have been available after the Cretaceous–Paleocene mass extinction, chameleons did not start to diversify in earnest until nearly 15 Myr after this catastrophic event. Hot and dry conditions in Africa and Madagascar at the time probably prevented angiosperm forests from becoming established, and this presumably hampered diversification of chameleon lineages at the end of the Cretaceous. Forests likely spread southward from Laurasia (Couvreur et al., 2011), first into north Africa (Axelrod and Raven, 1978; Upchurch et al., 1998, 1999) and later became widespread on the continent and Madagascar around the Eocene (Wells, 2003; Jacobs, 2004). It is apparent that both tropical rainforest and open-canopy forest were present (Jacobs, 2004; Secord et al., 2008; Couvreur et al., 2011), which would have provided multiple opportunities for chameleons to become established and diversify. Thus, it is likely that chameleons took advantage of this expanding arboreal niche as the forest vegetation became dominant, particularly in the tropics (Axelrod and Raven, 1978; Jacobs, 2004). The two main clades initially diverged much earlier than the Eocene, and these earliest branching lineages are primarily terrestrial (i.e., living on top of the leaf litter). Because all chameleons have adaptations that are beneficial in arboreal habitats (fused opposable digits, lateral compression, independently rotating eyes, projectile tongue) (Herrel et al., 2011; Chapter 2), they would be predisposed to an arboreal lifestyle but did not take full advantage of that habitat until it became dominant over the landscape. It is likely that early lineages utilized some aspect of the forest floor and leaf litter, in which their specializations were beneficial. For example, terrestrial forest chameleons occupy the surface of the leaf litter as well as low sticks and branches. In contrast, other reptiles that utilize leaf litter (e.g., skinks) typically push through the litter to forage or escape predation. Early chameleons may have found a new, empty niche on top of the litter and extending some centimeters into the low vegetation. In this case, morphological adaptations such as grasping hands and feet would make negotiation over the leaf litter and upward several centimeters more feasible.

Within genera, lineage diversification began in the Oligocene, presumably due to forest fragmentation associated with a global climatic cooling trend (Zachos et al., 2001; Couvreur et al., 2008; Tolley et al., 2011, 2013). It is thought that lineage diversification was prompted by vicariance as forests fragmented (Tolley et al., 2011), confining lineages to isolated patches of forest from the Oligocene through the Miocene (Tolley et al., 2006, 2008; Menegon et al., 2009; Tilbury and Tolley, 2009b; Townsend et al., 2009; Tolley et al., 2011). Because most chameleons (about 75%) are forest specialists, and forest (whether for terrestrial or arboreal use) is considered the ancestral habitat type (Tolley et al., 2013), the fragmentation of this habitat would have been a strong factor for promoting species-level diversification through vicariance.

Some extant lineages, however, are not forest endemics. In the genera *Bradypodion* and *Trioceros*, 35% and 45% of species, respectively, inhabit open heathland/grassland habitats, as do some *Calumma* (Raxworthy and Nussbaum, 2006; Glaw and Vences, 2007; Tolley and Burger, 2007; Tilbury, 2010; Chapter 5), while *Chamaeleo* and *Furcifer* occupy an assortment of diverse habitats (Glaw and Vences, 2007; Tilbury, 2010). Many of these lineages are typically part of much younger radiations (Fig. 7.1 in the color insert) (Nagy et al., 2012; Tolley et al. 2006, 2008), presumably associated with expansion of open habitats since the Miocene/Pliocene. In contrast, some genera are nearly exclusive to forests (*Brookesia, Kinyongia, Rhampholeon*), and very few of these species have made the transition to open heath/grassland (Townsend et al., 2009; Measey and Tolley, 2011; Tolley et al., 2011). These genera tend to lack recent radiations and instead are composed mainly of paleo-endemics (Townsend et al., 2009; Tolley et al., 2011). Aside from a few exceptions, such as *Brookesia ambreensis* complex (Townsend et al., 2009), all lineages within these genera are older than 5 Myr, with high proportion of lineages are older than 20 Myr (Fig. 7.1) (*Brookesia*, >40%; *Rhampholeon*, >65%; *Kinyongia*, >30%).

The recent radiation of lineages into the open habitats of heath/grasslands seems to be accompanied by convergence in morphological features that are adaptive to that vegetation structure (Tolley et al., 2006, 2008). Although only a few lineages have been studied to date, these have comparatively short tails and smaller hands and feet than their strictly forest/arboreal counterparts (Herrel et al., 2011; Herrel et al., 2013; Hopkins and Tolley, 2011; daSilva and Tolley, 2013; Potgieter, 2012). While these lineages are not terrestrial, the vegetation structure of these open environments consists of much finer perches than those in forest or woodland. Comparison of these habitats with forest in South African systems shows that on average the available perches are twice as large in forested canopy vegetation.

Some clades also occur primarily in open-canopy systems (e.g. woodland savanna in Africa and xeric forests in Madagascar), but these clades (e.g., *Chamaeleo* and *Furcifer*) have not radiated to the extent of open heath/grass habitat lineages in the last 6 Myr. Most lineages in these clades diverged more than 10 Mya (Tolley et al., 2013). While savanna and xeric forests are younger systems than forest, the perch structure (diameter) available to chameleons may be similar to that of primary forests. Thus, these open-canopy chameleons may not have undergone diversification when these habitats emerged, with chameleons relying on the trees and bushes despite the more open aspect of the macrohabitat. Regardless, these are typically very widespread species, suggesting they are generalists that can tolerate a diverse range of vegetation structure. Unlike species of most other genera, *Chamaeleo dilepis*, in particular, is often observed running across the ground as it heads for the next available tree or bush. This species performs equally well as the terrestrial congener, *C. namaquensis*, in running trials (Herrel et al., 2013), suggesting that it is highly vagile and prone to dispersal when necessary. Such enhanced performance may be the case with most *Chamaeleo* as compared with range-restricted taxa. Given their widespread distribution and the apparent ability of these species to tolerate a spectrum of habitats, it is possible that certain traits characteristic to this genus resulted in a phenotype that is suitable for range expansion similar to that found in the global radiation of toads (i.e., Van Bocxlaer et al., 2010).

7.2 DIVERSITY AND DISTRIBUTION

There are at least 196 described species of chameleons (Chapter 8, Appendix) within 11 genera, 7 of which are in Africa, 3 are in Madagascar, and 1 is endemic to the Seychelles. Fifty-five percent of species are found in Africa and 44% in Madagascar, with the remaining few species in Asia and Europe. The most widespread genus is *Chamaeleo*, with only 14 species but a near pan-African distribution for the genus. Most *Chamaeleo* species are found in savanna and occasionally forests, but are essentially excluded from the true rainforest and montane forest, the Sahara, and the xeric regions of Southern Africa. An exception is *C. namaquensis*, which is the only chameleon to live entirely as a terrestrial desert specialist. This species diverged early from other *Chamaeleo* (ca. 40 Mya), probably because of isolation and adaptation in its unique desert habitat, which has persisted in some form since the early Oligocene (van Zinderen Bakker, 1975; Tolley et al., 2013). Two species of *Chamaeleo* are found in Africa and the Mediterranean region (*C. chamaeleon*, and *C. africanus*). Although there are population-level differences consistent with migration from Africa to Europe, the time frame of dispersal has not been established but appears to be recent (Kosuch et al., 1999; Paulo et al., 2002; Dimaki et al., 2008). Three species are found in Asia (*C. arabicus* in Yemen and Oman, *C. calyptratus* in Yemen, and *C. zeylanicus* in India and Sri Lanka) and are related to north African/European species (*C. chamaeleon*, *C. africanus*, and *C. calcaricarens*), diverging from a common ancestor in the Miocene (Tolley et al., 2013). One species is endemic to Socotra Island (*C. monachus*) and this may correspond with vicariance, as Socotra fragmented from Africa about 20 Mya, although overseas dispersal cannot be ruled out (Macey et al., 2008; Fournier et al., 2010; Tolley et al., 2013). The north African, European, and Asian species are a monophyletic clade, and they diverged from other African species in the early Miocene (Tolley et al., 2013). The distribution of this clade is consistent with biogeographic regions of Africa, the strong biotic links between North Africa and

Europe (Linder et al., 2012), and an exchange of biotic elements between Africa and Asia particularly in the Miocene (Chevret and Dobigny, 2005; Gaubert and Cordeiro-Estrela, 2006).

African Genera

The *Bradypodion,* with 17 species, are limited to South Africa, with one species (*B. setaroi*) ranging into Mozambique. They occur around the southern margin of the continent, and are excluded from the arid interior. Most species have narrow distributions in coastal or montane forests, some of which are found in a single forest patch (e.g., *B. ngomeense* and *B. damaranum*) or in a few small forest fragments (e.g., *B. nemorale, B. caeruleogula,* and *B. caffer*). Several species are widespread (e.g., *B. occidentale, B. ventrale,* and *B. gutturale*), inhabiting low bushes in treeless environments, particularly in the succulent Karoo and Nama-Karoo biomes. A number of species have radiated into the fynbos and grassland biomes, presumably within the past 5 Myr as these vegetation types became dominant along the southern margin of Africa (e.g., *B. atromontanum, B. taeniabronchum,* and *B. melanocephalum*). These species utilize open habitats with fine perch structure, and have apparently adapted morphologically to habitat structure dissimilar from forest (Measey et al., 2009; Herrel et al., 2011; da Silva and Tolley, 2013). Forest-restricted species are typically much older lineages (mid-Miocene) than fynbos/grassland lineages, and they lack extant sister species, suggesting that extinction filtering through the loss of forest habitat since the Miocene has occurred (Tolley et al., 2008).

Chamaeleo is the most widespread genus of chameleons, ranging from South Africa to Europe and Asia, spanning a distance of nearly 8000 km. It is the only genus found on more than one landmass—that is, across Africa, coastal regions of Europe, Socotra Island, and parts of the near East and India. Aside from a few atypical records (Tilbury, 2010), in Africa, *Chamaeleo* is absent only from the driest areas, particularly the Sahara and the Namib regions and the true rainforests. Despite its wide distribution, the genus is not particularly specious, with only 14 species. Members of the genus are archetypical, almost always resembling the concept of what a chameleon should look like (e.g., *C. dilepis* and *C. chamaeleon*). Most species are considered generalists because they are often found in diverse habitats, on a wide range of vegetation, and often crossing the ground at a surprisingly quick pace. Perhaps the most radical member of the genus is *C. namaquensis,* a fully terrestrial chameleon found primarily in the xeric regions of Namibia and South Africa, where it inhabits gravel plains and low dunes (Burrage, 1973; Herrel et al., 2013).

Kinyongia, with 18 species, are basically restricted to Afromontane forests of the Eastern Arc Mountains, the Albertine Rift, and the Kenyan Highlands. They are arboreal, forest specialists, usually found deep in the canopy, although a few species can tolerate altered habitats (e.g., *K. boehmei*) and at least one species has been observed in high-altitude shrubby habitats (Greenbaum et al., 2012). Because of their strong habitat preferences, their evolutionary history and present-day fragmented distribution is the result of forest dynamics since the Oligocene (Tolley et al., 2011). There are three main clades of *Kinyongia,* and these are allopatric in distribution. The Eastern Arc Mountains are divided, with one clade in the south (e.g., Nguru, Uluguru, and Udzungwa) and the other to the north (e.g., the Taita, Kilimanjaro, Meru, Usambara, and Pare mountains). The Nguru and East Usambara mountains, however, contain members of both clades (*K. fischeri* and *K. tenuis*, respectively), indicating some form of contact between the northern and southern mountain blocks during the Miocene (Tolley et al., 2011). Several species have colonized Neogene volcanic mountains (e.g., *K. uthmoelleri* and *K. tavetana*), but none of these lineages are endemic to these volcanic mountains, pointing to recent dispersal events rather than long-term isolation (Tolley et al., 2011). The remaining clade contains species found in both the Albertine Rift and the Kenyan Highlands, which is consistent with the strong biogeographic association between these two regions (Lovett, 1993; Fjeldså and Lovett, 1997).

The genus *Nadzikambia* has only two known species, *N. mlanjensis* and *N. baylissi*, both of which are restricted to high-altitude forest, above elevations of 900 m on Mount Mulanje (Malawi) and Mount Mabu (Mozambique), respectively. While the Mulanje chameleon has been known for a number of years (Broadley, 1965), and was thought to be monotypic (Tilbury et al., 2006), recent exploration on nearby Mabu revealed its sister taxon (Branch and Tolley, 2010). Undeniably, not only Mabu but also other nearby inselbergs, such as Chiperone, Namuli, and Inago, hold promise for exciting new discoveries of chameleons. These inselbergs are neglected in terms of biodiversity inventories, yet they consistently turn up with new taxa, including herpetofauna—for example, vipers (*Atheris;* Branch and Bayliss, 2009), freshwater crabs (*Potamonautes;* Daniels and Bayliss, 2012), and bats (*Rhinolophus;* Monadjem et al., 2010).

The genera *Rhampholeon* and *Rieppeleon*, along with the Malagasy genus *Brookesia* are sister to all the other chameleons. These genera comprise fairly similar-looking lizards: diurnal, leaf-litter dwellers, very small in size, with a short tail and, especially in the African species, often a soft snout appendage. Most of the species are associated with montane wet forests or coastal forests. The genus *Rhampholeon* comprises 14 described extant species arranged into three subgenera—*Rhampholeon* (*Rhampholeon*), *Rhampholeon* (*Rhinodigitum*) and *Rhampholeon* (*Bicuspis*)—and all species but one occur in east Africa and are often restricted to isolated forest patches or single mountain blocks (Tilbury, 2010). The diversification of the *Rhampholeon* lineages broadly coincides with landscape uplifts and associated periods of forest desiccation; in fact, sister taxa within *Rhampholeon* tend to be geographic neighbors, inhabiting close, isolated montane forests, strongly suggesting regional aridification and the resulting vicariance as the driving force behind speciation in *Rhampholeon* (Matthee et al., 2004).

The only West African species, *R. spectrum*, belongs to the subgenus *Rhampholeon* and is widespread in the closed-canopy forests of Cameroon, Central African Republic, Equatorial Guinea, and Democratic Republic of the Congo, with an isolated population on Bioko Island. The other species in the subgenus *Rhampholeon* (*R. spinosus, R. temporalis,* and *R. viridis*) occur on the other side of a huge gap encompassing most of the Congo River, basin in the forest fragments of the Pare and Usambara mountains, in the northern part of the Eastern Arc Mountains of Tanzania (Matthee et al., 2004; Mariaux and Tilbury, 2006; Tilbury, 2010). The subgenus *Rhinodigitum* is composed of eight forest-associated species (*R. acuminatus, R. beraduccii, R. boulengeri, R. chapmanorum, R. moyeri, R. nchisiensis,*

R. platyceps, and *R. uluguruensis*), which are widespread across the mountainous regions of Tanzania, the Albertine Rift and the Southern Rift, south to southern Malawi, with a major radiation of species in the Central and southern Eastern Arc Mountains. Several populations in the region still need taxonomic assessment (Tilbury, 2010). The subgenus *Bicuspis,* which is composed of only two species (*R. gorongosae* and *R. marshalli*), occurs in the Highlands of Eastern Zimbabwe and on Mount Gorongosa in central Mozambique.

The genus *Rieppeleon* is composed of three species with an east African distribution; they occur only in the eastern branch of the Great Rift Valley and are almost completely confined to coastal habitats. Molecular analyses suggest that a single lineage persisted for about 20 Myr and subsequently diversified into three distinct lineages (Matthee et al., 2004). *Rieppeleon brachyurus* and *R. brevicaudatus* are widespread across coastal forest and moist miombo of southern Kenya, Tanzania, and southern Malawi, with isolated populations in some Eastern Arc mountains (both species) and Shire highlands and Zomba Mountain (*R. brachyurus*). *Rieppeleon kerstenii* has successfully adapted to a variety of habitats, including arid, semidesert areas, unlike the other members of *Rieppeleon* and *Rhampholeon* (Tilbury, 2010).

The *Trioceros,* with 40 species, is the most speciose genus of chameleons and is widespread across tropical Africa, from Ethiopia on the north to Mozambique in the south. Along with its sister genus *Kinyongia,* the genus *Trioceros* represents a radiation of arboreal chameleons and likely originated during the Eocene, when up to 60% of the African continent had an equatorial climate dominated by forest habitat (Scotese, 2002). The majority of species are still associated with forest habitat, especially on mountains. Some species however, occur in areas dominated by other types of vegetation, but always with relatively high rainfall (e.g., *T. rudis, T. sternfeldi, T. schubotzi,* and *T. tempeli* on montane and upper montane forest/grassland ecotones and scrubby areas, *T. goetzei* on montane grasslands, and *T. melleri* on coastal semidecidous forests of East Africa). They are virtually absent from the lowland hot and dry savannas, and this could be interpreted as a high degree of niche conservatism for montane habitats. *Trioceros* are typically range-restricted species, occurring on a single mountains range (*T. narraioca, T. laterispinins, T. marsabitensis,* and *T. balebicornutus*) or a few forest fragments (*T. werneri, T. harennae,* and *T. incornutus*). *Trioceros* is the only genus in which the development of annulated bony horns is present (Tilbury, 2010), often in both sexes. This feature can be striking, with up to six bony horns in *T. quadricornis,* although the more common condition is two or three in males and one or none in females (*T. montium, T. werneri, T. jacksonii,* and *T. deremensis*).

Malagasy Genera

There are three chameleon genera in Madagascar (*Brookesia, Calumma,* and *Furcifer*) and one in the Seychelles Islands (*Archaius*) which lie 800 km northeast of Madagascar. Chameleons are also found on some of the coastal islands, like Nosy Be, which has at least six different species in sympatry from all Malagasy genera (Andreone et al., 2003). The bulk of species are confined to northern and eastern rainforests, with few species found in the dry southern regions. The predominantly terrestrial living *Brookesia* (commonly known as leaf or pygmy chameleons) are extremely diverse, with 32 species. They are exceptionally cryptic, and by means of coloration and spinous scalation, they easily resemble leaves, moss, lichen, or bark. Although all are small in body size, one group (*B. minima*) shows extreme miniaturization, and includes one of the smallest amniote vertebrates on Earth (*B. micra,* with a body length of 16 mm; Glaw et al., 2012). *Brookesia* are primarily distributed in the north, in wet tropical rainforests and karstic dry forests, with most species being extremely range-restricted (Glaw and Vences, 2007). Fewer species occur in the rainforests of the east and the dry forests of the west. This genus is also distinctive among chameleons because they form a separate, ancient radiation from all other genera and species (Townsend et al., 2009; Tolley et al., 2013).

The diversity within the genus *Calumma* is also extraordinary, with 33 species described (Raxworthy and Nussbaum, 2006; Glaw and Vences, 2007; Gehring et al., 2010) and incredibly another 26 highly divergent lineages that have been termed "candidate species," owing to both their genetic and morphological distinctiveness (Gehring et al., 2012). Indeed, this genus contains a bewildering array of diverse morphological forms, with colorful body patterns and outrageous-looking rostral processes. Like the *Brookesia,* they are distributed mainly in the northern and eastern rainforests (Glaw and Vences, 2007); however, they are predominantly arboreal, living deep in the canopy.

Twenty-two species are known from the genus *Furcifer,* all of which tend to be large-bodied chameleons (e.g., body length >100 mm). The genus is widely distributed across Madagascar, from the dry, open forests in the south to the rainforests in the north and east. The majority of species are found in the arid south and west regions (e.g., *F. tuzetae, F. angeli, F. antimena,* and *F. labordi;* Glaw and Vences, 2007). Individual species can be range-restricted (e.g., *F. belalandaensis,* and *F. timoni*), but other species are tolerant of different habitats, elevations, and even transformed landscapes (e.g., *F. pardalis, F. lateralis,* and *F. oustaleti*). Some of the largest chameleons are in this genus, *F. oustaleti* (body length, 280 mm), *F. pardalis* (body length, 250 mm for males), and *F. verrucosus* (body length, 265 mm for males), although the largest chameleon recorded is *Calumma parsonii* (body length, 295 mm for males).

There is a single species in the genus *Archaius* (*A. tigris*), which is endemic to the Seychelles Islands, making this the most range-restricted chameleon genus. It is a monotypic genus, having diverged from its sister group, the African genus *Rieppeleon,* some 35 Mya (Townsend et al., 2011b; Tolley et al., 2013). Until recently, it was thought to be within the genus *Calumma,* with its closest relatives on Madagascar (Klaver and Böhme, 1986; Raxworthy et al., 2002; Glaw and Vences, 2007), but inclusive molecular phylogenies have shown that this is a case of oceanic dispersal from Africa.

7.3 REGIONAL DIVERSITY

Madagascar is one of the most biologically diverse regions on Earth, despite its small geographic size (comparable in area to Kenya or France). Its rich biotic assemblage evolved in near isolation, since its initial separation from Africa about 140 Mya and India about

88 Mya (Seward et al. 2004; Yoder and Nowak, 2006). This isolation has allowed the fauna and flora of the island to diversify to such a degree that it is recognized as one of the top 10 most diverse regions on the planet in terms of species richness and endemism (Myers et al., 2000; Mittermeier et al., 2004). Madagascar has a totally unique native amphibian fauna, and all native terrestrial mammals are endemic, as are most reptiles (Goodman and Benstead, 2003, 2005). Much of its biota is microendemic, often isolated in small forest patches, particularly in the north and east of the island or associated with montane regions (Goodman and Benstead, 2005; Wilmé et al., 2006; Wollenberg et al., 2008; Vences et al., 2009; Glaw et al., 2012). Despite its isolation, oceanic dispersal to and from Madagascar has occurred for multiple taxa (e.g., Vences et al. 2003, 2004; Rocha et al., 2005; Yoder and Nowak, 2006), including chameleons (Raxworthy et al., 2002; Tolley et al., 2013). For its geographic size, the diversity of chameleons in Madagascar is exceptional (Raselimanana and Rakotomalala, 2003), rivaled only by that of the montane forests in the Eastern Arc. Two genera are endemic to the island (*Brookesia* and *Calumma*), with a third genus that is near-endemic (*Furcifer*) of which two species are on the Comoro Islands (*F. pollen* and *F. cephalolepis*) (Fig. 7.2 in the color insert). Nearly half of all known chameleon species (44%) occur in Madagascar, and studies suggest there are multiple as yet undescribed species that will be added to the inventory of this island (Gehring et al., 2012).

The *Brookesia* are found primarily in the north and eastern side of the island, particularly in rainforest patches, although a few species can be found in dry forest (Raxworthy and Nussbaum, 1995; Raselimanana and Rakotomalala, 2003) (Fig. 7.2). They are predominantly range-restricted, and several species (including *B. desperata,* and *B. micra*) even have some of the smallest known ranges for chameleons, possibly less than 10 km^2 and certainly less than 50 km^2 (Raxworthy and Nussbaum, 1995; Raselimanana and Rakotomalala, 2003; Glaw et al., 2012; F. Glaw, personal communication). Northern and eastern Madagascar is also home to numerous species of *Calumma,* particularly in the humid rainforest, and in some cases in high-altitude heathland up to 2800 m (i.e., *C. tsaratananense;* Glaw and Vences, 2007; Raxworthy and Nussbaum, 1996). The strong overlap in distribution of these two genera (Fig. 7.2) certainly contributes strongly to the high overall species richness in the north and east, which in turn correspond with the distribution of montane and lowland rainforest. *Furcifer* is the most widespread genus, found in virtually all parts of the island (Fig. 7.2) and in all habitat types (Glaw and Vences, 2007). The majority of species are not range-restricted, and some are distributed widely. Only 3 of the 20 *Furcifer* species on Madagascar have ranges smaller than 5000 km^2 (Table 7.1), of which one (*F. timoni*) can be considered extremely range-restricted (ca. 385 km^2), probably occurring only at Montagne d'Ambre National Park. Two species (*F. polleni* and *F. cephalolepis*) occur on the Comoros Islands (Mayotte and Grand Comoro, respectively) located in the Mozambique Channel, some 300 to 500 km from Madagascar (Hawlitschek et al., 2011).

East Africa is a broad geographic region encompassing several countries: Ethiopia, Kenya, Uganda, Rwanda, Burundi, Tanzania, Malawi, and northern Mozambique. The major

TABLE 7.1 Areas of High Chameleon Diversity

Distribution Area	No. of Species	Amount of Range Restricted	Range-Restricted Species
< = 200 km²			
Madagascar	67	12 (18%)	*Brookesia bekolosy, B. bonsi, B. confidens, B. desperata, B. micra, B. tristis, B. tuberculata, Calumma capuroni, C. hafahafa, C. jejy, C. peyrierasi, C. tarzan, C. tsaratananense*
Eastern Arc and Southern Rift	40	8 (19%) (29%)	*Kinyongia magomberae, Nadzikambia baylissi, N. mlanjensis, Rhampholeon acuminatus, R. chapmanorum, R. beraducii, R. viridis, Trioceros hanangensis*
Albertine Rift	13	0	none
Congolian highlands	9	0	none
Kenyan and south Sudan highlands	12	5 (42%)	*Kinyongia asheorum, Trioceros kinangopensis, T. marsabitensis, T. narraioca, T. ntunte*
Ethiopian highlands	5	0	none
Southern Africa	21	5 (24%) (21%)	*Bradypodion caeruleogula, B. caffer, B. nemorale, B. ngomeense, Rhampholeon gorongosae*
< = 5000 km²			
Madagascar	67	38 (57%)	*Brookesia ambreensis, B. antakarana, B. bekolosy , B. bonsi, B. confidens, B. decaryi, B. dentata, B. desperata, B. exarmata, B. karchei, B. lineata, B. lolontany, B. micra, B. perarmata, B. peyrierasi, B. ramanantsoai, B. tristis, B. tuberculata, B. valerieae, Calumma amber, C. ambreense, C. andringitraense, C. capuroni, C. fallax, C. furcifer, C. gallus, C. glawi, C. guibei, C. hafahafa, C. hilleniusi, C. jejy, C. peyrierasi, C. tarzan, C. tsaratananense, C. vatosoa, C. vencesi, C. vohibola, Furcifer balteatus, F. nicosiai, F. timoni*

TABLE 7.1 *(Continued)*

Distribution Area	No. of Species	Amount of Range Restricted	Range-Restricted Species
5000 km^2			
Eastern Arc and Southern Rift	40	30 (75%)	*Kinyongia boehmei, K. fischeri, K. magomberae, K. matschiei, K. multituberculata, K. oxyrhina, K. tavetana, K. tenuis, K. uluguruensis, K. uthmoelleri, K. vanheygeni, K. vosseleri, Nadzikambia baylissi, N. mlanjensis, Rhampholeon acuminatus, R. beraduccii, R. chapmanorum, R. gorongosae, R. moyeri, R. nchisiensis, R. platyceps, R. spinosus, R. temporalis, R. uluguruensis, R. viridis, Trioceros deremensis, T. fuelleborni, T. hanangensis, T. incornutus, T. laterispinis*
Albertine Rift	13	3 (23%)	*Kinyongia carpenteri, K. xenorhina, Trioceros schoutedeni*
Congolian highlands	9	1 (11%)	*T. eisentrauti*
Kenyan and south Sudan highlands	12	10 (83%)	*Kinyongia asheorum, K. excubitor, Trioceros conirostratus, T. kinangopensis, T. kinetensis, T. marsabitensis, T. narraioca, T. ntunte, T. nyirit, T. schubotzi*
Ethiopian highlands	5	2 (40%)	*Trioceros balebicornutus, T. harennae*
Southern Africa	21	14 (67%)	*Bradypodion atromontanum, B. caeruleogula, B. caffer, B. damaranum, B. kentanicum, B. nemorale, B. ngomeense, B. pumilum, B. setaroi, B. taeniabronchum, B. thamnobates, B. transvaalense, Rhampholeon gorongosae Rhampholeon marshalli*

geomorphological feature that has greatly shaped the current pattern of biological diversity across this entire ecoregion is the presence of the so-called Eastern Afromontane Biodiversity Hotspot (Mittermeier et al., 2004). It covers more than 1 million km^2 and includes three main ancient mountain ranges: (1) the Eastern Arc Mountains and Southern Rift Highlands, which stretch from southeastern Kenya to southern Tanzania, Malawi, and northern Mozambique, (2) the Albertine Rift, which includes parts of Rwanda, Burundi, Uganda, Tanzania, and the Kivu region in the Democratic Republic of Congo, and (3) the Ethiopian Highlands, which cover much of Ethiopia, as well as small parts of Eritrea, Djibouti, and Sudan, and are bisected by the Great Rift Valley (Mittermeier et al., 2004). In addition to these three principal massifs, a number of outlying montane isolates are part of this hotspot, including the Neogene volcanic highlands of the Kenyan Highlands (e.g., Mount Kenya, Mount Elgon, Aberdares Range, and Marsabit Range) and Tanzanian volcanic northern Highlands (e.g., Mount Kilimanjaro, Mount Meru, Mount Kitumbeine, and Mount Hanang). These mountains are diverse in age, with a basement of several hundred million years in age, and the most recent uplift occurring approximately 7 Mya, but with major volcanic building events only a few million years ago (Griffiths, 1993; Chorowicz, 2005; Le Gall et al., 2008; Nonnotte et al., 2008).

The extremely complex geomorphological history of the region has produced an array of habitats that certainly contribute to building the extraordinary biodiversity of the region (Brooks et al., 2002). First, the Eastern Arc Mountains and Southern Highlands of Tanzania are remarkable for their high species richness and levels of endemism across numerous taxa, chameleons included (Tolley et al., 2011; Lovett and Wasser 1993; Brooks et al., 2002; Burgess et al., 2007; Davenport et al., 2006; Menegon and Davenport, 2008; Menegon et al., 2008). Patterns of diversity and distribution within the forest-dwelling fauna are intricate, probably reflecting countless fluctuations in climate and long-term persistence of rainforests, the latter being critical in structuring current biodiversity in Eastern Arc montane isolates. This is undoubtedly demonstrated by the prevalence of deeply divergent, endemic monophyletic Eastern Arc montane taxa (Wilkinson et al., 2003; Loader et al., 2004; Tolley et al., 2011). The archipelago stretches across 1000 km, from the Livingstone Mountains in southern Tanzania to the Taita Hills in southern Kenya (Lovett and Wasser, 1993; GEF, 2002), and its proximity to the Indian Ocean ensures high rainfall (e.g., 3000 mm/yr on the eastern slopes of the Uluguru Mountains, falling to 600 mm/year in the western rain shadow) (GEF, 2002). This orographic effect produced relatively stable climatic conditions for at least 30 million years (Axelrod and Raven, 1978). The high rainfall and long-term climatic stability, together with the fragmentation of the mountain blocks, resulted in montane forests that are both ancient and biologically diverse. An intriguing consequence is that several chameleon genera have evolved en masse as forest specialists and are rarely found outside these isolated montane patches. That is, most species of *Kinyongia* and *Rhampholeon* are strictly endemic to montane forest, as are the species in the viviparous clade of *Trioceros* (Fig. 7.2 in the color insert).

The Albertine Rift is also a region of exceptionally high species diversity, with more endemic and threatened vertebrates than elsewhere in Africa (Plumptre et al., 2007). It also contains at least 16% of Africa's plant species, with over 6400 species identified from the region. Despite this, chameleon richness is not notably high, with only 13 species, of which 3 are range-restricted (Table 7.1). Range-restricted taxa, however, are all montane specialists, again highlighting the importance of montane isolates for harboring diversity. Because this region is less explored than the Eastern Arc in terms of chameleons, it could potentially harbor more diversity than is currently known.

In the Ethiopian Highlands, there are two main forested areas, which are notable for their exceptional diversity. On the eastern side of the Great Rift Valley, the Bale Mountains soar to heights of nearly 4400 m and form an extensive massif that harbors an impressive assemblage of biodiversity, with many unique, rare, and spectacular but also threatened taxa. The region incorporates the largest single area (>4,000 km^2) in Africa over 3000 m in altitude: consequently, it is a crucially important reservoir of Afromontane and Afroalpine habitats, both of which are globally threatened (Largen and Spawls, 2010). The highest part of the Bale Mountains, the Sanetti Plateau, and associated peaks comprise Afroalpine grasslands, with their own highly endemic biota. To the South, the Sanetti Plateau terminates at the steep Harenna escarpment that rapidly drops from 3200 to 2000 m in just 8 km. This escarpment exhibits a dramatic vegetation gradient from the plateau grasslands through low heather scrub, giant heather forest (*Erica* spp.), Hagenia, bamboo (*Arundinaria*) and *Podocarpus* forest, high-altitude woodland, and finally dropping into medium-altitude rainforest (Friis, 2010; Miehe and Miehe, 1994). Although the region is considered diverse in its overall fauna and flora, the chameleon diversity is not high and is limited to only five species in the genus *Trioceros* and *Chamaeleo* (Table 7.1). Outside these varied and diverse mountain ranges of East Africa, a lengthy stretch of coastal forest once covered the region from Kenya through Mozambique, but this is now highly fragmented because of human impact. The vast expanse of Miombo woodlands and dry savanna is exceptionally low for chameleon richness, with a paucity of species, all of which are widespread (e.g., *Chamaeleo dilepis, C. africanus,* and *C. gracilis*).

Southern Africa, the region south of the Zambezi River in the east, and the Cunene River in the west (ca. 17°S), is primarily a mesic to xeric environment, and lacks extensive patches of forest that characterize the tropics. The area is best characterized by grassland and savanna, particularly in the interior, which leaves little chameleon habitat. The western margin is dominated by xeric habitats, including the Namib desert, which has persisted in some form since the early Oligocene (ca. 35 Mya), in part because of the South Atlantic anticyclonic system and the cold Benguela current, which have an aridifying effect along the coast. The southern margin of the continent is home to the Cape Floristic Region, which consists of two unique vegetation types, the fynbos and the succulent Karoo, plus patches of Afromontane forest that are fed by moisture from the warm Aghulas Current. The eastern margin is tropical to semitropical, and it is here that the landscape becomes a mosaic of forest patches and grasslands within which chameleons thrive. Afromontane forests are scattered from the Drakensberg Mountains northward to the Eastern Highlands of Zimbabwe. The mountains fall away to the Maputo-Pondo-Albany (MPA) region, an eastward-lying escarpment that hosts a unique form of Scarp Forest, and in turn this area descends to the coastal region with a distinct coastal belt forest that is floristically disjunct from Afromontane. The extensive habitat heterogeneity in the MPA is most certainly a contributing factor to the high species richness and phylogenetic diversity of that region (Tolley et al., 2008).

Although not as species rich as Madagascar and East Africa, Southern Africa is home to many endemic species. This includes 17 species of *Bradypodion*, 2 of *Rhampholeon* (*R. marshalli* and *R. gorongosae*), and the only fully terrestrial desert chameleon, *Chamaeleo namaquensis*. While *C. namaquensis* has a large range, along the eastern margins of South Africa and Namibia, it is excluded from regions with rocky terrain and highlands. It is a specialist in open terrestrial habitats and therefore does not overlap in distribution with any other chameleon species through most of its range, except perhaps marginally with *C. dilepis* (Tilbury, 2010). Only 1 of the 21 species found in Southern Africa is not endemic, and not surprisingly, this is the pan-African species, *C. dilepis*. Most species are narrow endemics, often confined to Afromontane forest patches (e.g., *B. damaranum* in the Knysna Forest), coastal forests (e.g., *B. caffer* and *B. setaroi* along the Indian Ocean coastal belt forests), or fynbos (e.g., *B. taeniabronchum* and *B. atromontanum*).

West and Central Africa are both dominated by a vast expanse of tropical rainforest interspersed with woodland and savanna, and three main broad biogeographic regions are recognized: Congolian, Guinean, and Shaba (Linder et al., 2012). The lowland Congolian forest encompasses most of the Congo River valley from the western coast of the continent, east to the Albertine Rift, representing the second biggest forest on the planet, although it is considered a rather homogeneous habitat with low species diversity (Linder et al., 2012). The Guinean region is divided into two distinct forest ecoregions: (1) the Upper Guinea forest ecoregion which stretches along the Atlantic coast from southern Guinea into eastern Sierra Leone, through Liberia, Côte d'Ivoire, and Ghana into western Togo, and (2) the Lower Guinea forest ecoregion, which extends along the Atlantic coast from western Nigeria to the Sanaga River in southwestern Cameroon. The two subregions are separated by the Dahomey Gap in Benin, a mosaic of savanna woodland and dry forest, although it is primarily transformed for agriculture at present. The Dahomey Gap is the only portion of the savanna–forest mosaic to breach the coastal forest belt, which otherwise runs from Western Senegal to Eastern Nigeria (Salzmann and Hoelzmann, 2005). The Guinean forests consist of a range of distinct vegetation zones varying from moist forests along the coast, freshwater swamp forests, and semideciduous forests inland with prolonged dry seasons. Of all West African countries, only Liberia lies entirely within the moist forest zone, although a substantial portion of Sierra Leone also falls within this zone. The area includes several important forest refugia created by the retraction and fragmentation of forests during the Pleistocene, particularly on montane regions, including the Cameroon and Nimba Highlands (Stuart and Adams, 1990). The Shaba region separates the Guinea–Congolian forests from the Zambezian savannas and consists of a broad transitional strip that stretches from the Angolan coast northeastward to the southern Albertine Rift

(Linder et al., 2012). Only the widespread genus *Chamaeleo* is found throughout west and central Africa across all three ecoregions. However, there are three genera (*Chamaeleo, Trioceros,* and *Rhampholeon*) that are localized at the interface of the Guinea–Congolian ecoregions (Fig. 7.2 in the color insert). In particular, there are numerous *Trioceros* species in Cameroon highlands (lower Guinea forest ecoregion), making this area the most species-rich within these three bioregions.

Chameleons are not well represented in North Africa, Europe, or Asia, with just four species of *Chamaeleo* found in the region (*C. africanus, C. chameleon, C. calyptratus,* and *C. zeylanicus*) (Fig. 7.2). Aside from *C. zeylanicus* in India, distributions do not extend far from the continental margin probably because of unsuitable climatic conditions. In Europe and Africa, chameleons are restricted to the fringe of the Mediterranean and are presumably prevented from dispersing northward because of the cooler climates or southward (in Africa) because of extreme xeric conditions across the Sahara. Similarly, chameleons do not range into the interior of the Arabian Peninsula, although are reasonably abundant along the west and south coasts. Chameleons were more widely distributed in Europe during the Miocene (Chapter 9), with a number of *Chamaeleo* fossils found in Europe (primarily in Germany, Czech Republic, and Switzerland). Chameleons, therefore, were more widespread in these northern regions during warmer global climate conditions, but presumably were excluded as the climate cooled in the late Miocene into the Pliocene (Zachos et al., 2001).

Chameleons cannot be considered excellent dispersers over water, nevertheless they are found on many coastal and oceanic islands, particularly in the western Indian Ocean. Given that most of these islands have never been connected to larger landmasses, oceanic dispersal is indeed achievable for chameleons. The most extreme case of oceanic dispersal is of *Archaius tigris,* which is endemic to the Seychelles Islands and whose closest relatives are the *Rieppeleon* from East Africa (Townsend et al., 2011b). These islands have been isolated for ca. 65 Myr and at present, they are more than 1200 km from the closest point in Africa (Somalia) but at least 1700 km from Tanzania, where *Rieppeleon brevicaudatus* occurs. This is the only chameleon that has managed to disperse such a large distance over water, presumably carried by prevailing paleo-currents on floating vegetation during the Oligocene (Townsend et al., 2011b). Other examples of long-distance oceanic dispersal are less extreme, but still remarkable. Two species of *Furcifer* have probably dispersed independently from Madagascar across the Mozambique Channel to colonize Grand Comoro (*F. cephalolepis,* 550 km from Madagascar) and Mayotte (*F. polleni,* 300 km from Madagascar) (Rocha et al., 2005). Approximately 250 km from the Horn of Africa lays Socotra Island, which has a single endemic species, *Chamaeleo monachus,* which diverged from its African sister taxa approximately 20 (±5) Mya. This island is a Gondwanan fragment, and the timing of divergence for *C. monachus* overlaps with the timing of separation between Socotra and Africa (ca. 20 Mya) (Girdler and Styles, 1978; Hébert et al., 2001; Bosworth et al., 2005; Fournier et al., 2010) suggesting vicariance, although oceanic dispersal cannot be entirely ruled out (Tolley et al., 2013). Off the east coast of Africa, populations of the widespread species *C. dilepis* are found on the coastal islands of Zanzibar and Pemba. Although Pemba is isolated from Africa by deep waters isolating it from the continent, for Zanzibar this could be the result of either overwater dispersal, or overland dispersal during the last ice age, when this island was connected to Africa (Agnarsson and Kuntner, 2012). In a similar example, *C. zeylanicus*, which is relatively widespread in India, is also found on the island of Sri Lanka, which lies just 50 km off the Indian coast but is separated from India by shallow waters that probably would have provided subaerial connections during ice ages, allowing overland dispersal of chameleons. Although *F. pardalis* is found on Réunion Island (700 km east of Madagascar) and Mauritius (200 km northeast of Réunion), these are known to be human-mediated introductions rather than natural dispersal (Vinson and Vinson, 1969; Cheke, 1987; Cheke and Hume, 2008; Cole, 2009).

While oceanic dispersal in the Indian Ocean appears relatively common for chameleons, the same is not true in the Atlantic Ocean, although there are some examples. Four species of chameleon (*Trioceros cristatus, T. feae, T. oweni,* and *Rhampholeon spectrum*) are found on Bioko Island, which lies in the Gulf of Guinea. Only 30 km from Cameroon, and across shallow waters, this island has been periodically connected to the African continent (Lee et al., 1994) possibly allowing chameleons to disperse over land rather than over water. Chameleons are not found on São Tomé, Príncipe, or Annobón, all of which lie further out into the Gulf of Guinea (230 to 350 km) in waters up to 1800 m in depth. Terrestrial dispersal can be ruled out for these islands, as they have never been connected to land, and oceanic dispersal seems to have not occurred (unless chameleons have become extinct on those islands), despite the distances being less than some of the dispersals that chameleons have made to Indian Ocean islands. The widespread *Chamaeleo chamaeleon* has dispersed to a few Mediterranean islands (Cadiz, Chios, Crete, Malaga, Samos, and Sicily) (Tilbury, 2010). Although it occurs on the Canary Islands (100 km from Morocco), this is probably the result of a human-mediated introduction (Lever, 2003).

7.4 PATTERNS OF ALPHA DIVERSITY

For chameleons, two regions stand out as remarkable for diversity and endemism: northeastern Madagascar and the Afromontane forests of the Eastern Arc and Southern Rift Mountains (extending from the Taita Hills in Kenya to the Livingstone Mountains in southern Tanzania). These areas are characterized by species with restricted distributions, particularly in montane areas, and this contributes to the high richness and endemism (Fig. 7.3 in the color insert). While some other regions also show elevated diversity (i.e., high-altitude West African Congolian forests and Kenyan and Ethiopian highlands), none rival Madagascar and the Eastern Arc.

Although Madagascar is home to some 87 species of chameleons (44% of the total), most of the diversity is centered in the rainforests of the north and east. In particular, many species of *Brookesia* and *Calumma* are found in isolated forest patches in the north of the island (Raselimanana and Rakotomalala, 2003). Indeed, northern Madagascar has high species richness and endemism for a number of other taxa (Andreone, 2004; Andreone et al., 2009), suggesting a common cause at some level for the elevated diversity. In particular, strong habitat heterogeneity, climatic gradients, and complex topographic structure may contribute to both low levels of extinction and high speciation rates, which have produced the elevated species richness and endemism in this region (Andreone et al.,

2001; Andreone, 2004; Pearson and Raxworthy, 2009). Southern and western Madagascar are less diverse in terms of species numbers, possibly because of a dry climate resulting in more arid and sparse vegetation.

East Africa (Ethiopia, Kenya, Tanzania, Malawi, northern Mozambique, Rwanda, and Uganda) holds a high percentage (38%) of the world's chameleons. In particular, diversity is elevated in the archipelago-like system of mountain blocks of the Eastern Arc and Southern Rift, primarily because of the high turnover of range-restricted species across and within the submontane and montane forest fragments and the co-occurrence of forest-associated species belonging to different genera. The Udzungwa Mountains for example, are home to seven sympatric species of chameleons from various genera (*Kinyongia oxyrhina, K. magomberae, Rieppeleon brevicaudatus, Rhampholeon moyeri, Trioceros werneri, T. tempeli,* and *T. laterispinins*), as are the Usambara Mountains (*K. multituberculata, K. matschiei, K. vosseleri, K. tenuis, Rh. spinosus, Rh. temporalis,* and *Ri. brevicaudatus*) (Tilbury, 2010). While species diversity is high, phylogenetic diversity also shows elevated diversity in these regions, in part because of the presence of paleo-endemic lineages, which are retained in forest refugia (Tolley et al., 2011). Species turnover between mountain blocks is high because of the presence of multiple range-restricted species, which contrasts patterns in the Albertine Rift, Kenyan, Ethiopian, and Cameroonian highlands. These areas collectively represent a third hotspot of chameleon diversity at global level, but montane- and forest-associated chameleon species have comparatively wider ranges, encompassing more than one mountain block (*Rh. boulengeri, T. rudis, K. adolfifriderici, T. johnstoni, T. montinum,* and *T. quadricornis*) with fewer hyper-endemics (e.g., *K. asheorum*). Overall, this results in lower richness and a reduced rate of both spatial and altitudinal species turnover. The decreased level of species richness outside the Eastern Arc Mountains can also be attributed to a lack of comparable lineage radiations outside this region. For example, the Eastern Arc Mountains holds several near-endemic clades of chameleons, which have radiated in situ (i.e., *Kinyongia,* and *Rhampholeon*), whereas other regions lack these near-endemic genera.

The escarpment of Southern Africa has a high number of species, many of which are range-restricted, but these are spread over a large area with little or no sympatry. The result is low species richness at a local scale but a large number of species regionally. Species richness and phylogenetic diversity are highest in areas that are known biodiversity hotspots (Tolley et al., 2008), particularly the MPA region (7 species of *Bradypodion,* 1 species of *Chamaeleo*). The high diversity is primarily due to the presence of patches of Afromontane, Scarp, and coastal forests, to which chameleons are strongly tied. The lack of connectivity between forests has resulted in isolation and produced a number of endemic lineages in the region.

7.5 PATTERNS OF BETA DIVERSITY

The aggregation of geographic grid cells, based on species composition across the distribution of all chameleons species produced by cluster analysis of distributional data, reveals extensive areas of high congruence between geography and species occurrence

BOX 7.1 Biogeographic Analysis

Spatial analysis of species richness (based on number of species) per 1-degree grid cells across the entire distribution of chameleons was used to map out areas of high diversity (Fig. 7.3 in the color insert), using 4300 records of chameleon occurrences. A neighborhood method was applied, where the richness for each cell is scored as the number of species in that cell plus the number of species in the neighboring eight cells. Every grid cell is scored in this way, which minimizes the influence of arbitrary boundaries between the grid cells (Slatyer et al., 2007). Weighted endemism was estimated by applying a continuous weighting function to species range size, assigning high weights to species with small ranges and progressively smaller weights to species with larger ranges (Slatyer et al., 2007). For biogeographic clustering (Fig. 7.4 in the color insert) the β_{sim} dissimilarity metric was used with unweighted pair-group methods. Both the spatial neighborhood and cluster analyses were conducted in Biodiverse 0.17 (Laffan et al., 2010).

(Fig. 7.4 in the color insert) (Box 7.1). On the broadest scale, seven main clusters congruent with specific geographic areas are identified: Madagascar, India, Mediterranean/southwest Arabia, southeast Arabia, sub-Saharan Africa, Southern Africa, and a portion of the Ethiopian Highlands (although this may be due to a lack of records from Ethiopia). Madagascar shares no genera with any of the other regions (except the Comoros) in which chameleons occur (Mediterranean, Africa, near East, and India) setting it strongly apart. Genera are shared across all other regions, although India has no species in common with other areas.

If Madagascar is considered separately, three discrete biogeographic partitions are retrieved for chameleons within the island: northern, eastern, and southwestern (Fig. 7.4 in the color insert, inset). The regions retrieved for chameleons correspond with major biomes in Madagascar (i.e., the forest mosaic in the north, the humid high altitude forests in the east, and the dry deciduous and arid spiny forests of the southwest) (Yoder and Nowak, 2006). In fact, these chameleon bioregions correspond well with the large-scale biogeographic regions for Malagasy fauna (Angel, 1942; Glaw and Vences, 2007). The resolution in this current analysis, however, is too broad to determine whether chameleons correspond with fine-scale centers of endemism that are defined by watersheds in Madagascar (Wilmé et al., 2006).

Biogeographic clustering that excludes Madagascar and India better defines the biogeographic partitions of chameleons on the African continent and Mediterranean/near East, and reveals an intriguing pattern with 10 recognizable biogeographic regions for chameleons (Fig. 7.4 in the color insert). The previously identified Southern African biogeographic region is partitioned into: (1) a western region (South Africa and Namibia) that runs to Angola, and (2) the southeastern coast, which terminates near the transition from subtropical to tropical environments. This transition is consistent with broad-scale biogeographic zonation that is statistically defined on the basis of combined vegetational and faunal datasets (Linder et al., 2012). Congruent with the Zambesian region, an extensive homogeneous area extends from northern Namibia, encompassing Angola, Zambia, Zimbabwe,

Botswana, Mozambique, Tanzania, and southern Democratic Republic of the Congo. Interestingly, the north Kivu area, including Ituri, southern Uganda, and Western Kenya, form a well-defined zoogeographic zone due to the occurrence of several species with relatively similar ranges (e.g., *Rhampholeon boulengeri, Trioceros johnstoni,* and *Kinyongia adolfifriderici*). A narrow, elongated biogeographic partition encompasses the whole central African belt from the coast of Guinea in the west to Somalia. However, this is superimposed with two less-well-defined areas, one covering most of West Africa, from Guinea to Central African Republic, and the other on the Ethiopian Highlands. The remaining three partitions are quite well defined, covering: (1) the Sahelian belt with scattered outliers along the upper river Nile, up to the delta, (2) the Mediterranean region with western Arabia, and (3) the southern part of the Arabian Peninsula across Yemen and Oman.

As in other attempts to identify cohesive biogeographic units in Africa, chameleons fail to retrieve separate entities for some of the areas with highest species richness and endemism. This is particularly apparent for the Eastern Arc Mountains and the Cameroonian Highlands, but is similar to the broad-scale biogeographic divisions for Africa (Linder et al., 2012). This contrasts to the regions defined by White (1983) and Wickens (1976); they considered the Eastern Afromontane region as distinct. Arguably, the comparatively small distribution ranges of species in these mountainous regions results in high turnover throughout the area. The result is that the region stands out for its uniqueness, rather than its similarity to other adjacent biogeographic assemblages. This intriguing pattern is shared with other groups of montane-associated vertebrates (Poynton and Boycott, 1996; Menegon and Salvidio, 2005; Poynton et al., 2006), and suggests that there is little shared biota across the many mountain blocks that make up the entire Eastern Arc region.

Although chameleons range across Africa, Madagascar, and parts of Europe and Asia, the overwhelming majority of species are range-restricted endemics. In Africa, most endemics are isolated in forest fragments or in heathlands at high elevation, whereas in Madagascar, a large number of endemics also occur at low altitudes and mid-altitudes. Chameleons should be considered conservative in the niche that they occupy (i.e., forest floor or arboreal) and any reduction in the occurrence of this niche will strongly influence their range size. Certainly, for forest endemics, this conservatism has impacted their evolutionary history. In general, these montane forest lineages are paleo-endemic, with deep divergences between taxa (Fig. 7.1 in the color insert) (Tolley et al., 2013; Matthee et al., 2004; Townsend et al., 2009), although some high-altitude species are comparatively young (Tolley et al., 2006; Nagy et al., 2012). Lineage diversification was likely initiated and driven by fragmentation of forests, beginning in the Oligocene and many of these lineages are retained in stable forest refugia, although some genera have radiated to habitats such as savanna, heathlands, and grasslands.

Overall, patterns of elevated species richness are clearly associated with the montane regions of Madagascar and Africa and are heavily influenced by species turnover associated with montane regions, rather than by similarities in assemblages. In both areas, this may be due to the enormous habitat diversity that is afforded through the broad altitudinal zonation, which provides diverse climates and habitats. Madagascar, however, has a number of low-altitude endemics, whereas these are generally lacking in Africa. This could be the result of Madagascar's climate being less arid throughout the history of chameleons (Wells, 2003), allowing endemics to persist in low-altitude areas that remained moist enough to support suitable forested habitat. Undeniably, the evolution of chameleons is entwined with the dynamics of forests, and together they have traveled through time to produce the abundant assortment of species that adorns our planet today.

ACKNOWLEDGMENTS

We would like to thank Miguel Vences and Colin Tilbury for giving us access to species distributional data, Nik Cole for useful information about Mauritius and Réunion Islands, John Poynton for helpful comments on African biogeography, and particularly Frank Glaw for numerous discussions and helpful advice regarding Madagascar.

EIGHT

Overview of the Systematics of the Chamaeleonidae

COLIN R. TILBURY

Much of the current tertiary taxonomy of the Chamaeleonidae is based on the works of Charles Klaver and Wolfgang Böhme who produced an organized phylogeny of the family using a multifaceted organismal approach to better define relationships (Klaver and Böhme, 1986). Beyond some refining, their original phylogeny currently remains the defining work for the tertiary systematics of the family. This chapter will draw largely from their work and explore the evolution of the expanded taxonomic landscape since their 1986 publication.

Chameleons share with their sister family the Agamidae the synapomorphies of acrodont dentition, the absence of caudal autotomy, and the presence of a ramus duodenalis branch of the cecal artery. Macey et al. (1997a) also identified three structural features of the mitochondrial genome that were shared by all acrodont lizards. Attempts to include the other acrodontine family (Agamidae) within a broadened family concept of the Chamaeleonidae (Frost and Etheridge, 1989; De Quieroz, 1995) have not met with much subsequent support. The monophyly of the Chameleonidae is strongly supported by many distinctive morphological peculiarities that are shared by all members of the family such as independently mobile turreted eyes, prehensile tail, "ballistic" tongue, opposable toes bound in uneven bundles, compressed body habitus, nonoverlapping scales and lack of osteoderms (Chapter 2). Many of these unusual characteristics are considered adaptations to an arboreal lifestyle.

Beyond various attempts to recognize several genera (Gray, 1865; Fitzinger, 1843; Angel, 1933, 1942) there were few early attempts to arrange the many various species of chameleons into an organized phylogeny. Hillenius (1959) divided the typical chameleons into "groups" based on shared external characteristics but did not attempt to derive a phylogeny. Previous works on skeletal and cranial structure covered only a small range of species (Vrolik, 1827; Siebenrock, 1893; Brock, 1941; Visser, 1972; Rieppel, 1981; and others), while several investigations into chromosome morphology (Brink, 1957; Matthey and van Brink, 1956, 1960; Matthey, 1957; Bourgat, 1973) covered around only 40 different taxa. Matthey (1957) identified two main groups of chromosomes—an Insular (Malagasy) group with 12 different chromosomal patterns, and a Continental (African) group with 4 patterns. Thereafter began a period of research that focused on the soft anatomy of lungs (Klaver 1973, 1977, 1979, 1981) and hemipenes (Böhme and Klaver, 1980). Klaver (1981) developed a familial reconstruction rooted on lung morphological characteristics but decided not to formally revise the classification at that time pending further work. Subsequently, Klaver and Böhme (1986) presented a formal revision of the family that utilized a combination of all the character sets including karyological, osteological, hemipenal, and lung morphology. The combined evidence suggested that the two genera generally recognized at that time (*Chamaeleo* and *Brookesia*) were paraphyletic. Although they were unable to investigate all the extant taxa, based on the amalgamated evidence they provisionally divided the family into two subfamilies and six genera. The two subfamilies consisted of the largely terrestrial pygmy group Brookesiinae and the more "typical" longer-tailed and largely arboreal Chamaeleoninae. The former encompassed the Malagasy genus *Brookesia* and the African genus *Rhampholeon,* and the latter included the remaining genera. Three genera were recognized in Madagascar (*Furcifer, Calumma,* and *Brookesia*) and three in Africa (*Rhampholeon, Bradypodion,* and *Chamaeleo,* with the latter further divided into two subgenera—*Chamaeleo (Chamaeleo)* and *Chamaeleo (Trioceros)*). The genus *Bradypodion* combined a number of East African species with a southern African group that had similar lung and hemipenal morphologies but differed with respect to other character states and reproductive strategy.

Understanding that the data sets were incomplete, Klaver and Böhme recognized that their phylogeny could only be provisional and would need to be tested by other comparative studies. Since 1986, their phylogeny has largely stood the test of time and has been widely accepted and used in almost all subsequent systematic papers. That is not to say that there has been no change. A number of papers have examined aspects of the higher systematics of the Chamaeleonidae (Rieppel and Crumly, 1997; Townsend and Larson, 2002; Raxworthy et al., 2002; Matthee et al., 2004; Tilbury et al., 2006; Tilbury and Tolley, 2009a; Townsend et al., 2011b; Tolley et al., 2013; Chapter 7). The major tertiary taxonomic changes that have occurred since 1986 have all involved African taxa and include the splitting of *Rhampholeon (sensu lato)* into two genera (*Rhampholeon* and *Rieppeleon;* Matthee et al., 2004), the splitting of *Bradypodion* (sensu Klaver and Böhme, 1986) into three genera (*Bradypodion, Kinyongia* and *Nadzikambia*; Tilbury et al., 2006) and the elevation of *Trioceros* from subgeneric to generic rank (Tilbury and Tolley, 2009a). All recent literature concerning the African Chamaeleonidae appears to have adopted the proposed changes. In 2010, the Seychelles endemic species *tigris* was allocated to the resurrected monotypic genus *Archaius* (Townsend et al., 2011b). Unless there is a reversal of some or all of the above taxonomic changes, it might be anticipated that the current genera allocation should stabilize.

8.1 EVOLUTION OF METHODOLOGY IN CHAMELEON TAXONOMY

The taxonomic system introduced by Linnaeus was initially based on describing the physical morphology (phenotype) of species. As time lapsed, it became apparent that within the Chamaeleonidae (as with many other families of vertebrates), phenotype is a relatively plastic phenomenon. The external morphology within the family shows strong ontogenetic, sexually dimorphic, sexually dichromatic and intraspecific variation. Many of the morphological characteristics used to supposedly define taxa were shown to be rather variable (Hillenius, 1959). Since most currently recognized genera were, at the time, considered to be synonyms of one or the other of two broad genera, assumptions of relationships within and between Continental and Insular species were often based simply on external similarities (Hillenius, 1959). With numerical taxonomy, attempts to derive hypotheses of relationships from simply comparing characteristics that matched each other were eventually found to be an "excursion into futility" (Ross, 1964). However, in the Chamaeleonidae, it was not only the gross morphology that was found to be variable. The chromosomes from a spectrum of Malagasy and African species (Matthey and van Brink, 1956, 1960; Brink, 1957; Matthey, 1957; Bourgat 1973) showed rather confusing and widespread variation in absolute numbers and morphology between species. Beyond broadly grouping them as Continental and Insular groups (and subgroups within each), karyology added little to revealing phylogenetic relationships. While it was relatively easy to describe the highly derived and often uniquely adorned forest species, populations with more conservative morphologies were usually recognized as sub-species, particularly the East African "*bitaeniatus* group" and the so-called dwarf chameleons of the *pumilum* group from South Africa.

Klaver (1981) and Rieppel (1981) were the first to introduce a cladistic interpretation to resolving relationships within Chamaeleonidae. Phylogenies constructed on the basis of phenetic or cladistic datasets are of course based on the hypotheses that the characteristics examined in such analyses reflect evolutionary relationships, whereas in fact they may not. Many characteristics identified within the Chamaeleonidae are plesiomorphic, and while it is generally easy to put the various species into groups based on external morphology, lung characteristics, reproductive strategies and cranioskeletal structure, it is often difficult to find common morphological synapomorphies that unite species into clearly defined groups. Cladistic methodology depends on identifying sufficient accurately polarizable characteristics to allow meaningful support of nodes. Robust phylogenetic reconstruction is rarely feasible when only limited character suites are available. In recent years, biologists are more aware of just how labile morphology is and how much conflict there can be between morphology and hypothesized evolutionary relationships. It is increasingly evident that the earlier focus on macromorphology with its unique structural derivations, has generally underrated the complexity of relationships.

The use of molecular markers to infer evolutionary relationships within Chamaeleonidae was first employed by Hofman, Maxon, and Arntzen (1991), who used starch gel electrophoresis to examine different enzyme loci and microcomplement fixation of albumin to infer immunological distance (ID). Then, following on the lead of molecular analyses in other families of squamates, the first researchers to use gene sequence analysis in the Chamaeleonidae (Pook and Wild, 1997) used the conserved 12S rRNA gene to investigate relationships within the West African "*Chamaeleo cristatus* group." Subsequently, other authors began to use other genes or combinations of molecular markers (e.g., nuclear genes RAG-1, mitochondrial genes ND1 and ND4, 16S rRNA, COI, flanking tRNA regions, and other potentially informative markers). In the past decade or so, molecular phylogenetics has blossomed rapidly, and has become the primary method behind current phylogenetic analysis. However, like any "new" method, it is not without its pitfalls, some of which are briefly mentioned below for the sake of perspective.

One concern involved with molecular phylogenetics is that it is based on the assumption that the degree of overall similarity between segments of base pair sequences of DNA reflects the degree of relatedness in the context of a Darwinian model of continual and gradual change—the so-called Molecular Assumption (Schwartz and Maresca, 2006). Molecular phylogenetics has been criticized for utilizing intrinsically the same methods as those used in simple phenetics (i.e., searching for similarities albeit comparing a much greater number of "characteristics"), questioning the wisdom of replacing a system that is widely acknowledged to have failed with another that uses essentially the same methods. Molecular phylogenetics is not based on cladistic principles, so does not provide distinctions between homoplasies and plesiomorph and apomorph characteristics. Topologies produced are supported at nodes by mathematical algorithms and statistical optimization, which will tend to obscure character conflicts (Mooi and Gill, 2010).

The ability of molecular phylogenetics to discriminate accurately at the level of deeper divergences is also controversial (Rocha et al., 2005; Heads, 2005; Nelson and Ladiges, 2009). At the deeper and by inference older nodes, the resolution provided blurs considerably. However, in the absence of a suitable fossil record for the Chamaeleonidae, molecular-based phylogenetic reconstructions are frequently being utilized to inform the biogeographic history of the Chamaeleonidae (Raxworthy et al., 2002; Townsend and Larson, 2002; Matthee et al., 2004; Tolley et al., 2006; Tolley et al., 2008; Townsend et al., 2011b; Tolley et al., 2011; Tolley et al., 2013; Chapter 7). In order to try and improve resolution, researchers now commonly employ wider ranges of genes to assemble larger arrays of base pair sequences for comparison.

8.2 CURRENT STATUS OF TAXONOMY OF THE CHAMAELEONIDAE

Since the description of the first chameleon *Lacerta chamaeleon* by Linneaus in 1758, some 300 specific names have been given to chameleons. At present, 196 species are currently recognized with an additional 13 named sub-species. Approximately 55% of these originate from continental Africa, 44% from Madagascar and the remainder from Asia, the Middle East, Europe and Africa's other offshore islands. Thirty distinct genera and subgenera have been erected since 1758, including 14 genera erected by Gray (1865) in one publication. Twenty two of these genera were erected before the turn of the 18th century, but only 8 since 1900.

Following Linneaus's description of the first species, progress in the next 140 years was rapid. Between 1758 and 1899, 77 currently recognized species were described. With the development of European naval competencies, exploration boomed and with the new ability to penetrate previously inaccessible regions, vast areas of Africa and Madagascar became available for exploration and colonization. Collections of fauna and flora made by scientific and occasional military expeditions and by missionaries stationed at remote outposts were repatriated to museums in Europe and included many chameleons. German colonists in East Africa (Tanzania), the British in Kenya and Uganda, and the French in Madagascar and parts of West Africa provided the grist for the main bulk of new descriptions. As a consequence, in the 20th century, 93 additional taxa were described. The rate of new species descriptions continues to accelerate, and more than 40 new taxa have already been described in the 21st century.

The family Chamaeleonidae currently consists of 11 extant genera, and 3 named subgenera within *Rhampholeon*. Three genera are restricted to Madagascar, the Comoro Islands and the Mascarene island of Réunion, and the remainder to the African block (including those species extant in the Middle East, Europe, the Indian subcontinent, and the Seychelles).

8.3 SUBFAMILIAL GROUPINGS WITHIN CHAMAELEONIDAE

In Klaver and Böhme's 1986 phylogeny, the family was divided into subfamilies, namely, Brookesiinae and Chamaeleoninae. The Brookesiinae (or pygmy chameleons) were characterized by the hemipenal synapomorphies—absence of truncal calyces and by the presence of dual apical ornamentation of the hemipenes, whereas the Chamaeleoninae (or ordinary chameleons) were characterized by the presence of truncal calyces and a quadruple apical ornamentation. The two genera *Brookesia* and *Rhampholeon* were differentiated by having different forms of apical ornamentation, namely, crests in *Brookesia* and horns in *Rhampholeon*. At the time, the hemipenal morphology of most of the known Malagasy *Brookesia* had been characterized, whereas only five of the nine then known African species of *Rhampholeon* had been available for examination. It was specifically noted by Klaver and Böhme (1986) that the characterization of the subfamily Brookesiinae assumed that the unexamined species also lacked hemipenal calyces. As time passed and more material came to light, the situation indeed changed. By 1997, it was shown that *Rieppeleon brevicaudatus* (previously *Rhampholeon brevicaudatus*) appeared to lack obvious apical structures on the hemipenis, and that *Rieppeleon brachyurus* and *Rieppeleon kerstenii* both had a quadruple apical ornamentation. In addition, *Rhampholeon temporalis* and *R. spectrum* were found to possess definite hemipenal calyces. These facts effectively nullified the phylogeny of the Chamaeleonidae at the subfamilial level, which placed undue emphasis on hemipenal characteristics. This was acknowledged by Klaver and Böhme (1997), who elected to suppress the subfamilial classification.

The generic relationships within the Chamaeleonidae have been addressed, in part, on several occasions. Methuen and Hewitt (1914) stated "The genus *Brookesia* approximates to the normal lacertilian type in so many respects that we may safely regard it (and *Rhampholeon* probably) as the most generalized of living chameleons." Hofman et al. (1991), analyzing starch gel electrophoretic patterns and albumin immunological distance, concluded that they could find no evidence for the Brookesiinae (represented by *Brookesia stumpffi* and *Rhampholeon* [now *Rieppeleon*] *kerstenii*) being the sister group of the Chamaeleoninae. They also concluded that the ID between these two species in their study indicated that they could not be placed in the same genus—indeed that the albumin of *Brookesia stumpffi* was more similar to that of *Furcifer pardalis* than it was to *Rhampholeon* (*Rieppeleon*) *kerstenii*. Rieppel and Crumly (1997), using cladistic analysis combining skull characteristics and hemipenal and lung morphology, placed *Brookesia* first and then *Rhampholeon* (*sensu lato* [*Rhampholeon* + *Rieppeleon*]) as 'basal' (i.e., sister to) to all the other typical chameleons. They also concluded that a monophyletic Brookesiinae could not be the sister taxon to a monophyletic Chamaeleoninae. More recently, Townsend and Larson (2002) placed *Rieppeleon* and then *Rhampholeon* as sister to *Brookesia*, while Raxworthy et al. (2002) placed *Brookesia* as sister to *Rhampholeon*/*Rieppeleon*. Tilbury and Tolley (2009a), using representatives of all the currently recognized extant chameleon lineages in Africa (but with only three species of *Furcifer* and notably absent *Calumma*), again placed *Brookesia* as sister to all other chameleons, followed by *Rhampholeon* and *Rieppeleon*. The most recent family-wide phylogeny (Tolley et al., 2013) places *Brookesia* as sister to all other genera, with *Rieppeleon*, and *Rhampholeon* diverging first from the other genera which form an unresolved polytomy. Müller and Hildenhagen (2009) investigated subdigital and subcaudal lamellar structures of a broad range of chamaeleonid genera and could not find support for grouping the three pygmy genera in one clade. Based on all this there is not much evidence to validate the existence of the combined pygmy genera as a subfamily within Chamaeleonidae.

Apart from their basic external phenotype, perhaps the only unifying morphological characteristic of the pygmy chameleons is the simple sac-like lung (lung type A; Klaver, 1981) common to all genera of pygmy chameleon and not seen in the more typical chameleons. This type of lung however is considered to be a primitive plesiomorphic character distributed widely in other lizard families and as such cannot be used to define a subgroup. The fact that all the pygmy chameleons retained this lung type may be due to something that had already been suggested in the past (Beddard, 1907; Werner, 1911; Klaver, 1973, 1977)—that is, that simple lungs are related to the small size and physiological requirements of these chameleons, with there being no selective advantage to have developed a more derived lung type. This notion is based on the assumption that the simple lung morphology of the pygmy chameleons is a true primitive retained feature and not a secondary reversal from a more derived structure. Aside from their lungs, one of the few other characteristics seen in almost all the pygmy chameleons is the rather unique squamation of sheets of stellate-edged granular tubercles that clad their bodies and limbs (Hillenius, 1986). Its taxonomic value has not yet been assessed.

Studies by several authors (Werner, 1902b; Methuen and Hewitt, 1914; Hillenius 1986) have shown that the pygmy chameleons have fewer ribs than typical chameleons—*Rhampholeon* with an average of 12 pairs, and *Brookesia* with 9 or less pairs—as compared with the typical chameleons, which have 13 or more pairs. This may be simply a function of size, with larger species having more pairs of ribs than the smaller species. Zippel et al. (1999) showed that about 50 vertebrae were usually found in the tails of "typical chameleons," while only 20 were present in *Brookesia thieli* (Boistel et al., 2010). Unfortunately, skeletal studies on a wider range of species within the Chamaeleonidae are lacking, and it would be unwise to infer generalized morphological attributes on the basis of a few observations.

Almost all phylogenetic studies support the notion that the pygmies (Brookesiinae, represented by *Brookesia, Rieppeleon,* and *Rhampholeon*) are not a monophyletic group. They are all largely terrestrial and have retained all the symplesiomorphies of an apparently arboreal ancestor. Whether the shortened tails of the pygmy chameleons represent the symplesiomorphic condition, or like their lung structure, perhaps represent an anatomical simplification whereby tiny lightweight creatures do not need a long fully prehensile tail, is debatable.

Klaver and Böhme (1997) abandoned their subfamilial classification because the synapomorphies that they had earlier identified were falsified by subsequent data. Tilbury (2010), however, preferred to use the same subfamilial classification as Klaver and Böhme (1986), on the basis of the absence of evidence principal—that is, that the absence of evidence does not necessarily equate to evidence of absence. As emphasized by Rieppel and Crumly (1997), the characteristics examined to derive phylograms can only be assumed to reflect the phylogenetic history of chameleons, while this may not necessarily be the case. Since neither phenetics, cladistics, nor molecular phylogenetics have been able to give an accurate account of relationships at the deeper end of the phylogram, this leads one to conclude that the jury is actually still out on this issue. But while it may be difficult to prove hypotheses of relationship, it is possible at least to make them unlikely on the basis of that evidence at hand. Since naive conviction alone will not suffice under scrutiny, it seems rational that persistence with the subfamilial division of the Chamaeleonidae should be abandoned. Notwithstanding this, the use of the two phenetic assemblages (pygmy chameleons and typical chameleons) to informally group the various species within the Chamaeleonidae still seems worthwhile for general descriptive purposes.

8.4 OVERVIEW OF EXTANT GENERA

Genus *Archaius* Gray 1865

TYPE SPECIES: *Chamaeleo tigris* Kuhl 1820

GENERIC SYNONYMS: None

CONTENT: *tigris*

ETYMOLOGY: Gr. *archaios*, ancient, primitive. The author gave no further explanation.

Figure 8.1 (see color insert)

The recent resurrection of the monotypic genus *Archaius* (Fig. 8.1) for the Seychelles chameleon (previously, *Calumma tigris* Kuhl 1820) was made after the finding (based on a multilocus phylogenetic reconstruction using mitochondrial and nuclear genes that the Seychelles endemic was the sister taxon of the east African genus *Rieppeleon* (Townsend et al., 2011b). Using a dated phylogeny, it was estimated that the mean divergence of the two clades occurred around the Mid Eocene to early Oligocene (ca. 38±10 Mya). The authors concede that the only external morphologicalcharacter that connects *Archaius* and *Rieppeleon* is the small scaly mental lobe seen in *tigris* and in one of the species of *Rieppeleon* (*brevicaudatus*). Previous studies that examined anatomical features of *tigris* (Bourgat and Domergue, 1971; Hillenius, 1986; Klaver and Böhme, 1986; Rieppel and Crumly, 1997; Raxworthy et al., 2002) loaned support to its inclusion within *Calumma*, while biochemical evidence (Hofman et al., 1991) based on immunological distance had suggested a possible African connection for *A. tigris*.

This case demonstrates the dilemma of how to reconcile conflicting genetic and anatomical datasets. In this particularly interesting situation, the molecular evidence appears to underline that genealogical ancestry may play little or no role in the development or retention of physical character attributes.

Genus *Bradypodion* Fitzinger 1843

TYPE SPECIES: *Lacerta pumila* Gmelin 1789

GENERIC SYNONYMS: *Microsaura* Gray 1865; type species *Microsaura melanocephala* Gray 1865, *Lophosaura* Gray 1865; type species (by subsequent designation) Loveridge 1957 = *Lacerta pumila* Gmelin 1789

CONTENT: *atromontanum, caeruleogula, caffer, damaranum, dracomontanum, gutturale, kentanicum, melanocephalum, nemorale, ngomeense, occidentale, pumilum, setaroi, taeniabronchum, thamnobates, transvaalense, ventrale*

ETYMOLOGY: Gr. *bradys*, slow; *podion*, diminutive of *podos*, foot: slow small-foot, referring to the slow movements generally and the small size of the type species

Figure 8.2 (see color insert)

The higher systematics of the genus *Bradypodion* (Fig. 8.2) has a rather convoluted history and is closely linked with the genus *Kinyongia* (below) to which section the reader is referred for a fuller discussion. Although long recognized to comprise a unique group of species, attempts to recognize the South African "dwarf" chameleons as a genus (Methuen and Hewitt, 1914; Fitzsimons, 1943; Raw, 1976) did not garner much support in the general literature. Hillenius (1959) and Mertens (1966) preferred to keep them within *Chamaeleo,* with most taxa designated as subspecies of *B. pumilum.* Klaver and Böhme (1986, 1997) included in *Bradypodion* (as resurrected by Raw [1976]) a group of East African species (the "*fischeri*" group), to form a heterogeneous group of species grouped more by the lack of synapomorphies that would ally them to other recognized genera. For this reason Branch (1998) specifically rejected this move, but final resolution of the expanded group was obtained only when various molecular studies subsequently demonstrated paraphyly of *Bradypodion* (*sensu* Klaver and Böhme, 1986; Hofman et al., 1991; Townsend and Larson, 2002; Tolley et al., 2004), and finally resulted in the division of *Bradypodion* into three genera (Tilbury et al., 2006).

As a group, these small to medium-sized species all share similar external morphological characteristics that in some species vary only in the degree of expression. The genus has developed independently derived viviparity with associated pigmentation of the parietal peritoneum of the abdominal cavity (Townsend and Larson, 2002). All species have a gular crest, composed in many species of prominent gular lobules with a tendency to form simple cones in the smaller species. A ventral crest is not present. None of the species possess any form of rostronasal projection or occipital lobes, and scalation is heterogeneous.

The structure of the cranial parietosquamosal complex suggests a retained plesiomorphic condition in contrast to all the other genera of typical chameleons, which share a derived cranium. The posteriorly projecting broad parietal bone forms a strongly recurved casque in all species. The temporal crest, formed along the postfrontal and squamosal bones, extends from the mid-postorbital rim, to divide the side of the head into superior and inferior temporal fossae (Parker, 1881; Brock, 1941; Rieppel, 1981). The lateral crests that arise as a posterior continuation of the supraorbital ridge delineate the lateral margin of the parietal crest, in contrast to other African genera, in which the lateral crest is formed along the postorbitofrontal and squamosal bones (homologous to the position of the temporal crest in *Bradypodion* [Raw, 1976]). In addition, the parietal bone possesses supratemporal processes that descend to meet the ascending ramus of the squamosals to enclose the supratemporal fossa. Apart from the pygmy genus *Brookesia, Bradypodion* is the only genus of chameleons to retain this plesiomorphic condition; however, Rieppel and Crumly (1997) consider this to be a secondary reversal in *Bradypodion.* The hemipenal morphology is characterized by a plesiomorph quadruple apical ornamentation of two pairs of denticulate rotulae that differs relatively little between species.

The lungs of *Bradypodion* are small; consist of a relatively simple sac with a number of small partial septae present on the anterior, dorsal, anteroventral walls; and have no external appendages or diverticulae. The larynx of *B. occidentale, B. damaranum,* and *B. ventrale* has been shown to have an associated gular pouch (Klaver, 1981). It is likely that this may be applicable to all the species of this genus. The karyotype $2n = 34 = 12M + 22m$ is shared by all species of *Bradypodion* examined to date (Matthey and van Brink, 1956; Gordon et al., 1987).

Other primitive characteristics, such as the retention of a well-developed Organ of Jacobson as described in *B. pumilum* (Brock, 1941; Malan, 1946) is generally absent in most other chameleons, but the extent of this and its phylogenetic implications have not been fully assessed in the Chamaeleonidae.

The taxonomy of dwarf chameleons of South Africa was previously problematic. Along with their allopatric pattern of distribution, all their similarities suggest that they are closely related, sharing a recent common ancestor from the Miocene (Tolley et al., 2006, 2008). Among species of *Bradypodion*, three broad phenetic assemblages can be identified. However, a study using mitochondrial DNA sequences (Tolley et al., 2004) showed these phenotypic groupings are not representative of evolutionary relationships. On the contrary, even some apparently well-defined species with vastly different phenotypes (e.g., *B. melanocephalum* and *B. thamnobates*) have little genetic divergence from each other, and others that appear to be closely related by virtue of phenotype (e.g., *B. occidentale* and *B. ventrale*) have markedly divergent genotypes. The various lineages within the clades that form the bulk of the species show a strong tendency to be linked within geographic regions.

Several species of *Bradypodion* (among others, *B. pumilum*) demonstrate considerable "ecomorph" variation between populations (Measey et al., 2009, 2011; Hopkins and Tolley, 2011; Herrel et al., 2011). This demonstrates the need for an integrated total evidence approach to alpha taxonomy, particularly in genera such as *Bradypodion*, in which the use of morphological characteristics alone has proven limitations.

Genus *Brookesia* Gray 1865

TYPE SPECIES: *Chamaeleo superciliaris* Kuhl 1820

GENERIC SYNONYMS: *Leandria* Angel 1933; type species *Leandria perarmata* Angel 1933, *Evoluticauda* Angel 1942; type species (by subsequent designation) Guibe 1954, *Brookesia tuberculata* Mocquard 1894

CONTENT: *ambreensis, antakarana, bekolosy, betschi, bonsi, brunoi, brygooi, confidens, decaryi, dentata, desperata, ebenaui, exarmata, griveaudi, karchei, lambertoni, lineata, lolontany, micra, minima, nasus, perarmata, peyrierasi, ramanantsoai, stumpffi, superciliaris, therezieni, thieli, tristis, tuberculata, vadoni, valerieae*

ETYMOLOGY: Named by Gray in honor of Joshua Brookes (1761–1833). Brookes was a British anatomist and founder of the Brookesian Museum of Comparative Anatomy. The first officially described species of *Brookesia* was part of his collection.

Figure 8.3 (see color insert)

This genus is currently comprised of 32 species, 20 of which have been described since 1970. *Brookesia antoetrae* was synonymized with *B. thieli* by Raxworthy and Nussbaum (1995), but its status should perhaps be reconsidered on account of the significant genetic divergence from *Br. thieli* demonstrated by Townsend and Larson (2002). At the time of this writing, there are several more species in the process of being described. *Brookesia* is the second most speciose genus among the Malagasy chameleons, and it is likely that there will be many more additions to this genus in the future (Fig. 8.3).

Following on the initial work of Siebenrock (1893), Rieppel (1981, 1987) and Rieppel and Crumly (1997) described the cranial structure of *Brookesia* as represented by *B. superciliaris* and *B. stumpffi*. The broad roughly trapezoidal parietal has no posterior sagittal spur, with the posterior margin being broad rather than pointed. The parietal drops posteroinferiorly to form a short posterolateral (supratemporal) process on each side and is considered to be a retained symplesiomorphy (Rieppel and Crumly, 1997). The nasals are fused into a single midline bone that prevents the premaxilla from making contact with the frontal and makes broad contact with the prefrontals on each side. Methuen and Hewitt (1914) were the first to suggest that *Brookesia* was the most primitive of all the chameleonine genera. Rieppel (1981, 1987) concurred with Methuen and Hewitt and suggested that a Malagasy origin for the family was likely, followed by dispersal to Africa, giving rise to *Rhampholeon*.

Brookesia (as represented by *B. superciliaris*) have a uniquely derived structure of the dorsal vertebrae (Siebenrock, 1893, Parker, 1942; Rieppel, 1987) and a unique structure of the pelvic girdle (synsacrum) (Werner 1902a,b), which may be considered as synapomorphies for the genus, assuming that the other members of the genus share the above features. In addition, the rather unique mode of walking with the tail tip used as a stabilizing rod (Glaw and Vences, 2007; Boistel et al., 2010) might also be viewed as a behavioral synapomorphy. Brygoo (1971) demonstrated that all members of *Brookesia* that were so examined had unpigmented testes and peritoneum, while in species of *Chamaeleo,* these were pigmented. Pigmentation of the testes was demonstrated in several (but not all) members of *Rhampholeon* and *Rieppeleon* (Klaver, 1979). Other random observations on skeletal characteristics such as the structure of the thoracic skeleton, numbers of rib-bearing vertebrae, and the configuration of the sternal rib attachments are of undetermined taxonomic significance (Werner, 1902b; Methuen and Hewitt, 1914). Because of the limited number of species examined in such studies, generalizations as to these attributes for the genus are unwise for the present and are mentioned here as areas that need further clarification.

There are three divergent clades currently nested within *Brookesia* (*nasus* c.s., *minima* c.s. and *superciliaris c.s.*), which can be distinguished by both morphological and genetic datasets. From a morphological perspective, *Brookesia nasus* and *Brookesia lolontany* do not possess lateral vertebral spines or pelvic shields, which could be viewed as synapomorphies for the other *Brookesia*. In *nasus/lolontany* the median dorsal ridge is narrow and in males has a well-defined series of midline crenulations, each occurring over a dorsal vertebral process. The body habitus and shape of *nasus/lolontany* is more akin to that seen in the African pygmy genera and also—uniquely among *Brookesia*—both *B. nasus* and *B. lolontany* possess a small rostronasal process. The head is not the broadly flattened structure seen in *Brookesia,* but has a rather longer snout and is more vertically compressed, in keeping with their body habitus. In *B. nasus,* the shape of the parietal appears to be similar to that seen in *B. superciliaris.* The dominant posterior head crest delineates the lateral edge of the parietal bone. Based purely on its external morphology, Klaver (1979) considered *Br. nasus* to be the intermediate taxon between typical *Rhampholeon* and typical *Brookesia*. Molecular studies that have included the species *nasus* and/or *lolontany* among the specimens examined have consistently shown them to form a deeply divergent clade within *Brookesia* (Raxworthy et al., 2002; Townsend and Larson, 2002; Townsend et al., 2009), estimated to have split at around 60 million years ago (Late Cretaceous) (Tolley et al., 2013) (Chapter 7). Raxworthy et al. (2002) placed *B. nasus*/*B. lolontany* as sister to the rest of *Brookesia* in their phylogram and Townsend and Larson (2002) suggested that *B. nasus* was sister to the rest of *Brookesia* but could not reject non-monophyly on the basis of statistical tests. The latter study, apart from identifying a relatively recent tandem duplication in the genome of *B. nasus* (that probably postdates the split of the two clades) demonstrated a net sequence divergence of 24% between *nasus* and the rest of *Brookesia*.

The remaining 29 species of *Brookesia* are divided into two groups: the so-called *minima* group (Brygoo, 1978; Raxworthy and Nussbaum, 1995; Townsend et al., 2009; Glaw et al., 2012) and the rest within a loosely termed "typical" group (Townsend et al., 2009). Both lateral vertebral spines and a pelvic shield are found in most of the species, although in some of the diminutive species within the *minima* group, these structures appear to be, at least macroscopically, absent.

The typical group consisting of 18 species is characterized by larger bodies, relatively large heads with truncated snouts, and in addition to the lateral parietal crest, a well-defined lateral crest (posterior extension of the supraorbital ridge), which delineates the superior edge of the postorbitofrontal and the squamosal. The *minima* group, on the other hand, currently consists of 11 described species and some candidate taxa (Glaw et al., 2012). The *minima* group contains some of the smallest amniote vertebrates in the world. Most species are less than 45 mm in total length. They tend to have somewhat smaller heads relative to their bodies, with generally indistinct head crests. This group, although having some remarkably similar phenetic attributes, shows marked genetic divergence, averaging around 17%, between taxa (Glaw et al., 2012).

Genus *Calumma* Gray 1865

TYPE SPECIES: *Chameleon cucullatus* Gray 1831

GENERIC SYNONYMS: *Crassonota* Gray 1865; type species *Chamaeleon nasutus* Dumeril and Bibron 1836

CONTENT: *amber, ambreense, andringitraense, boettgeri, brevicorne, capuroni, crypticum, cucullatum, fallax, furcifer, gallus, gastrotaenia, glawi, globifer, guibei, guillaumeti, hafahafa, hilleniusi, jejy, linotum, malthe, marojezense, nasutum, oshaughnessyi, parsonii, peltierorum, peyrierasi, tarzan, tsaratananense, tsycorne, vatosoa, vencesi, vohibola*

ETYMOLOGY: L. transcription of Gr. *kalymma*, veil or hood, referring to the large occipital lobes of the type species of this genus

Figure 8.4 (see color insert)

Excluding the species allied to the genus *Furcifer* (Fig. 8.4), the remainder of the species of "typical" Malagasy chameleons could be divided into at least two groups based on hemipenal apical morphology. Although no common apomorph features were identified, Klaver and Böhme (1986) placed both groups within the resurrected genus *Calumma*. In erecting *Calumma,* Klaver and Böhme (1986) allowed that the monophyly of the genus had yet to be demonstrated. As predicted, this has been questioned or paraphyly implied in several publications (Townsend and Larson 2002; Raxworthy et al., 2002; Townsend et al., 2011b). Townsend and Larson (2002) identified three distinct clades within *Calumma* among the 10 species sampled, which did not collectively form a monophyletic group. However statistical tests based on parsimony did not reject monophyly of *Calumma*. At this point in time although suggestive, the available molecular data does not reliably support or reject monophyly of the genus (Gehring et al., 2011).

Many species have been described in the last decade—the genus is currently composed of 33 species, and this seems likely to increase significantly in the future as cryptic taxa are identified and described. As a genus they appear to be restricted to humid forests and are the most commonly found genus in the cooler montane areas of Madagascar. They have not been recorded from the dry southwest of Madagascar (Glaw and Vences, 2007).

In Madagascar, high levels of microendemism with cryptic speciation has been demonstrated among several groups of vertebrates and in particular amphibians. The chameleons are proving to be no exception (Glaw et al., 2012). Raxworthy and Nussbaum (2006) identified several taxa within what had been previously regarded as *Calumma brevicorne*. Gehring et al. (2012) examined the genetic diversity within the *C. nasutum* group and demonstrated that in addition to the currently 7 named species, an additional 26 OTUs (operational taxonomic units) were identified, many of which will probably be elevated to species rank in the near future. This is the only genus of chameleons in Madagascar in which occipital lobes occur in some of the species. Several species bear a close appearance to some species within the African genus *Kinyongia*.

Genus *Chamaeleo* Laurenti 1768

TYPE SPECIES: *Chamaeleo parisiensium* Laurenti 1768 (= *Lacerta chamaeleon* Linnaeus 1758).
GENERIC SYNONYMS: *Phumanola* Gray 1865; type species *Chamaeleo namaquensis* Smith 1831. *Calyptosaura* Gray 1865; type species (by subsequent designation) Loveridge 1957, *Chamaeleo calyptratus* Dumeril and Bibron 1851. *Erizia* Gray 1865; type species (by subsequent designation) Loveridge 1957, *Chamaeleo senegalensis* Daudin 1802. *Dilepis* Gray 1865; type species *Chamaeleon dilepis* Leach 1819
CONTENT: *africanus, anchietae, arabicus, calcaricarens, calyptratus, chamaeleon, dilepis**, *gracilis, laevigatus, monachus, namaquensis, necasi, senegalensis, zeylanicus.* (*Pending a full review of the "*dilepis* complex." *C. dilepis* is considered here to include (as synonyms) the described forms *roperi, quilensis, isabellinus, petersii, ruspolii, martensi,* and *idjwiensis.*)
ETYMOLOGY: Gr. *khamai*, on the ground; *leon*, lion: ground lion
Figure 8.5 (see color insert)

The African genus *Chamaeleo* (Fig. 8.5) as construed by Klaver and Böhme (1986) united all chameleons with large lung septae that subdivided the intraluminal space, a condition that was supposed to have evolved from lung type B (Klaver, 1981). Two sub-genera were proposed based on fundamental differences in lung septation—*C. (Chamaeleo)* and

C. *(Trioceros)*—which differed in the number, origin, and attachments of these septae within the lung. In *C. (Chamaeleo)* (lung type C), two straight septae arise from near the bronchial orifice of the lung and terminate a variable distance caudal to their origin. In *C. (Trioceros)* (lung types D, E, and F) two to three usually anteriorly curved septae are connected to the ventral, medial and lateral walls of the lung and terminate a variable distance from the dorsal wall (Klaver, 1981).

In addition to the divergent lung morphology, several apomorphic characteristics unite the species grouped within each of the two sub-genera, but apart from an uninformative plesiomorphic similarity in the structure of the casque, there were no obvious synapomorphies to unite them within a single genus. Using a combination of sequences from mitochondrial and nuclear DNA (Tilbury and Tolley, 2009a), it was demonstrated that *Trioceros* was neither nested within nor was sister to the clade containing *Chamaeleo*, and as such it was proposed that *Trioceros* be elevated from within the genus *Chamaeleo* and instated as a separate genus.

The genus *Chamaeleo*, currently composed of 14 species and 5 subspecies, is the most widely spread of all chameleon genera, covering almost the whole of Africa and with all species represented in Europe, the Middle East and Asia belonging to this genus. The various species are united by the synapomorphy of the type C lung (Klaver, 1981). All species within *Chamaeleo* possess a gular pouch although this is also found sporadically in *Trioceros* and possibly in all *Bradypodion* (Klaver, 1981). Other potential synapomorphies for *Chamaeleo* are a particular chromosomal formula ($2n = 24 = 12M + 12m$—seen in all six species of *Chamaeleo* so examined to date) and the synapomorphy of a multirotulae ornamentation on the hemipenes—although the symplesiomorphic four rotulae condition is retained in both *C. namaquensis* and *C. arabicus*. In addition, the males (and occasionally females) of many species from this genus are adorned with tarsal spurs—not seen in any other genus—and which could be considered to be a synapomorphy for the genus. In all species, a lateral crest is present and none of the species have a temporal crest. The tail is unadorned with a crest in all species. Scalation is generally of homogeneous to finely heterogeneous granular tubercles.

The most wide-ranging African species, *Chamaeleo dilepis*, encompasses a number of described sub-species and other species treated as synonyms, often differing from the nominate form by morphological characteristics that show confusing intrapopulational and interpopulational variation. The forms *roperi* and *quilensis* have been demonstrated to show considerable divergences from typical *dilepis* and may represent full species (Ullenbruch et al., 2007). However, the "*dilepis* complex" as a whole is in urgent need of a full and exhaustive analysis to elucidate the relationships of its component forms using tissues derived from topotypic specimens.

Gray (1865) placed *Chamaeleo namaquensis* within the monotypic genus *Phumanola*, but this name was never recognized as valid or used subsequently and was consequently synonymized. Within the genus, *C. namaquensis* is indeed a unique species that, although possessing the characteristic genus chromosomal formula and lung type, has as mentioned above, also retained the symplesiomorphic quadruple hemipenal apical ornamentation. It has a uniquely constructed dorsal crest and lacks the prominent gular crests and tarsal spurs seen in most other species within the genus. Rieppel and Crumly (1997) identified several cranial structural characteristics in *namaquensis* that are not seen in other members of the genus. Townsend and Larson (2002) commented that *C. namaquensis* was an ancient lineage and demonstrated an uncorrected sequence divergence of 17% between *namaquensis* and other chameleons in the genus. Hofman et al. (1991) examined the albumin immunological distance between *C. namaquensis* and both *C. dilepis quilensis* and *Furcifer pardalis* and concluded that *C. namaquensis* was closer to *F. pardalis* than it was to *C. dilepis*. Although the above character divergences may support a future taxonomic move for *C. namaquensis*, given the fact that it possesses synapomorphies that align it to *Chamaeleo*, and that it forms part of a monophyletic *Chamaeleo* (Tolley et al., 2013), it seems appropriate to leave it within this genus.

Genus *Furcifer* Fitzinger 1843

TYPE SPECIES: *Chamaeleo bifidus* Brongniart 1800

GENERIC SYNONYMS: *Diceros* Swainson 1839; type species *Diceros bifurcatus* = *Chamaeleo bifidus* Brongniart 1800. *Apola* Gray 1865; type species *Chamaeleo lateralis* Gray 1831. *Sauroceras* Gray 1865; type species *Chameleo rhinoceratus* Gray 1845. *Dicranosaura* Gray 1865; type species (by subsequent designation) Loveridge 1957—*Chameleo bifurcus* Gray (= *Chamaeleo bifidus* Brongniart 1800). *Cyneosaura* Gray 1865; type species *Chamaeleo pardalis* Cuvier 1829

CONTENT: *angel, antimena, balteatus, belalandaensis, bifidus, campani, cephalolepis, labordi, lateralis, major, minor, monoceras, nicosiai, oustaleti, pardalis, petteri, polleni, rhinoceratus, timoni, tuzetae, verrucosus, viridis, willsii*

ETYMOLOGY: L. *furca*, a two-pronged fork; *-fer*, suffix meaning bearing or bearer (from *fer(r)e*, to bear, to carry): fork-bearer, denoting the forked rostral appendage in male specimens of the type species of this genus

Figure 8.6 (see color insert)

Currently composed of 23 taxa, the genus *Furcifer* (Fig. 8.6) has been relatively quiet in terms of alpha taxonomy with only 6 new species being described in the past 40 years. Nevertheless, among some species, the presence of molecular differences and color morphs may indicate the presence of further cryptic species within the currently described cohort (Glaw et al., 2009). *F. lateralis* has been recently split into three species (Florio et al., 2012). Based on morphological grounds only, *Furcifer monoceras* was tentatively synonymized with *F. rhinoceratus* (Glaw and Vences, 2007), but given the high levels of cryptic speciation observed in Malagasy chameleons, this may have been a premature move. Published phylogenies identify *Furcifer* as a monophyletic genus (Townsend and Larson, 2002; Raxworthy et al., 2002; Townsend et al., 2011b; Tolley et al., 2013).

In the phylogeny of Klaver and Böhme (1986), of the non-Brookesiinae species, the genus *Furcifer* was characterized by the hemipenal synapomorphies of apical pedunculi and auriculae. The lung and cranial morphologies of *Furcifer* show the same basic symplesiomorphic structure as in other genera such as *Calumma*, *Kinyongia* and *Bradypodion* (lung type B—Klaver, 1981). None of the species in the genus possess occipital lobes. Several species within *Furcifer* possess prominent helmet-like casques, a feature not seen in *Calumma*. The males of several species within *Furcifer* share convergent cranial ornamentation with species of the African genus *Kinyongia* such as dual blade-like horns (*F. bifidus*, *F. minor*, *F. balteatus*, *F. petteri* and *F. willsii*) or a single laterally compressed rostral process formed by the fusion of the rostral/preorbital crests (*F. angeli*, *F. antimena*, *F. labordi*, and *F. rhinoceratus*). As a group, most species are distributed in the more arid western and north-western parts of Madagascar, with only a few species adapted to rainforest (Chapter 7). Two species (*F. cephalolepis* and *F. polleni*) occur on the Comoro Islands.

Genus *Kinyongia* Tilbury, Tolley, and Branch 2006

TYPE SPECIES: *Chamaeleon fischeri* Reichenow 1887

GENERIC SYNONYMS: None

CONTENT: *adolfifriderici, asheorum, boehmei, carpenteri, excubitor, fischeri, gyrolepis, magomberae, matschiei, multituberculata, oxyrhina, tavetana, tenuis, uluguruensis, uthmoelleri, vanheygeni, vosseleri, xenorhina*

ETYMOLOGY: Derived from the KiSwahili word *Kinyonga* meaning "chameleon"

Figure 8.7 (see color insert)

When Klaver and Böhme (1986) proposed their systematic revision of the family Chamaeleonidae, the genus *Bradypodion* was retained although the generic characteristics of *Bradypodion* (*sensu* Raw, 1976) were not thought to be defining for a purely South African genus. The genus was thus expanded to include a group of east and central African chameleons the "*fischeri* group." the most southerly species of this group being *Nadzikambia mlanjense* from southern Malawi. The genus *Bradypodion* was now composed of a heterogeneous group of 21 species and subspecies of chameleons that did not comprise a well-defined cluster and was itself not diagnosable by synapomorphies.

In spite of the arguments by Hillenius (1959), Klaver (1973) and Klaver and Böhme (1986), comparison of the southern African species as a group with those species from further north suggested that there were at least two geographically well-defined groups that differed in reproductive physiology, cranial and external morphologicalcharacteristics and as a combined group were difficult to characterize. The character of viviparity exists in the South African *Bradypodion* and several groups of East African species allied within *Trioceros*. The presence of pigmentation of the parietal peritoneum in *Bradypodion* (first noted by Beddard [1907] and further elaborated on by Klaver [1973]) gave Bauer (1997) support to argue against the inclusion of the central and east African species within the genus *Bradypodion*, by emphasizing the lack of pigmentation of the parietal peritoneum in this group. Hofman et al. (1991) using immunological distance between the serum albumins of a range of taxa, found that the east African taxon *Bradypodion tavetanum* was closer to the Malagasy genus *Furcifer* than it was to the South African *Bradypodion* (represented by *B. thamnobates, B. setaroi* and *B. melanocephalum*). They concluded that there was no support to classify them in the same genus. Townsend and Larson (2002) who examined mitochondrial DNA also inferred that the genus *Bradypodion* (*sensu* Klaver and Böhme, 1986) was paraphyletic. More recently, Tolley et al. (2004) demonstrated large net sequence divergences for mitochondrial gene sequences in ribosomal RNA, and nuclear genes between *B. pumilum* and *B. mlanjense* from Malawi. Finally Tilbury et al. (2006) moved to separate the two groups and formally designated the east African "*fischeri* group" as a new genus *Kinyongia* and created a monotypic genus, *Nadzikambia*, for *mlanjense*.

The genus *Kinyongia* is currently composed of 18 described taxa (Fig. 8.7 in the color insert). Since the erection of the genus, there have been several new additions to *Kinyongia*—6 new taxa described since 2000 (Nečas et al., 2009; Nečas , 2009; Menegon et al., 2009; Greenbaum et al., 2012) and several species elevated from the synonymy of *K. fischeri* (Mariaux et al., 2008). Within *Kinyongia* itself, three divergent clades have been identified (Tolley et al., 2011), more or less confined to geographic regions consisting of the Albertine Rift/Kenyan highlands, the northeastern and southeastern Arc Mountains. The range of the most westerly species, *K. adolfifriderici*, extends to the eastern DRC and the most northerly species *K. asheorum* occurs on Mount Nyiro in northern Kenya. They are essentially confined to relict montane or submontane forest biomes.

Apart from lung similarities (lung type B—Klaver 1981), the species of this group also share several phenotypic characteristics in common with some species of the Malagasy genera *Calumma and Furcifer.* In some species, the cranial ornamentation of paired rostronasal blade-like horns is seen (*K. boehmei, K. fischeri, K. matschiei, K. multituberculata, K. tavetana, K. uluguruensis,* and *K. vosseleri*) and in others, a fusion of the preorbital canthal ridges on the snout into a single vertically flattened rostral process (*K. carpenteri, K. xenorhina,* and *K. oxyrhina*). None of the described species have gular or ventral crests, occipital dermal lobes, preorbital or annulated horns, although the recently described form *K. asheorum* has scattered gular cones and some prominent cones on the snout. Background scalation is generally of granular or flattened tubercles often clustered into "rosettes," especially on the flanks and separated by interstitial skin. In some species, a few tubercles may have the stellate-edge appearance approaching that seen in the pygmy chameleons. Plantar surfaces are smooth and claws are simple.

The cranial structure of the only species of this genus that has been analyzed to date (*K. fischeri*) demonstrates the same relatively apomorphic structure of the parietal bone and temporal fossa as seen in specimens of the genera *Trioceros, Chamaeleo, Furcifer* and *Calumma* (Hillenius, 1988). The lungs are generally simple with a varying number of small partial ridge-like septae that arise from the dorsal, ventral and cranial walls similar to that of *Nadzikambia, Bradypodion,* and the Malagasy genera *Calumma* and *Furcifer.* The septae protrude into the lung for only a short distance. There is no membranofibrous diaphragm at the dorsocranial end of the lung (cf. *Chamaeleo* and *Trioceros*). Several of the species have varying numbers of diverticulae appended to the inferior and caudal margins of the lung. None of the species so far examined possess a gular pouch.

In species in which the hemipenal morphology has been examined, the hemipenes are calyculate with a plesiomorphic four-rotulae apical ornamentation, and with little else on the apex apart from the rotulae. Apart from some differences in the size and occasionally the shape of the rotulae, there may be little to distinguish between them. All the species are oviparous. The karyotype of the only species to date that has been analyzed, *K. fischeri* is $2n = 36 = 12M + 24m$.

Genus *Nadzikambia* Tilbury, Tolley, and Branch, 2006

TYPE SPECIES: *Chamaeleo mlanjense* Broadley 1965

GENERIC SYNONYMS: None

CONTENT: *baylissi, mlanjensis*

ETYMOLOGY: Derived from the ChiChewa word *Nadzikambe,* meaning "chameleon"

Figure 8.8 (see color insert)

The two species of this genus (Fig. 8.8) are found on two separate, inselbergs near the southern tip of Malawi, and in central Mozambique where they inhabit submontane forest.

In external morphology, these chameleons differ little from that seen in the genus *Kinyongia.* There are no gular or ventral crests and occipital lobes are absent. All the other body crests are weakly developed. The head is unadorned by rostronasal processes and the casque is only weakly elevated. Scalation is weakly heterogeneous with rounded tubercles forming rosettes of tubercles on the lower flanks.

The skull morphology is undescribed but is anticipated to be similar to that of *Kinyongia, Trioceros* and *Chamaeleo.* The lungs of *Nadzikambia mlanjensis* are sac-like with a series of small partial septae found on the anterior dorsal and ventral walls. Two pairs of long diverticulae trail from the posterior and inferoposterior margins of the lung. There is no gular pouch (Klaver, 1977). The hemipenes of *Nadzikambia* are uniquely derived (Branch and Tolley, 2010). The pedicel is short with shallow truncal calyces. The spermatic sulcus drains into a shallow cup-like apex. The distal (asulcal) rim of the cup is smooth in *N. baylissi* and coarsely fringed in *N. mlanjensis.* The distal margins of the spermatic sulcus are thickened. There are no apical rotulae. Reproduction is oviparous.

The monophyly of this genus was essentially established on the basis of its evolutionary relationships which places it as far from *Bradypodion* and *Kinyongia* as it is from *Chamaeleo* and *Trioceros* and the pygmy genera *Rhampholeon* and *Rieppeleon* (Tilbury et al., 2006).

Genus *Rhampholeon* Günther 1874

TYPE SPECIES: *Chamaeleo spectrum* Bucholz 1874

GENERIC SYNONYMS: None

SUBGENUS: *Rhampholeon (Bicuspis)* Loveridge 1956

TYPE SPECIES *Rhampholeon marshalli* Boulenger 1906

CONTENT: *gorongosae, marshalli*

SUBGENUS: *Rhampholeon (Rhampholeon)* Günther 1874

TYPE SPECIES *Chamaeleo spectrum* Bucholz 1874

CONTENT: *spectrum, spinosus, temporalis, viridis*

SUBGENUS: *Rhampholeon* (*Rhinodigitum*) Matthee, Tilbury, and Townsend 2004

TYPE SPECIES *Rhampholeon platyceps* Günther 1893

CONTENT: *acuminatus, beraduccii, boulengeri, chapmanorum, moyeri, nchisiensis, platyceps, uluguruensis*

ETYMOLOGY: Gr. *rhamphos,* beak; *leon,* lion: beaked lion, referring to the soft granulated rostral appendage present in the type species of this genus.

Figure 8.9 (see color insert)

Following on the cranial studies of Rieppel (1987) and Werner (1902b), and the morphological assessment of Tilbury (1992), Matthee et al. (2004) split the African species of pygmy chameleons into two genera *Rhampholeon* and *Rieppeleon* and further divided *Rhampholeon* into three subgenera, namely, *Rhampholeon* (*Rhampholeon*); *Rhampholeon* (*Rhinodigitum*) and *Rhampholeon* (*Bicuspis*). The genus *Rhampholeon* currently comprises 14 species, 5 of which have been described since 2000 (Fig. 8.9). There are several more species in the process of being described. As a group, all species of *Rhampholeon* can be characterized by possessing, to variable degrees, a rostronasal projection. Apart from three species of the *Rhampholeon* (*Rhampholeon*) subgenus, all have strongly bicuspid claws and all have, to a varying degree of development, accessory palmar/plantar spines at the base of each toe or finger. The chromosomal morphology of two species of *Rhampholeon* (*R. spectrum* and *R. marshalli*) have in common a derived continuous karyotype of $2n = 20 = 18M + 2m$ (Matthey, 1957; Wright and Broadley, 1973). The three subgenera show a geographic distribution that suggests that vicariance associated with the development of the African Rift Valley system and climate changes were the main drivers to speciation within the genus (Matthee et al., 2004). Three of the four species of the *R.* (*Rhampholeon*) subgenus are spread out along the northern part of the Eastern Arc of montane refugia and the remaining species (*R. spectrum*) from the northwestern Congo basin as far west as Nigeria. *R. spectrum* occupies an isolated position in the subgenus in that it seems to be sister to a clade containing the rest of the lineages. This subgenus has retained the symplesiomorphic condition of a calyculate hemipenal truncus and a rather more complex apical ornamentation as compared with other subgenera within *Rhampholeon*. In addition, the subclade containing the other three species (*R. spinosus, R. temporalis* and *R. viridis*) has retained the simple nonbicuspid claws seen in all other species of "typical" chameleons. *R (Bicuspis)* comprises two species of pygmy chameleon that are found to the south of the Gregorian Rift. This subgenus is characterized by a specific hemipenal morphology of two bifid apical horns surmounting a short bag-like noncalyculate truncus. In addition, the parietal peritoneal membranes are pigmented with melanin, which, with their short gestation time from deposition to hatching (Broadley and Blake 1979, Humphreys 1990), indicates a partial adaptation to cool weather in apparently moving toward a viviparous mode of reproduction. The rest of the species in *Rhampholeon* are within the subgenus *R.* (*Rhinodigitum*). This subgenus has the hemipenal synapomorphy of two simple horn-like apical structures surmounted on a short bag-like acalyculate truncus. The presence or absence of axillary and/or inguinal pits (mite pockets) has been used in the past for descriptive purposes, but as characteristics in some species, these elements have proved to be rather variable and not consistent enough to be a reliable taxonomic character.

Genus *Rieppeleon* Matthee, Tilbury, and Townsend, 2004

TYPE SPECIES: *Chamaeleo kerstenii* Peters 1868

GENERIC SYNONYMS: NONE

CONTENT: *brachyurus, brevicaudatus, kerstenii*

ETYMOLOGY: Named for the comparative anatomist Olivier Rieppel who first identified the paraphyly of the African pygmy chameleons by virtue of differences in cranial anatomy

Figure 8.10 (see color insert)

Three described species are ascribed to this genus (Fig. 8.10). Among the African pygmy chameleons, *Rieppeleon* in general and *R. kerstenii* in particular have retained a superficial phenetic similarity to species within the genus *Brookesia*. All three species possess a cutaneous ridge that extends posteriorly from the temporal crest along the midflank. They also all have closely knit acuminate tubercles that form a spinous mat on their soles and palms. There are no accessory palmar spines. Claws are weakly bicuspid. *R. kerstenii* has a dual gular ornamentation very like that seen in some species of *Brookesia*. *R. brevicaudatus* has a small composite lobule on the mentum; the latter anatomical feature is purported to link *Rieppeleon* to the single species within its sister clade *Archaius tigris*.

The hemipenes of both *Rieppeleon brachyurus* and *R. kerstenii* are short and acalyculate and have a quadruple apical ornamentation, while *R. brevicaudatus* appears to have no apical ornamentation apart from a cluster of a few mucosal buttons on the asulcal aspect of the apex.

In the skulls of *Rieppeleon kerstenii* and *R. brachyurus* (Rieppel's *Rhampholeon* type I) the dorsal process of the squamosal is reduced and does not make contact with the parietal, leading to an incomplete closure of the posterior part of the temporal fossa (Rieppel, 1987), while in *R. brevicaudatus,* the ascending process of the squamosal does make contact with the parietal. The nasals are paired, but *R. brevicaudatus* has an otherwise unique arrangement in which the nasal forms a lateral projection that meets with an anterior projection of the prefrontal. In all three species, the prefrontals make direct and usually extensive contact with the nasal bones (unlike in *Rhampholeon,* in which there is no contact). The supratemporal bone is absent in all three members of the genus. The parietal bone forms a flattened roughly triangular skull table with the posterior angle elongated into a narrow sagittal spur of varying length in all cases (Rieppel, 1987). The phylogenetics of *Rieppeleon* suggest that *R. brevicaudatus* is sister to a clade that contains *R. brachyurus* and *R. kerstenii* (Matthee et al., 2004).

Genus *Trioceros* Swainson 1839

TYPE SPECIES: *Trioceros oweni* Gray 1831

GENERIC SYNONYMS: *Trioceros* Fitzinger 1843; type species *Chameleo oweni* Gray 1831. *Pterosaurus* Gray 1865; type species *Chameleon cristatus* Stutchbury 1837. *Ensirostris* Gray 1865; type species *Ensirostris melleri* Gray 1865

CONTENT: *affinis, balebicornutus, bitaeniatus, camerunensis, chapini, conirostratus, cristatus, deremensis, ellioti, feae, fuelleborni, goetzei, hanangensis, harennae, hoehnelii, incornutus, ituriensis, jacksonii, johnstoni, kinangopensis, kinetensis, laterispinis, marsabitensis, melleri, montium, narraioca, ntunte, nyirit, oweni, perreti, pfefferi, quadricornis, rudis, schoutedeni, schubotzi, serratus, sternfeldi, tempeli, werneri, wiedersheimi*

ETYMOLOGY: L. *tri-*, prefix meaning "three," Gr. *keros*, horn: three-horn, denoting the three cranial horns present in the males of the type species of this genus

Figure 8.11 (see color insert)

This is the most speciose African genus (Fig. 8.11 in the color insert) with currently 40 species and 6 subspecies recognized (including *Trioceros eisentrauti* recently classified as a subspecies of *T. quadricornis*). The genus has a tropical pan-African distribution extending from east to west, with the most southerly species occurring in Mozambique and Malawi (*melleri*) and the most northerly in Ethiopia (*affinis*). Most of the species within *Trioceros* are confined to wet evergreen forest or montane biotopes or their peripheries with only a few species found out of evergreen forest proper. (*T. melleri, T. bitaeniatus, T. goetzei, T. schubotzi,* and *T. rudis*)

This is the only genus in which the development of annulated bony horns is seen. This character does not occur in all species of this genus but is found in representatives across the various subgroups within the genus. Other head ornamentation may include such features as occipital lobes and dual gular crests although these are not found only within this genus. The scalation is variable from subhomogeneous and granular to markedly heterogeneous.

The bony structure of the cranial parietal complex of the few species in which it is known, conforms with that of all other genera of typical chameleons (except *Bradypodion*)—that is, the relatively apomorphic condition in which the parietal bone is usually small, triangular, and without descending supratemporal processes. A narrow sagittal spinous process or spur forms the midline support of the casque underlying the parietal crest. There are no identified bony features of the cranium specific to the genus *Trioceros* apart from the species that have cylindrical preorbital or rostral horns. These may be considered to be a synapomorphy within the genus (Klaver and Böhme, 1986).

The lungs of this genus are characterized by having one to three large septae that are attached to and possibly arise from the ventral, medial and lateral walls of the lung (Klaver, 1981). The origin and the nature of these septae are regarded as synapomorphic characteristics for the genus. The septae subdivide the lung into two to four chambers arranged from anterior to posterior (lung types D, E and F; Klaver, 1981). The septae do not reach the dorsal wall, allowing all chambers to communicate with each other via a common space in the dorsal zone. A membranofibrous diaphragm partially delimits a small dorsocranial compartment at the cranial end of the lung. Varying numbers of small partial septae arise from the anterodorsal and anteroventral walls of the lung. In all species examined so far, the inferior and posterior surfaces of the lungs are festooned with diverticulae of varying length, structure, and numbers.

The hemipenes are calyculate with a plesiomorphic four rotulae apical ornamentation (Böhme and Klaver, 1980; Klaver and Böhme, 1986). The species differ in the finer detail of calyceal structure, rotulae size and orientation, and the number and site of apical papillae.

The karyotype of representative species from two of the species groups within *Trioceros* differs from 2n = 36 = 12M + 24m in the *cristatus* group to 2n = 24 = 12M + 12m in the *bitaeniatus* group. Hofman et al. (1991) using immunological distance to assess the relationships within the Chamaeleonidae found that all of the several species of *Trioceros* they examined (*T. montium, T. werneri, T. tempeli, T. hoehnelii, T. jacksonii, T. bitaeniatus,* and *T. elliotii*) were all closer to *Furcifer pardalis* than to *Chamaeleo quilensis.*

This African genus comprises a somewhat heterogeneous collection of chameleons which encompasses at least four species complexes (*Trioceros affinis, T. cristatus, T. bitaeniatu, T. werneri*; Hillenius, 1959; Klaver, 1981; Koreny, 2006), which are currently loosely named "groups" as well as a single species that does not fit into any of the other groups, namely, *T. melleri*. Three of the five subgroups within this genus (*T. affinis, T. bitaeniatus,* and *T. werneri*) have probably independently developed a viviparous reproductive strategy, while the fourth (*T. cristatus*) and *T. melleri* have retained oviparous modes of reproduction. The viviparous groups have a dense melanotic infiltration of the parietal peritoneum—a condition usually associated with viviparity in the Chamaeleonidae. The species composition of the various groups based on morphological characteristics is largely congruent with molecular phylogenies that have incorporated sufficient taxa across the groups (Townsend and Larson, 2002; Koreny, 2006; Tilbury and Tolley, 2009a), although a comprehensive analysis of the evolutionary relationships with the genus as a whole is still as yet outstanding.

The "*cristatus* group" of West African species includes *Trioceros feae, T. camerunensis, T. peretti, T. pfefferi, T. quadricornis* and subspecies, *T. wiedersheimi, T. cristatus, T. serratus,* and *T. montium*. A study on the genetic affinities of the *cristatus* group (Pook and Wild, 1997) demonstrated two distinct lineages within the group, one including *T. camerunensis, T. feae, T. cristatus,* and *T. montium* and the other including *T. quadricornis, T. wiedersheimi,* and *T. pfefferi*. Barej et al. (2010) elevated *T. serratus* from the synonymy of *T. wiedersheimi* and elevated the subspecies *T. wiedersheimi peretti* to a full species (both moves based on an average uncorrected p-distance (genetic divergence) within *T. wiedersheimi sensu lato* for segments of

12S and 16S rRNA of <4%) and placed *T. eisentrauti* as a sub-species of *T. quadricornis* based on low genetic divergence of <1%. Four species of chameleon ranging from west to east Africa comprising *T. oweni, T. ituriensis, T. deremensis* and *T. johnstoni* are not easy to place in any group. These latter four species provide a link between the east and west African species of *Trioceros*. Barej et al. (2010) placed *T. oweni* as sister to the "*cristatus*" radiation. *T. johnstoni* has the same karyotype as *T. cristatus* and *T. wiedersheimi* (2n = 36 = 12M + 24m). Like the west African *cristatus* group, the lungs of these four species have two large septae arising from the ventral wall of the lung and a variable number of diverticulae (lung type E; Klaver, 1981). In addition, all these species are oviparous. For the meantime and until further resolution is obtained, *T. johnstoni, T. oweni, T. deremensis* and *T. ituriensis* can be loosely grouped together with affinities to the *cristatus* group.

Although the recently described Ethiopian form *Trioceros balebicornutus* shows several morphological characteristics in common with *T. deremensis* and the West African group, such as paired annulated horns and the absence of parietal and temporal crests, its genetic affinities clearly lie within the Ethiopian region and ally it to *T. affinis* and *T. harennae* (Koreny, 2006).

The "*bitaeniatus* group" includes *Trioceros bitaeniatus, T. chapini, T. conirostratus, T. ellioti, T. hanangensis, T. hoehnelii, T. jacksonii, T. kinangopensis, T. kinetensis, T. marsabitensis, T. narraioca, T. ntunte, T. nyirit, T. rudis, T. schoutedeni, T. schubotzi,* and *sternfeldi*. A study using mtDNA (Koreny, 2006) demonstrated four principal clades within the "*bitaeniatus* group" with *T. jacksonii* as sister to the rest of the group. These clades were estimated to have diverged about 4 Mya to 4.5 Mya. The species in which the karyotype is known at present (*T. bitaeniatus, T. hoehnelii,* and *T. jacksonii*) all have the same karyotype (2n + 24 + 20M + 4m) (Klaver and Böhme, 1986). Species of the *bitaeniatus* group are viviparous and have a melanotic infiltration of the parietal peritoneum. The lungs of all the species of the "*bitaeniatus* group" so far examined conform to the same structure, namely, lung type F (Klaver, 1981). The species *T. chapini*, which occurs in West Africa, below the Bulge of Africa, is included in the *bitaeniatus* group based on its lung morphology (Klaver, 1981). Neither its genetics nor its mode of reproduction is known at present. Within the *bitaeniatus* group, *T. rudis* forms a heterogeneous group comprising several populations of chameleons from diverse localities. This "*rudis* subgroup" is in need of a specific review of its inter-relationships. There are almost certainly a number of unresolved species captured within the current definition of *rudis*. Stipala et al. (2011, 2012) have described additional species from northern Kenya.

The "*werneri* group" includes *Trioceros werneri, T. incornutus, T. goetzei, T. tempeli, T. laterispinis,* and *T. fuelleborni*. This group has a viviparous reproductive strategy associated with pigmentation of the parietal peritoneum. The hemipenes of the members of this group are characterized by the presence of large sulcal rotulae and smaller asulcal rotulae. Their lung structure is generally similar to the species of the *cristatus* group—lung type E (Klaver, 1981).

Lastly, the African giant one-horned chameleon *Trioceros melleri* is placed by itself in a fifth group by virtue of its unique lung morphology (Klaver, 1981). This oviparous species has three large septae arising from the ventral, medial, and lateral walls of the lung that divide the lung into four chambers (lung type D—Klaver, 1981).

Within general taxonomy and with regard to the Chamaeleonidae in particular, the use of categories such as subspecies, variants, and other loose descriptive concepts will need to be resolved. The concept of the "ecomorph" (Measey et al., 2009; Herrel et al., 2011) appears to have a place within some of the more complex taxa, especially within species in which the use of variety has been applied in the past. The use of subgenera to provide nomenclatural stability and to recognize evolutionary relationships between clades that encompass subclades could be usefully employed. This was initiated by Klaver and Böhme (1986) and also followed by Matthee et al. (2004). More recently within another reptilian family, subgeneric allocations were given to clades within the genus *Naja* (Wallach et al., 2009).

There are very few recent studies on the comparative anatomy of the Chamaeleonidae (Glaw et al., 1999; Zippel et al., 1999; Herrel et al., 2001b; Bergmann et al., 2003; Müller and Hildenhagen, 2009; Boistel et al., 2010). With modern methods of imaging and three-dimensional computed tomography, the anatomical features of chameleons could be much better appreciated and may potentially provide a wealth of new structural apomorphies.

Molecular phylogenetics is not without its limitations. One area of concern is the appearance of partial molecular phylogenies based on limited or possibly inappropriate genes and with incomplete taxon sampling. This has led to conflicting and confusing phylogenies. There appears to be sufficient macrostructure in the Chamaeleonidae to enable the identification of supporting apomorphies for revealed genetic clades, and these somehow need to be incorporated into phylograms. More than that, it will remain a future challenge to find, identify and assign all the currently undescribed extant taxa to complete the alpha taxonomic inventory. Almost all recently published studies describing new chameleon taxa have followed the "integrated taxonomic" approach (e.g., Gehring et al., 2012), using a tandem combination of morphological and molecular data sets to provide a model approach to alpha-level taxonomy.

Given the lability of phenotypic expression in chameleons, morphological and genetic conflicts are to be expected. If there is one thing that phylogenetics has shown, it is that resolution of the evolutionary relationships within the Chamaeleonidae is by no means complete. It remains a challenge for taxonomists to identify the principal supporting synapomorphies for all phylogenies. Further studies are necessary to marry macromorphology and micromorphology with molecular phylogenies into an understandable and workable taxonomic framework. Taxonomists do not have all the answers yet, but going forward, they need to focus on obtaining the most meaningful and best-quality data sets to produce the most robust phylogenetic reconstructions.

ACKNOWLEDGEMENTS

The author wishes to thank Bill Branch and Charles Klaver for their sage advice and comments on this chapter.

NINE

Fossil History of Chameleons

ARNAU BOLET and SUSAN E. EVANS

Chameleons are a highly characteristic and morphologically specialized group of lizards, with more than 190 species in 11 accepted genera (Appendix). Apart from recent introductions, the group is mainly distributed across southern Europe, Africa, the Middle East, southern India, Sri Lanka, and Madagascar, reaching its greatest diversity in the latter (Chapter 7). Ingroup relationships of living chameleons are based strongly on molecular studies (e.g., Townsend and Larson, 2002, 2011a; Tolley et al., 2013), particularly at the species level, where morphological characters are less reliable (Tolley et al., 2004, 2011; Tilbury and Tolley, 2009; Townsend et al., 2009; Gehring et al., 2012). Uncertainties as to the relationships of chameleons with agamids, as well as conflicting ideas as to the position of Iguania as a whole, have hampered the study of their origin and early history and, again, much of the recent literature on this topic has also been based on molecular analyses (e.g., Macey et al., 2000a; Townsend et al., 2011a). Fossils have the potential to provide valuable information regarding the early evolution and paleobiogeography of the group but, unfortunately, the fossil record of chameleons is extremely limited (Tables 9.1, 9.2, Fig. 9.1), there are no complete specimens, and representatives of the group appear surprisingly late (Early Miocene, ca. 21 Mya) in contrast to other squamates.

9.1 PHYLOGENETIC RELATIONSHIPS OF IGUANIA AND ACRODONTA

Iguania is a large and diverse squamate group. Until recently, it was considered to be the sister taxon of all other squamates, which together constituted Scleroglossa (e.g., Estes et al., 1988), and this tree topology is still consistently obtained by phylogenetic analyses based on morphological characters (e.g., Conrad, 2008; Gauthier et al., 2012). However, molecular

TABLE 9.1 Key Events in the Geological Timescale Relevant to Chameleon Evolution

Period	Epoch	Age (Ma)	Fossil record	Geological event
Neogene	Pleistocene	2.6–0.01	Malagasy chameleon fossils	
	Pliocene	5.3–2.6	African chameleon fossils	
	Miocene	23–5.3	Earliest African chameleon fossils	East African uplift and changes in current flow across Mozambique Channel
Paleogene	Oligocene	34–23	First agamid records from Africa	
	Eocene	56–34	First records of extant agamid genera	India collides with Asia Opening of Iranian route between Eurasia and Africa
	Paleocene	65–56	Acrodont jaws reported from Asia	Late Cretaceous/Paleocene exchanges between Europe and Africa
Mesozoic	Late Cretaceous	100–65	Priscagamid records from Asia	India-Seychelles separate from Madagascar (c. 88 Ma) West to east current flow across Mozambique Channel
	Early Cretaceous	145–100	Stem-iguanian from Mexico and possible early iguanians reported from Mongolia	West Gondwana divides (Africa and South America) Madagascar, India, Seychelles separate from remaining East Gondwana Limited Eurasian-African exchange via western Trans-Tethyan route
Jurassic		200–145	Earliest records of squamates (Laurasia) and in Late Jurassic/Early Cretaceous earliest records of major modern lineages	Pangaea breaks up Gondwana begins to fragment into East and West Gondwana Madagascar separates from Africa Eastern Asia at least partly isolated

SOURCES: Dates based on Ogg et al. (2008). Geologic events are from Gheerbrandt and Rage (2006), Ali and Krause (2011), Townsend et al. (2011b).

TABLE 9.2 Known Chameleon Fossils, the Area in Which They Were Found, and the Ages of the Fossils

Epoch and Age (Ma)	Taxon	Region	Level	References	Map[a]
Holocene (0.01–present)	*Chamaeleo chamaeleon*	Lebanon	Palaeolithic	Hooijer, 1961; Haas, 1952	1
	Chamaeleo chamaeleon	Spain	Holocene	Talavera and Sanchiz, 1983	2
	?*Chamaeleo intermedius*	Madagascar	?Holocene	Estes, 1983a	3
Pleistocene (2.6–0.01)	*Chamaeleo* sp.	Israel	Middle Pleistocene	Maul et al., 2011	4
	?*Chamaeleo intermedius*	Madagascar	?Pleistocene	Estes, 1983a	5
	Trioceros jacksonii	Tanzania	Pleistocene	Leakey, 1965	6
Pliocene (5.3–2.6)	?*Bradypodion* sp.	South Africa	Pliocene	Tolley et al., 2006	7
Miocene (23–5.3)	*Rhampholeon* sp.	Kenya	Miocene	Rieppel et al., 1992	8
	Chamaeleo intermedius	Kenya	Late Miocene	Hillenius, 1978a	9
	?*Chamaeleo intermedius*	Kenya	Early Miocene	Estes, 1983a	10
	Chamaeleo sulcodentatus	Germany	MN5–MN6	Schleich, 1994; Böhme, 2003	11
	Chamaeleo andrusovi	Czech Republic	MN3–MN4	Čerňanský, 2010	12
	Chamaeleo caroliquarti	Germany	MN5–MN6	Schleich, 1994; Böhme, 2003, 2010	11
	Chamaeleo sp.	Germany	MN4–MN6	Böhme, 2003, 2010	11
	Chamaeleo simplex	Germany	MN5	Schleich, 1994	11
	Chamaeleonidae indeterminate	Germany	MN4–MN6	Čerňanský, 2011; Böhme, 2010	11
	Chamaeleo pfeili	Germany	MN4b	Schleich, 1984; Čerňanský, 2011	11
	Chamaeleo caroliquarti	Germany and Czech Republic	MN3–MN6	Moody and Roček, 1980; Böhme, 2003, 2010; Čerňanský, 2010	11, 12

(Continued)

TABLE 9.2 *(Continued)*

Epoch and Age (Ma)	Taxon	Region	Level	References	Map[a]
	Chamaeleo bavaricus	Germany	MN5–MN6	Schleich, 1983, 1994	11
	Chamaeleo spp.	Switzerland	MN5–MN6	Bolliger, 1992	13

a. Map numbers correspond with Fig. 9.1.

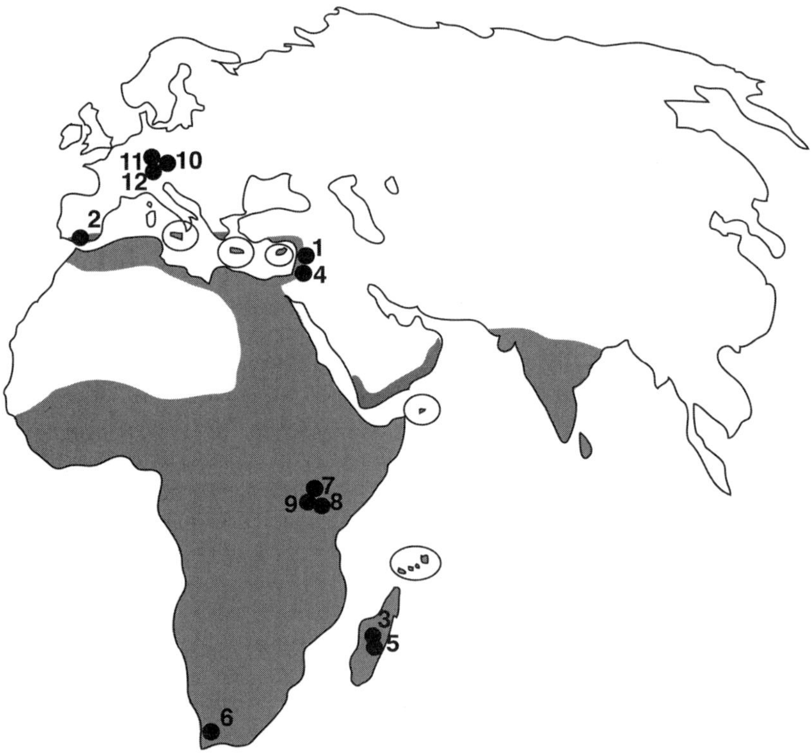

FIGURE 9.1. Map of localities that have yielded fossil chameleon remains. Numbers correspond to those in Table 9.2.

phylogenies (e.g., Townsend et al., 2004; Vidal and Hedges, 2005) and those based on combined evidence (e.g., Wiens et al., 2006; 2010), obtain a very different phylogeny in which Iguania is placed as the sister group of Anguimorpha (e.g., Townsend et al., 2004; Vidal and Hedges, 2005) or Anguimorpha + Serpentes, rendering Scleroglossa paraphyletic. This major difference between topologies makes the morphology of stem-iguanians difficult to reconstruct and has a major effect on estimated dates of origin.

In older classifications (e.g., Estes et al., 1988), Iguania was divided into Iguanidae (formed by all pleurodont iguanians) and Acrodonta (formed by all acrodont iguanians,

Chamaeleonidae, and Agamidae) (Estes et al., 1988). Subsequent classifications (e.g., Frost and Etheridge, 1989; Schulte et al., 2003; Conrad and Norell, 2007; Conrad, 2008) subdivided pleurodont iguanians into several distinct families (e.g., Opluridae, Crotaphytidae, and Polychrotidae). Under this scheme, Iguanidae is a far less inclusive clade than it was formerly (comprising the iguanines of earlier classifications), although, confusingly, many molecular trees retain Iguanidae for pleurodont iguanians as a whole (e.g., Wiens et al., 2006). Whether or not pleurodont iguanians form a monophyletic group in relation to acrodont iguanians (Pleurodonta of Frost et al., 2001; Conrad and Norell, 2007) remains uncertain (Conrad, 2008), but molecular trees generally support monophyly (e.g., Townsend et al., 2004, 2009, 2011a; Wiens et al., 2006, 2010).

Acrodontan monophyly has been corroborated by morphological (Moody, 1980; Estes et al., 1988; Frost and Etheridge, 1989; Gauthier, 2012) and molecular studies (e.g., Macey et al., 1997b, 2000a; Honda et al., 2000). Within Acrodonta, there is a consensus that Chamaeleonidae is monophyletic but a lack of agreement as to the relationships of chameleons to other acrodontan taxa. Camp (1923) argued that chameleons were an offshoot of the agamid stem, a view supported by some morphological (e.g., Moody, 1980; Estes et al., 1988; Frost and Etheridge, 1989) and molecular (e.g., Honda et al., 2000) studies that place the genera *Uromastyx* and *Leiolepis* (either together or in series) as the sister taxa of all other acrodontans, or of nonchamaeleonid acrodontans (i.e., the remaining Agamidae). In support of Moody's (1980) proposal, Gauthier et al. (2012) recovered a sister-taxon relationship between Leiolepidinae (*Leiolepis* + *Uromastyx*) and the taxon they named Chamaeleonoidea (Chamaeleonidae + Agaminae). Okajima and Kumazawa (2010), however, reported strong molecular evidence for agamid monophyly and argued that *Uromastyx* and *Leiolepis* are the sister taxa of other agamids; this topology was also recovered by Townsend et al. (2011a) and Hutchinson et al. (2012).

Here, Hutchinson et al. (2012) are followed, and the term *acrodontan* is therefore used to refer to taxa attributed to crown-group Acrodonta whereas *acrodont* refers to teeth that are fused to the crest of the jaw. A lizard may have acrodont jaws without being acrodontan, either because it lies on the stem of Acrodonta or because it has evolved the condition independently. Acrodonty has arisen at least three times within Lepidosauria—in Rhynchocephalia, in Acrodonta, and in trogonophid amphisbaenians; it may also have evolved in extinct lineages. Problems arise where fossil taxa are represented by isolated jaws and partial dentitions as, for example, with many of the taxa described from Paleogene deposits.

9.2 FOSSIL RECORD OF ACRODONTA

Mesozoic and Paleocene Fossil Record of Acrodont Jawed Lizards

The first undoubted fossil squamate assemblages are from the Middle Jurassic, ca. 165 Mya of Britain (Evans, 2003), Kyrgyzstan (Averianov, 2000), and Siberia (Averianov et al., 2005). The only earlier records are from the Late Triassic (*Tikiguania,* Datta and Ray, 2006) and Early Jurassic (*Bharatagama* and unnamed taxa, Evans et al., 2002) of India. In both *Tikiguania* and *Bharatagama*, a majority of the teeth have an acrodont implantation, and both taxa were originally referred to Acrodonta. *Tikiguania* comprises a single, well-preserved dentary from the Carnian Tiki Formation (ca. 225 Mya) and raised early suspicions because of its completeness and modern appearance as compared with the otherwise very fragmentary small vertebrate remains from the same beds (S.E. Evans, personal observation, 2010). One reanalysis (Hutchinson et al., 2012) has shown that the jaw is closely similar to that of extant draconine agamids (e.g., *Calotes, Gonocephalus*) that live in the same region today, and "*Tikiguania*" is almost certainly a recent or at least a Neogene inclusion in an older deposit. *Bharatagama*, from the Early Jurassic (Sinemurian–Pleinsbachian, ca. 190 Mya; Bandyopadhyay et al., 2010) Kota Formation, is represented by a number of partial specimens. It co-occurs with other lepidosaurs, including typical rhynchocephalians (e.g., *Godavarisaurus, Rebbanasaurus;* Evans et al., 2001) from which it differs in several respects, including the presence of recurved pleurodont teeth on the premaxilla, anterior maxilla, and anterior dentary. However, without skull or postcranial material, its phylogenetic position remains uncertain.

Aside from these early genera, the first putative record of an acrodont lizard is from the Early Cretaceous of China, in the form of the Yixian (Barremian) taxon *Xianglong* from China (Li et al., 2007). *Xianglong* is an exquisite specimen preserving traces of elongate gliding ribs analogous to those of the living draconine agamid, *Draco*. However, the type and only specimen is juvenile and the skull is extremely poorly preserved (S.E. Evans, personal observation, 2007). What was interpreted as an acrodont dentition may be broken bone along the edge of a rather amorphous crushed skull mass.

Priscagamids were first described from the Late Cretaceous of Mongolia (*Priscagama, Pleurodontagama;* Borsuk-Białynicka and Moody, 1984), although Gilmore's (1943) *Mimeosaurus* was later included in the group. Many more specimens, assigned to several genera, have been described by Alifanov (1989, 1996) and Gao and Norell (2000). Priscagamids have a mixed pleurodont (anteriorly) and acrodont (posteriorly) dentition. Their skulls are iguanian in character, having large eyes, constricted frontals, open temporal fenestrae and large triradiate postorbitals that contact the skull roof and form most or all of the posterodorsal orbital rim. Priscagamids resemble living acrodontans in having mainly acrodont marginal dentition, a reduced median premaxilla (Gao and Norell, 2000), and loss or extreme reduction of the postfrontal, but differ in the retention of a large splenial and pterygoid teeth (some taxa). Phylogenetic analyses using morphological characters (e.g., Frost and Etheridge, 1989; Conrad, 2008; Smith, 2009) place priscagamids on the acrodontan stem. Most genera are known from the Campanian–Maastrichtian (Late Cretaceous, ca. 80 to 65 Mya) of China and Mongolia, but Nessov (1988) mentions specimens from the Coniacian (ca. 88.6 to 85.8 Mya) of Uzbekhistan, and Alifanov (1993) reported them as present in the Early Cretaceous Mongolian locality of Höövör (dated as Aptian–Albian, ca. 110 Mya), although he omitted them from the list in a later account of that fauna (Alifanov, 2000). If priscagamids are stem-acrodontans, then the Coniacian record provides a minimum date of divergence from pleurodont iguanians, the jaws of which are also reported from Höövör (Alifanov, 2000). No priscagamid has been recorded from post-Cretaceous deposits.

Isodontosaurus was first described from the Late Cretaceous of Mongolia (Gilmore, 1943) on the basis of partial mandibles bearing uniform marginal teeth (hence the name) with a pleurodont or pleuroacrodont implantation. It has also been recovered from Chinese Inner Mongolia (Gao and Norell, 2000), along with a range of similar taxa from the Cretaceous and Paleogene of Mongolia (Alifanov, 2004). *Isodontosaurus* was classified variably as an anguid (Gilmore, 1943), a scincomorph (Estes, 1983a), and an agamid (e.g., Borsuk-Białynicka, 1991; Alifanov, 1993). Gao and Norell (2000) described new skull and postcranial specimens and attributed it to "Iguania *incertae sedis*," noting points of similarity with acrodontans, except for the tooth implantation. Conrad and Norell (2007) placed it in a similar position following a phylogenetic analysis, but in Conrad (2008), it emerged as the sister taxon of Priscagamidae + Acrodonta. If this is correct, then despite its relatively recent age, it may provide an indication of the primitive acrodontan morphology.

Similarly enigmatic are a number of fragmentary jaws and/or skulls from the Paleocene of China, some of which have been referred to Acrodonta or to Squamata *incertae sedis* (e.g., Wang and Li, 2008). These include indeterminate skull material of *Anhuisaurus huainanensis* (Hou, 1974), said to have acrodont teeth but this cannot be confirmed (S.E. Evans, personal observation, 2010). *Qianshanosaurus huangpuensis* (Hou, 1974) and *Changjiangosaurus huananensis* (Hou, 1976) are partial jaws with closely spaced pleurodont or pleuroacrodont teeth and peculiarly large angular processes. The dentition is somewhat like that of *Isodontosaurus*, but the position of all of these taxa remains equivocal pending more complete material. Further new genera from the Middle Eocene of Mongolia were referred to the Chanjiangosauridae by Alifanov (2009).

The Fossil Record of "Agamidae"

In addition to the problematic specimens mentioned above, the Paleocene of China has yielded rare lizard jaws with a more clearly acrodont dentition, at least in the posterior part of the jaw, although the anterior teeth may be pleurodont. Notable among these is *Tinosaurus doumuensis* (Hou, 1974) which, with *T. postremus* (Averianov, 2000) from the Paleocene of Kazakhstan, represents the first record of the apparently ubiquitous Paleocene–Eocene acrodont genus *Tinosaurus*. Specimens attributed to this taxon are also recorded from the Early Eocene of China (Gilmore, 1943; Dong, 1965; Li, 1991a,b), India (Prasad and Bajpai, 2008), and Europe (Hecht and Hoffstetter, 1962; Augé, 1990; Rage and Augé, 1993; Augé and Smith, 1997) and from the Middle Eocene of Kazakhstan (Chkhikvadze, 1985), Pakistan (Rage, 1987), and North America (Leidy, 1872, 1873; Marsh, 1872). Neither the monophyly of *Tinosaurus* nor its phylogenetic position is well established and it has been considered a "form-taxon" (Smith et al., 2011). Most attributed specimens share a suite of characteristics: the anterior pleurodont teeth, one of which is often a caniniform; well-spaced tricuspid posterior teeth (subacrodont with their bases above the prominent subdental shelf); the slender anteriorly tapering dentary; and maxillae with a strong palatal process and a horizontal premaxillary process (Augé and Smith, 1997). However, Smith et al. (2011) observed that many of these features are widespread among living agamids. *Tinosaurus* resembles priscagamids in having a Meckelian fossa that is open posteriorly and pleurodont anterior teeth that may include a caniniform, but *Tinosaurus* differs from priscagamids and resembles many crown-group acrodontans in lacking a splenial. Whether *Tinosaurus* is a primitive early agamid, a surviving member of a more geographically widespread stem-acrodontan lineage, or a collection of unrelated acrodont taxa remains unclear, pending the description of skull remains (Smith et al., 2011).

Other putative agamids or stem-agamids include *Vastanagama susani* from the early Eocene in India (Prasad and Bajpai, 2008); *Zephyrosaurus hypsochorosus, Talosaurus tribolosus, Mergenagama paurosa, Pseudotinosaurus asiaticus,* and *P. ascriptivus* from the middle Eocene of Mongolia (Alifanov, 1991; Gao and Dashzeveg, 1999); and *Brevidensilacerta xichuanensis, Huadiansaurus sunjiatunsis,* and unnamed taxa from the middle Eocene of China (Dong, 1965; Li, 1991a; Smith et al., 2011). Indeterminate acrodont lizards have also been reported from the Middle Eocene (Lutetian, Irdinmanhan) in the Shinzhaly locality of Kazakhstan (Zerova and Chkhikvadze, 1984). Rana (2005) reported the presence of "agamids" from Cretaceous–Paleocene sedimentary sequences in the Deccan Traps of India, but this was based on a misinterpretation (R. Rana, personal communication, February 2012). The jaws illustrated in the original paper are clearly not acrodont.

The earliest fossils attributed to living acrodontan genera (or at least as close relatives) include unnamed *Uromastyx*-like specimens from the Early Eocene (ca. 50 Mya) of Kyrgyzstan (Averianov and Danilov, 1996; Smith et al. 2011) and a reported *Leiolepis* from the middle Eocene (ca. 40 Mya) of China (Alifanov, 2009). Moody and Roček (1980) reported *Uromastyx* (*"Palaeochamaeleo"*) from the late Eocene of France (Phosphorites du Quercy), but the precise horizon was not known, and Estes (1983a) dated it as late Eocene or early Oligocene. Further material of *Uromastyx* (and a second agamid, *Quercygama*) has been recorded from Oligocene horizons in the Phosphorites du Quercy (Filhol, 1877; de Stefano, 1903; Augé and Smith, 1997), and it is likely that the original Palaeochamaeleo material also came from this level (Augé, 2005). Indeterminate acrodont lizards have been recovered from the Oligocene of Rigal-Jouet and Coderet also in France (Augé, 2005).

The Gondwanan record of acrodontan lizards is very poor. Two badly preserved jaw elements with acrodont teeth are known from the Late Paleocene of Morocco (Augé and Rage, 2006) but, based on their morphology, they could equally be rhynchocephalian. The Oligocene locality Quarry 5 of the Jebel Qatrani Formation in Egypt has yielded remains of cf. *Uromastyx* (Holmes et al., 2010), and Thomas et al. (1991) list acrodont jaw material from the Oligocene of Oman. The Australian record of acrodontans starts close to the Oligocene–Miocene boundary (Riversleigh), with representatives of the extant genus *Physignathus,* as well as *Sulcatidens* and further unnamed acrodontans (Covacevich et al., 1990).

Neogene records of acrodontan lizards in Europe are too abundant to be covered here in detail (see Delfino et al., 2008, for an extensive review). They have been recorded from European deposits ranging in age from Miocene to Pleistocene. Their geographical range includes both southern (Greece, Italy, Portugal, and Spain) and northern (Belgium, France, Germany, Romania, Hungary, Switzerland, and Ukraine) regions. Outside Europe, *Laudakia* is recorded from the Pleistocene of Israel (Maul et al., 2011), *Agama* from the Miocene and Pleistocene of India (Joshi and Kotlia, 2010), and both *Uromastyx* sp. and *Calotes* sp. from the Indian Pliocene (Patnaik and Schleich, 1998). In Australia, there are Miocene records of *Physignathus* (Archer et al., 2006) and Pleistocene records of the extant *Chlamydosaurus* (Bennett, 1875).

The Fossil Record of Chameleons

In contrast to the relatively widespread fossil record of acrodont jawed lizards reviewed above, that of chameleons is extremely poor. Whereas most groups of lizards have attributable representatives from at least the Late Cretaceous, the first undoubted occurrences of chameleons in the fossil record occur in the early Miocene of Europe (20 to 21 Mya; Moody and Roček, 1980) and Kenya (19 to 20 Mya; Pickford, 1986). As the Miocene specimens are very close to extant taxa, there is clearly a substantial gap in the fossil record during which recognizable chameleons evolved from more generalized acrodontan ancestors. Furthermore, there are no complete specimens.

One problem with the early record may be the difficulty of distinguishing fragmentary remains of early or stem-chameleons from those of other acrodontan lineages. The Cretaceous genus *Mimeosaurus* (Gilmore, 1943), the Eocene *Palaeochamaeleo* (de Stefano, 1903), and *Tinosaurus* (Marsh, 1872) were all originally referred to as Chamaeleonidae on the basis of their acrodont dentitions, but are no longer regarded as such (Moody and Roček, 1980; see above). The latter authors noted the following differences between chameleon jaws and those of agamids: the absence of anterior pleurodont teeth (*Uromastyx* is an exception); teeth exactly fused to the upper border of the dentary, whereas in agamids they tend to be slightly lingual in placement; teeth strongly compressed labially, with a flatter lingual surface (agamid teeth tend to bulge lingually); middle and posterior teeth separated at their bases by a gap measuring 15 to 20% of tooth length; and splenial always absent. However, these features can be affected by age and preservation (Moody and Roček, 1980).

With the exception of one dubious record from the Paleocene of China (*Anquingosaurus;* Hou, 1976), chamaeleonid fossils can be grouped as follows: fragmentary jaw material from several Early–Middle Miocene European (Molasse Basin: Germany, Switzerland, and Czech Republic; Čerňanský, 2010) and African (Pickford, 1986) localities; rare more complete early Miocene specimens from Kenya (Hillenius, 1978a; Rieppel et al., 1992); and scattered records from the Pliocene–Holocene, mainly from Lebanon, Israel, Madagascar, South Africa, Tanzania, and Spain.

Paleocene

A putative chamaeleonid, *Anquingosaurus brevicephalus,* was reported from the early Paleocene of Wang-Hu-Dun Series, Qian-Shan district, Anhui, China (Hou, 1976). This would represent the earliest record of the group, but the type and only specimen is represented by a poorly preserved skull of indeterminate affinity (S.E. Evans, personal observation, 2010). There is nothing to suggest that it is a chameleon.

Miocene

Six chameleon fossil species have been named from Europe: *Chamaeleo caroliquarti* (Moody and Roček, 1980), *Ch bavaricus* (Schleich, 1983), *Ch pfeili* (Schleich, 1984), *Ch simplex* (Schleich, 1994), *Ch sulcodentatus* (Schleich, 1994), and *Ch andrusovi* (Čerňanský, 2010). The status of many of these "species" is uncertain because of the fragmentary nature of the specimens and the absence of autapomorphies (Čerňanský, 2010, 2011), but attribution to the genus *Chamaeleo* is more secure, suggesting that it was present in the Miocene of Europe from at least MN3 to MN6 (ca. 21-13 Mya) (MN are mammalian standard levels for the Neogene, see van der Meulen et al., 2011).

Chamaeleo caroliquarti was a large chameleon (estimated total length, 0.5 M; Böhme, 2003) represented by partial dentaries, maxillae, and cranial bones. It has been reported from the following localities (data from Böhme [2003], except where indicated): Merkur Nord, Czech Republic (MN3a; Fejfar and Schleich, 1994; Vejvalka, 1997; Čerňanský, 2010), Wintershof West, Germany (MN3a; Moody and Roček, 1980); Petersbuch 28, 36, and 36 II, Germany (MN4); Erkertshofen 1, Germany (MN4b); Dolnice, Czech Republic (MN4b; Moody and Roček, 1980; Roček, 1984; Čerňanský, 2010); Gisseltshausen 1b, Germany (MN5); Griesbeckerzell 1b, Germany (MN5); Untereichen-Altenstadt 565 m, Germany (MN5; Prieto et al., 2009); Petersbuch 39, Germany (MN6); Laimering 2a and 3, Germany (MN6), and Wannenwaldtobel 2, Germany (Von Volker, 1999; Böhme, 2010). Material from Sandelzhausen B and C3/D1, Germany, has been recently referred to *Ch* aff. *cariloquarti* (Böhme, 2010).

Chamaeleo bavaricus is recorded from Germany at Unterempfenbach (MN5), Aresing (MN5), Arth 1a (MN5), Sandelzhausen C3/D1, and Laimering 2a (MN6) (Schleich, 1983, 1994; Böhme, 2003, 2010).

Chamaeleo pfeili has been described only from the German locality of Rauscheröd (MN4b; Schleich, 1984), with closely similar material from the Bavarian site of Langenau (Čerňanský, 2011).

Chamaeleo sulcodentatus is another species reported from the German localities of Massendorf (MN5), Göttschlag 1b (MN6), and Laimering 3 (MN6), and from Rümikon, Switzerland (MN6) (Schleich, 1994; Böhme, 2003), although material of this species is extremely fragmentary and makes comparisons with other taxa difficult (Čerňanský, 2010). The same is true for the poorly known species *Ch simplex* (MN5, Germany) described by Schleich (1984).

Material from the German locality of Puttenhausen (MN5) probably corresponds to either *Chamaeleo sulcodentatus* or *Ch bavaricus* (Böhme, 2003), and material from the following German (and Swiss, as noted) localities has been referred to *Chamaeleo* sp. (data from Böhme, 2003, except where indicated): Petersbuch 2 (MN4a), Eiboden (MN4), Rembach (MN4b), Schiessen (MN5), Niederaichbach and Niederaichbach (also known as "links") (MN5), Puttenhausen 2 (MN5), Walda 2 (Oben) (MN5), Altenstadt (MN5), Hambach 6C

(MN5; Mörs, 2002; Mörs et al., 2000), Steinberg (Ries) (MN6), Uzwil-Nutzenbuech, Switzerland (MN6), and Ornberg, Switzerland (MN6; Bolliger, 1992). Böhme (2010) reported a second, yet undescribed, large chamaeleonid species from Gisseltshausen 1b (MN6), and Čerňanský (2011) referred chamaeleonid material from Langenau to Chamaeleonidae *indeterminate*. and from Petersbuch 2 to a second taxon of chamaeleonid.

Chamaeleo andrusovi was erected by Čerňanský (2010) for new material from the Czech locality of Dolnice (MN4) and for part of the material originally referred to *Ch. caroliquarti* by Roček (1984); Čerňanský (2010) considered *Ch. caroliquarti* a *nomen dubium* because the holotype dentary cannot be differentiated from that of the recent *Ch. calyptratus, Calumma globifer,* or *Furcifer pardalis*. Moreover, the paratypic material of *Ch. caroliquarti* may represent a second morphotype as it appears identical to the dentary of the recent *Ch. chamaeleon* (Čerňanský, 2010). Unlike the type material of previously named Miocene species (mostly fragmentary jaws), the holotype and paratypes of *Ch. andrusovi* are skull bones bearing characters that differentiate it from the known extant species. This is commendable, but it makes it difficult to compare *Ch. andrusovi* with the other Miocene species, and more complete material may lead to synonomy.

Miocene European chamaeleonids thus occur in many localities across Germany, Switzerland, and the Czech Republic, with the oldest at ca. 20 to 21 Mya (Merkur-North, Czech Republic, MN3a). Most specimens consist of incomplete maxillae and dentaries that are of uncertain specific position but are probably attributable to *Chamaeleo,* which is the only chameleon genus living in Europe today (Spain, Greece, Turkey, and Mediterranean Islands). The more northern distribution of this genus in Miocene times can be correlated to the Miocene climatic optimum, a period that allowed the migration of thermophilic ectotherms toward central Europe (Böhme, 2010). With other reptiles, such as turtles and alligatorids, lizard groups including varanids, chamaeleonids, and cordylids reached their northernmost distribution at this time. This situation probably favored dispersion from Africa to Asia, either directly or via Europe.

In the Miocene of Africa, chameleons are represented by several specimens from Kenya, although few have been described. The only named species is *Chamaeleo intermedius* (Hillenius, 1978a), based on a specimen from Fort Ternan (Upper Miocene, 13.7±0.3 Mya; Pickford et al., 2006). This specimen, the holotype (KMN-FT 3833, Nairobi), is a cast in calcite, preserving the head and anterior part of the body. It provides relatively little morphological information, but its attribution to *Chamaeleo* seems justified, and the species probably belongs to the *Ch chamaeleon* group (Estes, 1983a), although Hillenius (1978a) suggested that it might be primitive with respect to the *Ch chamaeleon* and *Ch bitaeniatus* groups. Pickford (1986) noted that there was abundant chamaeleonid material from the primate localities of Koru and Songhor (Lower Miocene, 20 to 19 Mya; Pickford, 2001). Specimens from these localities have been less securely attributed to *Ch intermedius* (Hillenius, 1978a; Estes, 1983a), but the material has not been formally described.

A second chamaeleonid specimen was described from the Lower Miocene (18 Mya; Drake et al., 1988) of Rusinga Island, also Kenya (Rieppel et al., 1992). This well

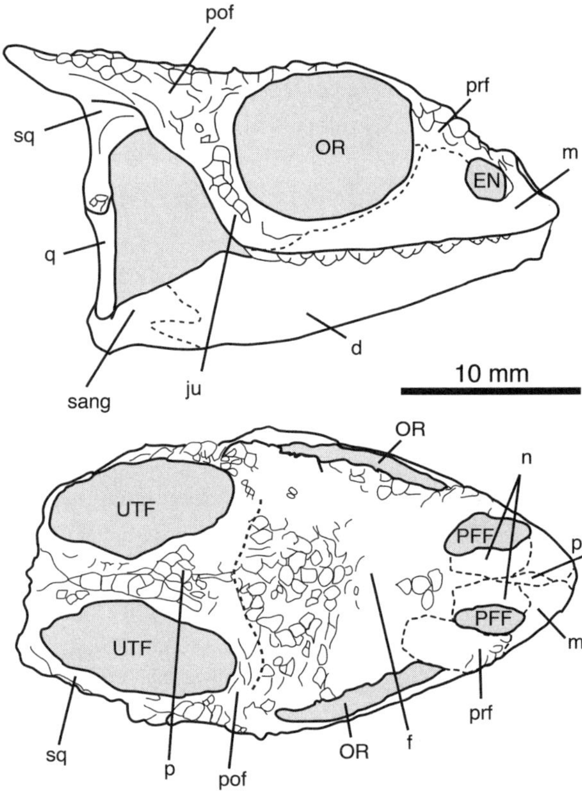

FIGURE 9.2. Drawing of the KNM-RU-18340 specimen (Rusinga Island, Kenya, Early Miocene) referred to *Rhampholeon,* in lateral (top) and dorsal (bottom) views.

LABELS: d = dentary, f = frontal, EN = external naris, ju = jugal, m = maxilla, n = nasals, OR = orbit, p = parietal, PFF = prefrontal fontanelle, pfr = prefrontal, pm = premaxilla, pof = postorbitofrontal, q = quadrate, sang = surangular, sq = squamosal, UTF = upper temporal fenestra.

preserved three-dimensional skull (KNM-RU-18340) (Fig. 9.2) is a little over 26 mm long, and is characterized by a complete separation of the prefrontal fontanelle from the external nares, paired nasals, and a parietal that narrows between the upper temporal fenestrae but widens posteriorly into a flat sculptured triangular plate. Rieppel et al. (1992) concluded that this was an early representative of the genus *Rhampholeon* (attributed to *Rhampholeon* type 2, now regarded as *Rhampholeon, sensu stricto;* Matthee et al., 2004), although no living species shows a fully separated prefrontal fontanelle. The age of this fossil is consistent with estimated divergence times of *Rhampholeo*n from its sister taxon *Rieppeleon* ("*Rhampholeon* type 1") (Matthee et al., 2004; ca. 26 to 28.3 Mya). The Rusinga specimen is both the only known complete fossil chameleon skull and the most complete chameleon fossil recovered. All of these Miocene specimens come from a relatively small area in the Rift Valley of Kenya and are from localities that generally represent vegetated areas on the slopes of Rift Valley volcanoes (Pickford, 1986). It may be that lower-lying regions provided a less suitable habitat for these specialized reptiles. Younger Neogene records of African chameleons are limited to reports of early Pliocene (5.2 Mya) fossil remains of *Bradypodion* from South Africa (Langebaan fossil Bed) (Tolley et al., 2006).

Pleistocene and Holocene

Unlike many other lizard groups, the fossil record of chameleons does not improve substantially in more recent deposits, although the distribution of fossils overlaps that of extant chameleons, namely southern Europe, Africa, Madagascar, and the Middle East. Despite the diversity of chameleons on Madagascar today, the fossil record of the group on the island is limited to two jaw elements, one from Children's Cave and the other from Cavern Ambatohomana, tentatively dated as Pleistocene and Holocene, respectively (Estes, 1983a). Similarly, the published record of Pleistocene specimens from Africa is limited to remains from Bed 1, Olduvai Gorge, Tanzania, dated to ca. 1.8 to 1.75 Mya (Walter et al., 1991), and identified by R. Hoffstetter (in Leakey [1965]) as belonging to a form close to *Chamaeleo jacksonii* (now *Trioceros jacksonii*). Disarticulated but abundant material of *Chamaeleo* sp. has been reported from the Middle Pleistocene of Israel (Maul et al., 2011) and may be identified at the species level in the future. *Chamaeleo chamaeleon* was tentatively identified at the Paleolithic site of Ksâr'akil in Lebanon (Hooijer, 1961) and was reported as abundant at the Abu Usba Cave (Mount Carmel) in Israel (Haas, 1952). It is also recorded from the Holocene of Málaga in Spain (Talavera and Sanchíz, 1983), where it is indistinguishable from the extant species that lives today in the south of Spain. This occurrence lends support to the view that the present distribution is natural and not the result of human introduction in historical times (Talavera and Sanchíz, 1983; Crespo and Oliveira, 1989; Blasco, 1997a,b), although Paulo et al. (2002) have proposed a double human introduction of African chameleons into the Iberian Peninsula based on molecular studies. *Chamaeleo chamaeleon* has recently been identified from the Moroccan Holocene locality of Guenfouda (Aouraghe et al., 2010). Rage (1972) reported chameleon remains from the Pleistocene of France, but these specimens were reattributed to agamids (Estes, 1983a).

9.3 ORIGINS OF ACRODONTA

The basal lepidosaurian dichotomy between Rhynchocephalia (the living *Sphenodon* and its fossil relatives) and Squamata must have occurred by at least the Middle Triassic (ca. 240 to 230 Mya) (Fig. 9.3, Table 9.1), given rhynchocephalian diversity in the Late Triassic, (e.g., Evans and Jones, 2010). As noted above, the first undoubted fossil squamate assemblages are from the Middle Jurassic, ca. 165 Mya of Europe (Evans, 1998, 2003) and Asia (Averianov, 2000, Averianov et al., 2005), leaving a considerable gap in the early record during which the group appears to have diversified. If Iguania is the sister taxon to all other squamates (morphological tree; e.g., Estes et al., 1988; Gauthier et al., 2012), then its origin would be predicted to be deep. *Tikiguania* (Datta and Ray, 2006) and *Bharatagama* (Evans et al., 2002) seemed to support this conclusion but, as explained above, *Tikiguania* is a recent or Neogene intrusion with draconine agamid affinities (Hutchinson et al., 2012) and the position of *Bharatagama* is uncertain.

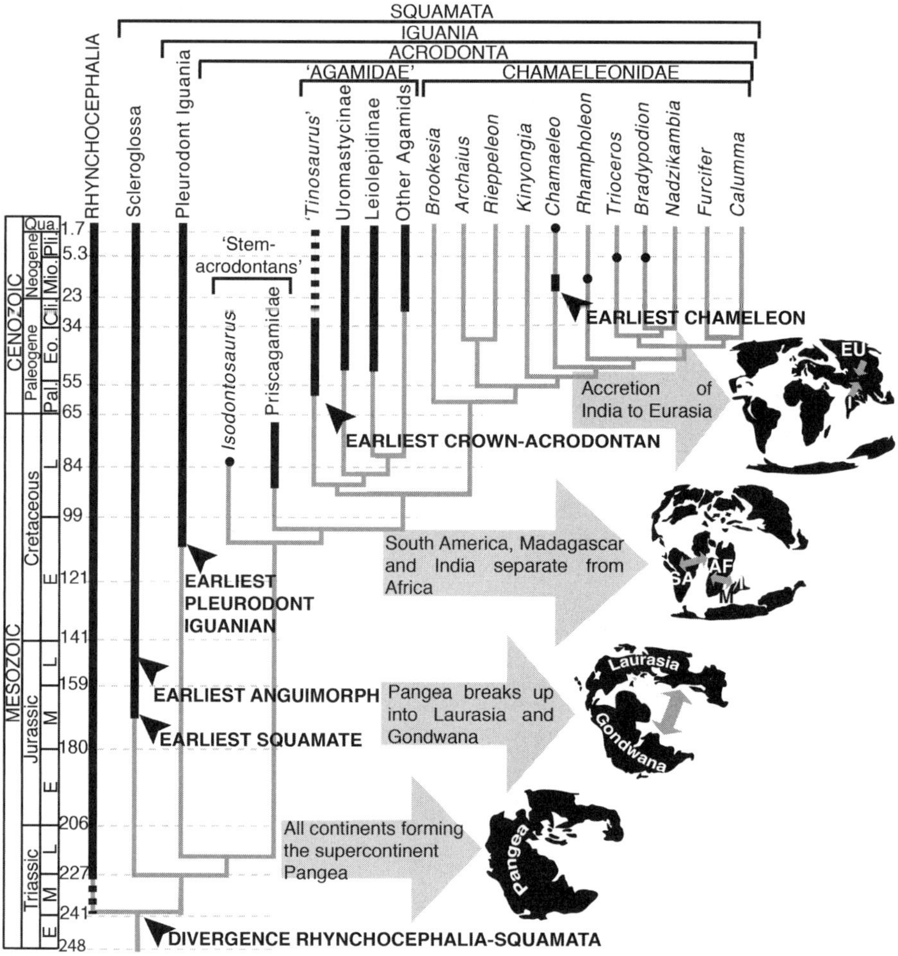

FIGURE 9.3. Temporally calibrated hypothesis of squamate interrelationships based on traditional morphologic analyses, with emphasis on acrodontan iguanians. Main geologic events are shown in paleogeographical maps.

LABELS: AF = Africa; EU = Eurasia; I = India; M = Madagascar; SA = South-America. In time scale, abbreviations correspond to E = Early, Eo = Eocene, L = Late, M = Middle, Mio = Miocene, Oli = Oligocene, Pal = Paleocene, Pli = Pliocene, Qua = Quaternary.

Based on the molecular tree topology (e.g., Townsend et al., 2004, 2011a; Vidal and Hedges, 2005; Wiens et al., 2010; Hutchinson et al., 2012), which places Iguania with Anguimorpha, iguanian origins would be shallower, but not markedly so, as the earliest currently accepted anguimorph (e.g., Conrad, 2008) is the Early Cretaceous (Berriasian, ca. 145 Mya) genus *Dorsetisaurus* from the Purbeck Limestone Formation of England (Hoffstetter, 1967), and dorsetisaur jaws and vertebrae have been described from the Late Jurassic (Kimmeridgian, ca. 150 Mya) of Portugal (from Guimarota; Seiffert, 1973; Broschinski, 2000) and North America (Morrison Formation; Prothero and Estes, 1980). This would be consistent with a recent molecular estimate for the anguimorph–iguanian divergence at 162.2 Mya (Wiens et al., 2006). The latter authors obtained a date of 146.4 Mya (Jurassic–Cretaceous boundary) for the basal iguanian dichotomy into pleurodont and acrodont iguanian lineages, although Townsend et al. (2011a) placed this split in the Aptian (late Early Cretaceous) at ca. 123 Mya. The oldest generally accepted stem-iguanian (based on both morphological [Conrad, 2008] and molecular [Wiens et al., 2006] analyses) is the mid-Aptian to late Aptian Mexican *Huehuecuetzpalli* (Reynoso, 1998), ca. 120 to 115 Mya, whereas Alifanov (2000) recorded "iguanid" remains from the Aptian–Albian (late Early Cretaceous, ca.110 Mya) locality of Höövör, Mongolia, and stem-acrodontans (priscagamids) are first recorded from the Coniacian (early Late Cretaceous, ca. 88.6 to 85.8 Mya ago) of Central Asia (Nessov, 1988). These records are compatible with the molecular dates, although, again, they indicate that a substantial part of the record is still missing.

Townsend et al. (2011a) dated the basal acrodontan dichotomy (chameleons + a monophyletic Agamidae) at 93 Mya (as compared with 78.5 Mya, dated by Wiens et al., 2006; and 47 to 90 Mya, dated by Raxworthy et al., 2002), the origin of *Uromastyx* at 87 Mya, that of *Leiolepis* at 82 Mya, and the diversification of "advanced" agamids at 80-70 Mya. Hugall et al. (2008) got similar results) The Early Jurassic *Bharatagama* is clearly incompatible with these dates, suggesting that its morphology is convergent, but there is no trace in the record of the early agamids and chameleons that should be present in Late Cretaceous deposits. Paleocene and Early–Middle Eocene records are dominated by the problematic *Tinosaurus* and other equally enigmatic taxa. The first fossils referred to modern genera, or their close relatives (cf. *Uromastyx;* Averianov and Danilov, 1996), are from the Early Eocene (ca. 50 Mya), and they remain rare until well into the Oligocene.

In terms of biogeography, Estes (1983b) suggested a Gondwanan origin for Iguania as a whole, an interpretation supported by Macey et al. (1997b, 2000b, 2006) based on molecular data. The latter works argued that the fragmentation of Gondwana contributed to early cladogenic events that led to the origin of major extant acrodontan clades. The different acrodont lineages would subsequently have entered Laurasia, assembling a complex Asian acrodont fauna. However, the conclusions of Macey et al. (2000b) and those of Schulte et al. (2003) are based on extremely old divergence estimates for some lineages of agamids, which would push the origins of Squamata well into the Paleozoic (Hugall and Lee, 2004). The latter authors modified the divergence times and suggested that the radiation of Australian agamids could no longer be correlated with Gondwanan fragmentation. Melville et al. (2011) also concluded that a Gondwanan origin was not possible, and they proposed that Australian agamids entered from Asia (where they have their closest relatives) and then underwent an in situ diversification through the Miocene. The hypothesis of Macey et al. (2000b) is also incongruent with the presence of the priscagamids and pleurodont iguanians in Asia from at least the early Late Cretaceous (ca. 89 Mya). Leaving aside the problematic *Bharatagama,* there is currently no substantial record of acrodontans on Gondwanan landmasses prior to the accretion of the corresponding landmasses to Laurasia (e.g., Prasad and Bajpai's 2008 account of early Eocene Indian agamids), although allowance must be made for the very poor record of nonserpentian squamates in Gondwana generally throughout the Mesozoic and Paleogene (Evans, 2003; Krause et al., 2003). There are, however, precollision records of numerous acrodont-jawed lizards (putative agamids) in Asia, and by the time of the full accretion of India to Asia (Early Eocene), lizards of this type are already recorded across Laurasia from China, Europe, and North America. Nonetheless, there is controversy as to the timing and sequence of the India–Asia contact (e.g., Briggs, 2003) and the potential for faunal exchange in the Late Cretaceous/Early Paleocene. Faunas from the Deccan Traps sedimentary sequence are said to contain elements from both regions (e.g., Briggs, 2003; Sahni, 2010). It could be argued that the accretion of the Southeast Asian plate that took place much earlier (ca. 120 Mya; Richter and Fuller, 1996) could have carried early Gondwanan acrodontans to Laurasia, but this plate had separated from the rest of Gondwana by at least the Late Jurassic (ca. 150 Mya; Metcalfe, 1996a,b), which predates molecular-based estimates of acrodontan origin.

On current evidence, therefore, a Laurasian (possibly Asian) origin of Acrodonta as a whole, and perhaps also of at least stem-agamids, is plausible. The modern distribution of acrodontans could be the result of dispersion into plates of Gondwanan origin after the separate collision of such plates with Laurasia, with subsequent diversification of modern clades, a view supported by Honda et al. (2000). It would also explain the absence of agamids from South America and Madagascar if these regions separated from Africa before the entry of agamids into Africa. Gondwana as a whole split from Laurasia around 180 Mya, although connections between southern Europe and Africa and through what is now the Middle East may have allowed sporadic faunal interchange in the Early and Late Cretaceous (Gheerbrandt and Rage, 2006; Zarcone et al., 2010). Madagascar separated from Africa in the Jurassic (ca. 160 Mya; Briggs, 2003), with Africa and South America separating in the mid-Cretaceous (ca. 110 Mya). With the exception of the acrodont dental fragments from the Paleocene of Morocco (Augé and Rage, 2006), the first secure evidence of African agamids is from the Oligocene. If Agamidae (including *Uromastyx* and *Leiolepis*; Okajima and Kumazawa, 2010; Hutchinson et al., 2012) and Chamaeleonidae are sister taxa, then the early history of agamids is important with respect to the center of the origin of chameleons.

9.4 ORIGINS OF CHAMAELEONIDAE

Living chameleons have a much more restricted geographical distribution than agamids, with their greatest diversity in Madagascar and East Africa (Chapter 7). Within the crown-group, most workers place the dichotomy between the Madagasy leaf chameleons of the genus *Brookesia* and all remaining taxa (Chapter 7). Raxworthy et al. (2002) dated this dichotomy at 68 to 35 Mya, but slightly different dates have been proposed more recently based on larger datasets—for example, 72 Mya (Townsend et al., 2009); 90 to 60 Mya (Townsend et al., 2011b), and 65 Mya (Tolley et al., 2013; Chapter 7). These dates range from the Late Cretaceous through to the Miocene, although many fall within the Paleogene. The oldest divergence dates are still 20 to 40 Myr after the estimated divergence of agamids and chameleons (e.g., 123 Mya; Townsend et al., 2011a) and the youngest roughly 100 Myr later. This would argue for the existence of a long stem prior to the divergence of crown-group taxa. On the basis of phylogenetic studies that placed *Brookesia* as the sister group to all other chameleons, Raxworthy et al. (2002) concluded that chameleons had evolved on Madagascar and radiated from there into Africa and the Seychelles. However, it is unlikely that the agamid–chameleon dichotomy occurred on Madagascar, given agamid distribution as outlined above, and therefore stem-chameleons would have to have dispersed to Madagascar from elsewhere. Moreover, it seems equally possible that the dichotomy within crown-group chameleons occurred in Africa with the ancestral stock of *Brookesia* dispersing to Madagascar and then radiating there (Hillenius, 1959; Blanc, 1972; Klaver, 1977; Hillenius,1978b; Tolley et al., 2013; Chapter 7). Townsend et al. (2011b) have demonstrated that the Seychelles chameleon *Archaius tigris* (formerly *Calumma tigris*) is more closely related to the East African *Rieppeleon* than to the Malagasy *Calumma*, giving a 38.4 Mya (27.8 to 48.5) age for the *Rieppeleon–Archaius* split and requiring an Africa-to-Seychelles dispersal during the Eocene–Oligocene (rather than a Madagascar–Seychelles one (Raxworthy et al., 2002). Current flow in the Late Cretaceous–Paleogene was predominantly west to east across the Mozambique Channel (e.g., Markwick and Valdes, 2004; Ali and Huber, 2010), and this was coupled with an extensive freshwater outflow due to drainage from major rivers that flowed out from East Africa (e.g., Markwick and Valdes, 2004; Townsend et al., 2011b). Mats of vegetation carried seaward by these flows provide the most plausible route by which chameleons (or stem-chameleons in the Raxworthy et al., 2002, model) reached Madagascar, but without a more complete fossil record (especially on Madagascar), these hypotheses are difficult to test (but see Tolley et al., 2013). The earliest records of African (Kenyan) chameleons are too recent (Early Miocene, ca. 21 Mya) to provide useful information on patterns of origin and dispersal.

If stem-acrodontans, in the form of priscagamids or their descendants, were limited to Asia, then the origins of chameleons (as of agamids) could plausibly be there also, with stem-chameleons entering Gondwana in the Late Cretaceous or Paleogene. Given estimated agamid-chamaeleonid divergence dates of 123 to 104 Mya (Townsend et al., 2011a; Chapter 7), it is clear that more than 100 Myr of chameleon history has yet to be recovered in the fossil record, and much of this is likely to be in Africa or Madagascar. Several factors may contribute to the poor record: the generally lower preservation potential of forested habitats; the poor record generally of Gondwanan squamates; the relative scarcity of Mesozoic and Paleogene horizons in Gondwana that have been sampled for small terrestrial tetrapods or are suitable for their preservation; and the potential difficulty of distinguishing between fragments of dentition from stem-agamids versus stem-chameleons. The Miocene Climate Optimum allowed the migration of African taxa to Europe, where sediments of this age have been quite well sampled, whereas most of the African fossils recovered to date have been a by-product of the search for human origins. Extensive work in the latest Cretaceous (ca. 66 to 70 Mya) of Madagascar (e.g., Krause et al., 1999) has revealed a very different fauna from that of today, but small tetrapods are rare and lizards are represented by a single incomplete specimen of a scincoid (Krause et al., 2003). There is then a complete hiatus in the record until the Pleistocene. Africa has a much greater potential to provide new agamid and chameleon fossils, as there are numerous localities in the Late Cretaceous, Paleogene, and Neogene, but relatively few have been sampled for microvertebrates.

Previous authors have suggested an Indian (or other Gondwanan landmass) origin for agamids (e.g., Estes, 1983b; Macey et al., 1997b; 2000b; 2006) and a Malagasy (e.g., Raxworthy et al., 2002) or African (Tolley et al., 2013) origin for chameleons. However, if the Asian priscagamids (and perhaps *Isodontosaurus*) represent stem-acrodontans, they are suggestive of a Laurasian origin for the group as a whole (Chapter 7). A diversity of acrodont-jawed lizards (mostly very fragmentary), whether agamid, stem-agamid or stem-acrodontan (or potentially even stem-chamaeleonid), are found widely across Laurasia in the Paleogene, and the earliest securely identified agamids (placed within or close to living taxa) are also Laurasian. Even allowing for the paucity of the Gondwanan lizard record, it is thus plausible that the immediate ancestors of derived agamids and of chameleons were also Laurasian, dispersing into the Gondwanan landmasses as they contacted Laurasia and then diversifying in situ. Apart from *Bharatagama,* the earliest secure records of acrodont-jawed lizards in India are from the Early Eocene, around the time of contact with Asia (Prasad and Bajpai, 2008). Although it has been hypothesized that crown-group chameleons arose in Madagascar (Raxworthy et al., 2002), it seems more plausible that they evolved in Africa (Tolley et al., 2013; Chapter 7). Either way, stem-chameleons or the ancestors of *Brookesia* reached Madagascar by dispersal across the Mozambique Channel, probably carried eastward with vegetation flowing out of major East African river systems. From Africa, chameleons later reached Europe and India, either through continental dispersion via what is now the Middle East or across temporary land bridges in the western Mediterranean (Tolley et al., 2013, Chapter 7, Gheerbrandt and Rage, 2006; Zarcone et al., 2010).

ACKNOWLEDGMENTS

Our thanks to A. Sahni, G.V.R. Prasad, and R.S. Rana for information (to S.E.) on Indian lizard faunas, and to J.M. Escribano for providing information (to A.B.) on extant chameleons. The manuscript benefited from comments by Krystal Tolley and an anonymous reviewer. A.B.'s work was supported by FPI grant (BES-2009-026731) and EEBB program associated with the project CGL2008-06533-C03-01/BTE, and by the project CGL2011-30069-C02-01 (Ministerio de Economía y Competitividad, Spain).

TEN

Chameleon Conservation

RICHARD JENKINS, G. JOHN MEASEY, CHRISTOPHER V. ANDERSON, and KRYSTAL A. TOLLEY

Across the globe, species are in decline, primarily because of chronic degradation of primary habitat from anthropogenic activities (Myers et al., 2000; Mittermeier et al., 2004; Forister et al., 2010). The result is extensive habitat loss, and populations of some species are now undoubtedly smaller than at any time in their history. When such species are range-restricted and the rate of decline is rapid, the effects are compounded, leaving a distinct chance of extinction in the wild (Brooks et al., 2002; Thomas et al., 2004; Butchart et al., 2010). Chameleons are not exempt from these pressures, and a number of species are faced with severe declines.

Although some chameleons occur over relatively wide geographic areas, inhabit degraded and anthropogenically modified habitats, including urban sites, and appear to have few active threats, many other species are less resilient and suffer from the negative impacts of global change. The challenges facing chameleon conservation are, therefore, to understand the biology and distribution of each species, to know where direct and/or indirect threats are taking place, and to know how to counter these threats with effective conservation actions. With this information, the most threatened chameleon species can be identified and appropriate conservation actions planned and implemented. It is necessary, therefore, to understand the biology and assess the conservation status of the world's chameleons and the types of threats they face to facilitate the identification of species that are in most need of conservationists' attention.

10.1 CONSERVATION STATUS OF CHAMELEONS

To prioritize conservation actions, conservation status of chameleons must first be assessed using a standard procedure that is both transparent and uses the best available scientific information (Rodrigues et al., 2006). The IUCN Red List of Threatened Species™

(hereafter referred to as "IUCN Red List") is the international standard for assessing the extent to which species are facing extinction (Rodrigues et al., 2006). By categorizing species into different threat categories, the IUCN Red List provides a compendium of knowledge on status and biology, and is often the starting point for conservation action. Also, by tracking the conservation status of species over time, trends are revealed that indicate whether conservation efforts are effective. Thorough assessments of the conservation status of the world's mammal and amphibian species have been completed (Stuart et al., 2004; Hoffmann et al., 2010), but a similar initiative for the reptiles has yet to be finished (but see, Böhm et al., 2013).

Chameleons on the IUCN Red List

In this section, the conservation status of the 103 chameleon species, assessed for the IUCN Red List, is summarized (www.iucnredlist.org), followed by a broader discussion of all 196 chameleon species (described as of August 1, 2012), including those species yet to be assessed for the IUCN Red List. Most of the chameleon species assessed to date are native and endemic to the Indian Ocean islands of Madagascar (n = 76), Seychelles Archipelago [Mahé, Silhouette, and Praslin] (n = 1), Comoros Archipelago (Mayotte and Grand Comore; n = 2) and Socotra (n = 1). Assessments for only 22 species from mainland Africa (which is biased toward the South African *Bradypodion*) and one from the Arabian Peninsula were available. The IUCN Red List categorizes extant species into six categories of descending threat status from Critically Endangered, Endangered, Vulnerable, Near Threatened, and Least Concern, while the category of Data Deficient is used for taxa for which insufficient information is available to determine whether or not they are threatened (IUCN, 2012).

Four chameleon species are classed as Critically Endangered (Table 10.1), all of which are endemic to Madagascar, making them, to our knowledge, currently the most threatened chameleons in the world. These species typically occur over a very small geographic range and are in decline because of active threats that are reducing the size and suitability of the remaining habitats (Andreone et al., 2011,a–d). Two of the species (*Calumma tarzan* and *C. hafahafa*) inhabit humid forest in the east of Madagascar, while the others occur in drier areas in the west (*Brookesia bonsi*) and southwest (*Furcifer belalandaensis*).

There are 23 chameleon species in the Endangered category: 19 from Madagascar, one from the Seychelles, and three from mainland Africa. These face similar threats to Critically Endangered species but are considered less prone to extinction because they occur over a wider geographic area, or occur in more sites. All of the Endangered chameleons in Madagascar (*Brookesia*, 9 spp.; *Calumma*, 7 spp.; and *Furcifer*, 3 spp.) and can be broadly divided into those that are restricted to either the eastern humid or western dry deciduous forests. Species in the east can be categorized as associated with low elevation (*Calumma gallus* and *C. furcifer*), middle elevation (*B. ramanantsoai*, *C. globifer*, *C. glawi*, and *F. balteatus*), or montane humid forests (*B. karchei*, *B. bekolosy*, *C. hilleniusi*, and *C. andringitraense*). Many Endangered chameleons are also restricted to relatively isolated forest blocks, including all the western Endangered species (*B. dentata*, *B. decaryi*, *B. exarmarta*, *B. perarmata*, and *F. nicosiai*). In mainland Africa, *Rhampholeon spinosus* of the Usambara Mountains in eastern Tanzania inhabits humid forest, while the two Endangered South African species (*Bradypodion caffer* and *B. taeniabronchum*) occur in coastal forest and montane fynbos, respectively. *Archaius tigris,* which is also Endangered, occurs naturally on two islands in the Seychelles, where it inhabits humid forests.

There are also 21 chameleons on the IUCN Red List categorized as Vulnerable, 17 as Near Threatened, and 4 as Data Deficient. It is important to note that an additional 34 chameleon species on the Red List are widespread and not considered of conservation concern (Least Concern). Some of these taxa are also able to survive in human-modified landscapes, including gardens and farmland.

Perhaps the most startling information from this assessment is that 63% of the known chameleon species, for which assessments have been completed, are categorized as Threatened or Near Threatened (i.e., Critically Endangered, Endangered, Vulnerable, or Near Threatened). This compares with a global average of 23% for 867 species of lizards (Böhm et al., 2013). This dramatic, nearly threefold, difference may be because many of the species already assessed are from Madagascar (many of which have relatively small distributions). However, it appears likely that once all chameleons are assessed, their level of threat may be greater than that of other lizards. This may be because so many species inhabit tropical forests, which themselves are threatened habitats. Also, a number of species are endemic to small geographic areas, often single mountains, and this increases their likely extinction in the face of threats. Illegal, or unsustainable, collection is also a current or potential threat for many species.

During the assessment of all species for the IUCN Red List, assessors are asked to list all actual and perceived threats to each species; these were then placed into a categorization scheme, with a broad category encompassing major threats (Table 10.2). Although very few mainland African chameleons have been assessed, and these are mostly southern African *Bradypodion*, agriculture and the exploitation of biological resources are a threat to nearly all species. These threats mostly relate to the continued degradation of forest systems for agriculture (both subsistence and commercial) and harvesting (selective logging and general removal) of timber, which together with natural system modifications impact as habitat loss/modifications to chameleons. The impact of mining is prominent in Madagascar, while invasive species appear to affect more species in mainland Africa (with a bias to southern Africa). While the loss or modification of forests is clearly the main threat to chameleons, additional research is needed to determine the way in which the threat operates. The susceptibly of different chameleon species will depend on whether they occupy the canopy or ground layer, their feeding habits, and their reproductive behavior. For example, species that favor naturally open areas (e.g., rivers, edges and gaps) may tolerate modest levels of disturbance (e.g., Jenkins et al., 1999; Chapter 5).

To date, almost all chameleon assessments (90%) have used the Red List's B criterion, which means that the threat level depends on the known extent of occurrence (or area of occupancy—see IUCN, 2012, for detailed definitions) and ongoing threats to the habitat

TABLE 10.1 Summary of Chameleons on The IUCN Red List of Threatened Species as of 2011

Red List Category	No. of Species
Critically Endangered	4
Endangered	23
Vulnerable	21
Near Threatened	17
Least Concern	34
Data Deficient	4
Not Evaluated	93

TABLE 10.2 Major Threats to Chameleon Species for Which There Were Recorded Threats, Obtained from the IUCN Red List of Threatened Species.

The 103 species assessed are divided into mainland Africa (African) and Madagascar (Malagasy).

Major Threats	All Species	%	African	%	Malagasy	%
Residential and commercial development	5	5.2	2	9.1	3	4.1
Energy production and mining	11	11.5	2	9.1	9	12.2
Agriculture and aquaculture	82	85.4	14	63.6	68	91.9
Biological resource use	79	82.3	15	68.2	64	86.5
Invasive and other problematic species, genes and diseases	5	5.2	3	13.6	2	2.7
Natural system modifications	25	26.0	3	13.6	22	29.7
Total species with threats	96		22		74	

within that area. Although these same criteria have been used for most species of reptiles and amphibians on the Red List, they are based on minimal information of the actual species concerned. All other criteria (except D2) are based on population data, for which we are ignorant for most, if not all, species of chameleon. We do know whether the density of different species of chameleons varies within the same habitats (see Chapter 5), and for those that are rare, or at least seldom encountered, it is imperative that we start making estimates of population size and population viability in habitats that are under threat. To make meaningful assessments for species that are traded (see the section on "Trade in

Chameleons," below), we also require population-level data. Perhaps the most appropriate assessment would be based on C criteria from population viability analysis (PVA), which determines the probability that a species will go extinct in a given period of time, and as such is extremely useful in conservation. Most PVAs are based on demographic models of populations over a number of years (e.g., using annual Capture–Mark–Recapture studies), but these need to be carefully tailored to the unusual life cycle of some chameleons (e.g., Karsten et al. 2008; Chapter 5).

Chameleons Restricted to Single Sites

While a full assessment of extinction threats to chameleons is not possible until The IUCN Red List of Threatened Species includes all known species and appropriate supporting information, it is useful to identify taxa that are restricted to single localities, probably occur in small populations, and may be sensitive to external threats such as habitat loss. This information provides a provisional indication across all chameleons of species that may warrant greater conservation attention. There is considerable value in identifying the species that are "narrow endemics"—that is, those that occur at a single locality. This approach is exemplified by the Alliance for Zero Extinction (AZE; www.zeroextinction.org), which uses three criteria in site selection: (1) endangerment, (2) irreplaceability, and (3) discreteness. Because the first criterion requires a species to be listed as Endangered to Critically Endangered on the IUCN Red List, this criterion has not been applied here because many of the species have yet to be assessed. However, the latter two criteria can be applied to chameleons by identification of sites that have a discrete boundary and are the sole areas where the chameleon species occur. All chameleons that can be considered "trigger species"—that is, those restricted to single management area/site are evaluated here.

Information on the distribution of each chameleon species was obtained from the IUCN Red List of Threatened Species, and other reference materials (Spawls et al., 2004; Tolley and Burger, 2007; Menegon et al., 2009; Branch and Tolley, 2010; Glaw et al., 2012; Stipala et al., 2012). There are 42 chameleon species that are endemic to discrete, small, geographical areas. We found 22 from Madagascar and 20 from mainland Africa, from 31 different sites (Table 10.3). Most sites were in Madagascar (n = 13), followed by Kenya (n = 6), Tanzania (n = 5), Mozambique (n = 2), South Africa (n = 1), Cameroon (n = 1), Ethiopia (n = 1), Malawi (n = 1), and Sudan (n = 1). Eight of these sites are listed by AZE, based on the presence of other taxa (Table 10.3) but an additional three from Madagascar meet current criteria for chameleons and should be incorporated into future AZE updates. With the exception of three karst outcrops at Bemaraha, Namoroka, and Ankarafantsika National Parks, all the sites are in montane areas.

Only two of these AZE-based sites are not known to be formally protected (Table 10.3). However, as sites managed traditionally by communities, informally, or privately are not well presented in our analysis, it is important not to assume that these sites are less well protected, or managed, than those included in nationally recognized reserves. It is imperative that sites with the entire known global population of a chameleon species are

TABLE 10.3 List of Sites (Alphabetical by Country) That Contain the Entire Known Population of Individual Chameleon Species

Country	Site Name	Protection Level	Trigger Species (= species restricted to discrete management area)
Cameroon	Manengouba Mountains	None at present	*Trioceros perreti* (NE)
Ethiopia	Bale Mountains*	National Park	*Trioceros balebicornutus* (NE), *Trioceros harennae* (NE)
Kenya	Marasabit Mountain	National Park, National Reserve and Forest Reserve	*Trioceros marsabitensis* (NE)
	Mount Kenya*	National Park, Forest Reserve, UNESCO-MAB Biosphere Reserve	*Trioceros schubotzi* (NE)
	Mount Kulal	Forest Reserve, UNESCO-MAB Biosphere Reserve	*Trioceros narraioca* (NE)
	Kinangop Peak	National Park	*Trioceros kinangopensis* (NE)
	Mount Nyiro	Private Reserve	*Kinyongia asheorum* (NE)
	Mount Nyiru	Forest Reserve	*Trioceros ntunte* (NE)
Madagascar	Andohahela*	National Park	*Calumma capuroni* (VU)
	Ankarana	Special Reserve	*Brookesia confidens* (NE)
	Ankarafantsika*	National Park	*Brookesia decaryi* (EN), *Brookesia dentata* (EN)
	Bemanevika*	Protected area under creation	*Calumma hafahafa* (CR)
	Tsingy de Bemaraha*	National Park, Strict Nature Reserve, World Heritage Site	*Brookesia perarmata* (EN), *Furcifer nicosiai* (EN)
	Forêt d'Ambre*	Special Reserve	*Brookesia desperata* (NE)

TABLE 10.3 *(Continued)*

Country	Site Name	Protection Level	Trigger Species (= species restricted to discrete management area)
	Manongarivo**	Special Reserve	*Brookesia bekolosy* (EN)
	Marojejy**	National Park	*Brookesia karchei* (EN), *Calumma jejy* (VU) *Calumma peyrierasi* (VU)
	Montagne d'Ambre	National Park	*Brookesia ambreensis* (NT), *Brookesia antakarana* (NT), *Brookesia tuberculata* (VU), *Calumma amber* (NT), *Calumma ambreense* (NT)
	Montagne des Français	Protected area under creation	*Brookesia tristis* (NE)
	Nosy Hara	Protected area under creation	*Brookesia micra* (NE)
	Tsingy de Namoroka**	National Park	*Brookesia bonsi* (CR)
	Tsaratanana*	Strict Nature Reserve	*Brookesia lolontany* (NT), *Calumma tsaratananense* (VU)
Malawi	Mount Mulanje	Forest Reserve, UNESCO-MAB Biosphere Reserve	*Nadzikambia mlanjensis* (NE)
Mozambique	Gorongosa Mountain	National Park	*Rhampholeon gorongosae* (NE)
	Mount Mabu	None	*Nadzikambia baylissi* (NE)
South Africa	Ngome Forest	Wilderness Area	*Bradypodion ngomeense* (NE)
Sudan	Imatong Mountains	Nature Conservation Area	*Trioceros kinetensis* (NE)

(Continued)

TABLE 10.3 *(Continued)*

Country	Site Name	Protection Level	Trigger Species (= species restricted to discrete management area)
Tanzania	Mount Hanang	Forest Reserve	*Trioceros hanangensis* (NE)
	Nguru South Forest	Catchment Reserve	*Rhampholeon acuminatus* (NE)
	Sali	Forest Reserve	*Rhampholeon beraduccii* (NE)
	Uluguru Mountains*	Forest Reserve	*Kinyongia uluguruensis* (NE)
	Udzungwa Mountains	Forest Reserve, National Park	*Kinyongia magomberae* (NE)

SOURCE: Information on Protection Level was taken from the World Database on Protected Areas (www.protectedplanet.net in August 2012) and, for Madagascar only, the Atlas Numerique du Systeme des Aires Protégées de Madagascar (http://atlas.rebioma.net).

Note: Some sites are already included on the Alliance for Zero Extinction (AZE) because of the presence of other species (*), while others qualify under the three criteria AZE uses (**).

ABBREVIATIONS: CR = Critically Endangered; EN = Endangered; VU = Vulnerable; NT = Near Threatened; NE = Not Evaluated.

communicated to relevant stakeholders, including management authorities and community leaders. In many cases, it may be sufficient to communicate the information because the chameleons may not require any specific monitoring or management but would benefit from, or be threatened by, the ongoing conservation and disturbance at the site. In a few instances, which need to be identified, it is possible that the chameleons are subject to specific threats (e.g., wild harvesting or habitat loss) and require direct conservation action.

10.2 TRADE IN CHAMELEONS

All chameleon species belonging to the genera *Archaius, Bradypodion, Calumma, Chamaeleo, Furcifer, Kinyongia, Nadzikambia,* and *Trioceros* are currently included in the Appendix II of the Convention on International Trade in Endangered Species of Wild Flora and Fauna (CITES). Species of the genus *Brookesia* are also included in Appendix II of CITES, with the exception of *Brookesia perarmata,* which is in Appendix I (Carpenter and Robson, 2005). Of the *Rhampholeon* species, only *R. spinosus* is listed in Appendix II of CITES (because of its inclusion under its previous classification as a species of *Bradypodion*), while all other *Rhampholeon* species and all *Rieppeleon* species are not currently CITES listed. Species in Appendix II can be traded as long as it is nondetrimental to wild populations, while commercial trade is effectively prohibited for Appendix I species under normal circumstances.

Because chameleons are colorful and ornate, they attract considerable attention from hobbyists, and many species are traded across international borders in large quantities. The removal of chameleons from the wild can be illegal (e.g., inside a protected area or without permission) or legal but detrimental (unsustainable harvest levels), and CITES is an important mechanism for monitoring, which allows reporting that can influence trade and protection measures. Patterns in the quantity and composition of international chameleon trade, obtained from the CITES Trade Database at the United Nations Environment Programme—World Conservation Monitoring Centre (UNEP-WCMC), can provide important insights into commercial trends, infractions, and impacts of suspensions as well as highlighting potential cases of unsustainable trade (e.g., Carpenter et al., 2004, 2005).

To provide an up-to-date review of the trends in commercial chameleon trade, the data from the CITES Trade Database were downloaded (CITES, 2012a) and corrected to reflect current taxonomic nomenclature (e.g., *Archaius, Trioceros*). The trends were then summarized utilizing data on the quantity of live chameleons imported between 1977 (when CITES trade regulations were initiated) and 2010, from countries with native chameleon populations. Thus, these trends reflected the documented commercial trade in live chameleons using reported import quantities, rather than the quantity for which permits were requested, from countries where chameleons naturally occur.

These data show that between 1977 and 2010, more than 1.37 million chameleons have been exported and nearly one million of those were exported in the second half of that time range, from 1994 on (CITES, 2012a). In fact, since 1994, nearly 59,000 chameleons were

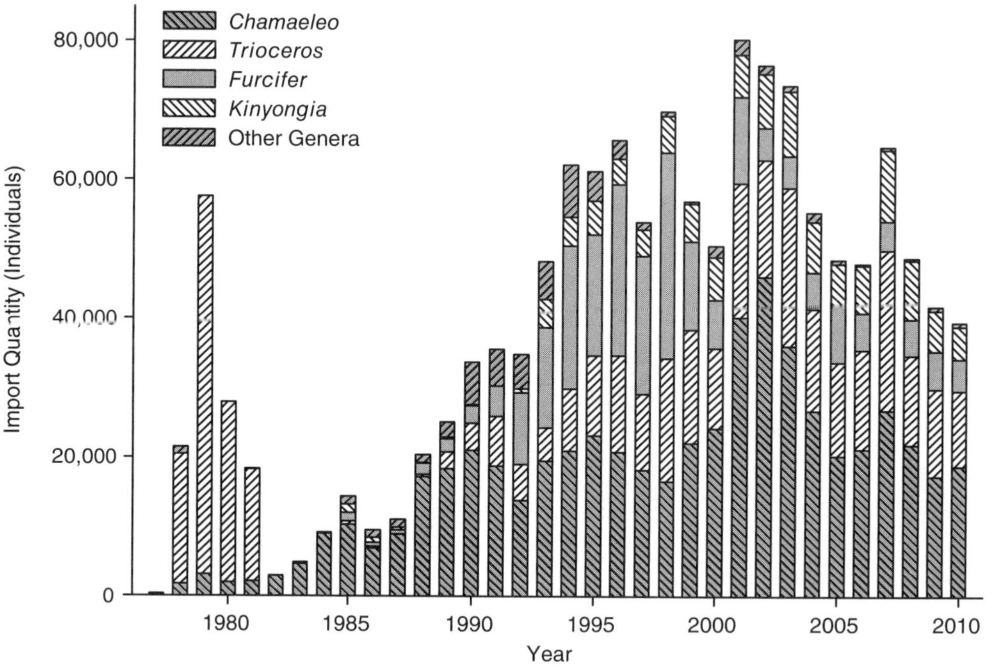

FIGURE 10.1. Annual chameleon exports from 1977 to 2010 by genus. Each bar represents the total export quantity for a year, subdivided to depict the proportion of the overall export quantity representing the top four chameleon genera exported over the period 1977 to 2010 (each genus designated by pattern).

exported annually from their countries of origin, with a high of more than 80,000 exported in 2001 alone.

The aggregated data show that the overall level of exportation of chameleons exhibits a bimodal distribution (Fig. 10.1) (CITES, 2012a), which is consistent with previously published studies (Carpenter et al., 2004, 2005). An initial spike in documented chameleon exports occurred between 1977 and 1981, when Kenya exported large numbers of *Trioceros* species prior to prohibiting the export of specimens caught in the wild. Following the sudden drop in exportation rates associated with the closure of Kenyan export of wild collected chameleons in 1981 (Carpenter et al., 2004), overall exportation rates climbed steadily until the mid-1990s when exportation rates began fluctuating year to year, but staying above 50,000 animals per year until 2005 (Fig. 10.1). Overall exportation rates remained relatively constant, from the mid-1990s through 2010, with a peak in annual export in 2001 (Fig. 10.1).

Chamaeleo species were exported in the highest numbers, followed by *Trioceros* and then *Furcifer* species (Table 10.4) (CITES, 2012a). Other genera, such as *Nadzikambia* and *Archaius*, were subject to little or no trade. It should be noted, however, that trade figures for *Rhampholeon* are limited to a single species, *Rhampholeon spinosus*, the only *Rhampholeon* listed by CITES, and thus no trade data are available for any other

TABLE 10.4 Total Live Chameleon Exports for Trade from 1977 to 2010 by Genus (Nonscientific/Zoological), by Absolute Quantity and Percentage of the Total Number Exported

Genus	Total Exported	%
Archaius	12	<0.001
Bradypodion	333	0.02
Brookesia	5,833[a]	0.4
Calumma	23,617	1.7
Chamaeleo	581,490	42.3
Furcifer	229,952	16.7
Kinyongia	113,367	8.3
Nadzikambia	0	0
Rhampholeon	?[b] (140[c])	? (0.01)
Rieppeleon	?[b]	?
Trioceros	392,509	28.6
"*Chamaeleo* spp." (Madagascar)	6,858	0.5
"*Chamaeleo* spp." (Mainland Africa)	19,525	1.4

a. Post-2003 data only.
b. CITES Trade Data lacking because of nonlisted status of genus.
c. CITES Trade Data for single species, *Rhampholeon spinosus*.

Rhampholeon or *Rieppeleon* species. Further, although trade data for *Brookesia* species are available for the period 1997 to 2010, trade data prior to the inclusion of this genus on CITES at the end of 2002 is incomplete (C. Anderson, personal observation). An example is *Brookesia perarmata,* a species endemic to Bemaraha National Park in western Madagascar, for which only 25 individuals were reportedly imported to the United States, all in 1997, but were observed for sale in far higher quantities during this period (C. Anderson, personal observation).

Accounting for more than a quarter of all imports from 1977 to 2010, *Chamaeleo senegalensis* remain the most traded chameleon species (Table 10.5), with sizable exports from Benin, Ghana, and Togo (CITES, 2012a; see also, Carpenter et al., 2004, 2005). This is followed by *Trioceros jacksonii* imports from countries where they are native (i.e., not including feral populations), which accounts for approximately 8.3% of total imports; however, more than 73% of that trade occurred prior to 1982, when Kenya was not restricting exports (Carpenter et al., 2004) and between 1990 and 2010, *T. jacksonii* was the ninth most traded species. From 1990 to 2010, *C. dilepis* was the second most exported chameleon, and it was third most since 1977. The majority of the 10 most traded species are the same when you consider the time period from 1977 to 2010 or from 1990 to 2010, however, *T. hoehnelii* is replaced on that list by *T. quadricornis* from 1990 to 2010.

Of the 10 most exported chameleon species from both aforementioned time periods (Table 10.5) (CITES, 2012a), 5 are listed as Least Concern on the IUCN Red List of Threatened Species (IUCN, 2012). *Chamaeleo gracilis, Trioceros jacksonii, T. melleri, Kinyongia*

TABLE 10.5 Top 10 Exported Chameleon Species from Native Countries of Origin from 1977 to 2010

Species	IUCN Red List	Total	%	Native Range
C. senegalensis	Least Concern	350,396	25.5	West Africa
T. jacksonii	Not Evaluated	114,511	8.3	Kenya and Tanzania
C. dilepis	Least Concern	108,214	7.9	Africa (Widespread)
C. gracilis	Not Evaluated	90,623	6.6	Africa (Widespread)
F. lateralis	Least Concern	82,957	6.0	Madagascar
F. pardalis	Least Concern	81,023	5.9	Madagascar
K. fischeri[a]	Not Evaluated	76,483	5.6	Tanzania
T. melleri	Not Evaluated	48,284	3.5	Tanzania, Mozambique and Malawi
T. hoehnelii	Least Concern	36,117	2.6	Kenya and Uganda
K. tavetana	Not Evaluated	33,904	2.5	Kenya and Tanzania

a. Includes *K. fischeri, K. matschiei, K. multituberculata, K. uluguruensis,* and *K. vosseleri.*

tavetana, T. quadricornis, and *K. fischeri* (for which CITES [2012a] Trade Data also includes *K. matschiei, K. multituberculata, K. uluguruensis,* and *K. vosseleri*); however, as with many other traded species, have not been assessed for The IUCN Red List (IUCN, 2012).

Chameleon Exports

Tanzania is the single largest exporter of chameleons, with over 345,000 recorded exports (Table 10.6, Fig. 10.2) (CITES, 2012a). Amazingly, almost 300,000 of those have been exported since 1996, and between 2004 and 2010, Tanzania alone exported an average 44% of the world's chameleons during that period (Fig. 10.2). Tanzania is followed by Togo (almost 284,000 individuals; Table 10.6) and Madagascar (almost 258,000 individuals; Table 10.6) as the largest exporter countries for chameleons. The height of Togo's chameleon exportation occurred between 1988 and 2003, but Togo's highest market share of chameleon exports occurred between 1982 and 1990, during which they accounted for 48 to 99% of the annual export (Fig. 10.2). Exportation levels for Togo, Benin, and Ghana may not be independent, however, as individual suppliers have been known to collect and export across these countries (A. Carpenter, personal communication). Peak exports from Madagascar occurred between 1992 and 1999, during which time Madagascar alone exported an average of 38% of the world's chameleons (Fig. 10.2). In 1994, however, the CITES secretariat requested that importing Party members suspend imports of all but four Malagasy *Chamaeleo* (now *Calumma* and *Furcifer*) species (Carpenter et al., 2004, 2005) and in 1999, Madagascar drastically cut annual export quotas of those four species, thus resulting in a decline in overall exports (Carpenter et al., 2005).

TABLE 10.6 Top 10 Native Chameleon Exporting Nations from 1977 to 2010

Country	Total Exported	Percentage
Tanzania	345,471	25.2
Togo	283,882	20.7
Madagascar	257,895	18.8
Kenya	128,501	9.4
Benin	80,275	5.8
Ghana	66,363	4.8
Cameroon	53,205	3.9
Burundi	38,919	2.8
Uganda	31,217	2.3
Equatorial Guinea	21,182	1.5

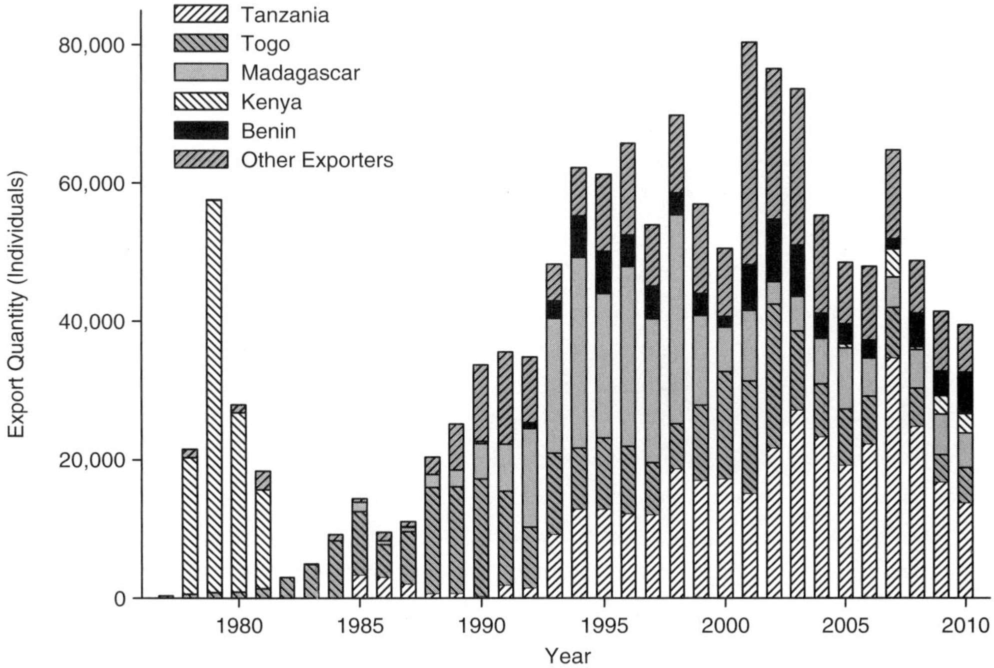

FIGURE 10.2. Annual chameleon exports from 1977 to 2010 based on actual import figures. Each bar represents the total export quantity for a year. Each bar is subdivided to depict the proportion of the overall export quantity exported by the top five chameleon-exporting nations over the period 1977 to 2010 and all other exporting nations.

Between 2001 and 2010, a number of other exporting countries notably increased their levels of exportation. Exports from Uganda, for instance, increased from less than 550 individuals exported between 1991 and 2000, to more than 30,600 individuals exported from 2001 to 2010. It should be noted, however, that 88% of that export volume occurred from 2001 to 2004 (CITES, 2012a). Other sizable increases in exportation include Benin (approximately 32,300 to 48,000 individuals), Jordan (70 individuals of one species to over 7200 of three species), Niger (no export to over 5250 individuals), and Mozambique (approximately 950 to 14,150 individuals). Kenya also saw an increase of exports, from 320 individuals to over 10,800, however, this increase was because of exports of chameleons reported on export documents to have been bred in captivity.

One increase in chameleon exportation is particularly noteworthy, however. Equatorial Guinea's export quantities increased from 4100 individuals from 1991 to 2000, to over 16,800 between 2001 and 2010 (CITES, 2012a). In fact, between 2004 and 2010, Equatorial Guinea was the sixth largest single exporter of chameleons, but 53% of the chameleons exported during that period were of species that are not known to occur within Equatorial Guinea (*Trioceros pfefferi, T. quadricornis,* and *T. wiedersheimi*). Despite the fact that these species do not occur in Equatorial Guinea, original CITES documents (as opposed to reexport documents) were issued for animals listed as caught in the wild, raising considerable concern about circumvention of international wildlife laws and management efforts.

Chameleon Imports

The United States is the single largest importer of chameleons, having imported more than 884,000 chameleons, or over 64% of all documented chameleon exports, between 1977 and 2010 (Table 10.7, Fig. 10.3) (CITES, 2012a). In fact, since 1978, the United States has imported less than 50% of all documented chameleon exports for a given year only four times, in 1984, 2003, 2009, and 2010 (Fig. 10.3). Europe as a whole has imported almost 26% of all documented chameleon exports with Asia as a whole accounting for only an additional 6.8% (Fig. 10.3). The overall global trends in chameleon exportation levels are, thus, highly correlated with that of the importation levels to the United States of America (Fig. 10.3).

Importation quantities between 2001 and 2010 to some countries have increased considerably as compared with quantities from 1991 to 2000. Ghana has emerged as a new import market, having no chameleon imports between 1991 and 2000 but almost 18,400 from 2001 to 2010 (CITES, 2012a). Germany also saw a considerable rise in imports (from approximately 16,000 to 56,400 individuals). Other sizable increases in importation include Canada (approximately 4200 to 8200 individuals), Japan (approximately 34,000 to 48,000), Mexico (20 to approximately 3600), and Thailand (approximately 100 to 3900). It should be noted, however, that importation to the United States between these two time periods declined from almost 361,000 to less than 332,000 individuals.

TABLE 10.7 Top 10 Chameleon-Importing Nations from 1977 to 2010

Country	Total Imported	Percentage
United States	884,276	64.4
Germany	111,324	8.1
Japan	84,607	6.2
Netherlands	61,087	4.4
Spain	43,061	3.1
France	37,588	2.7
Belgium	35,183	2.6
United Kingdom	32,019	2.3
Ghana	18,398	1.3
Canada	12,589	0.9

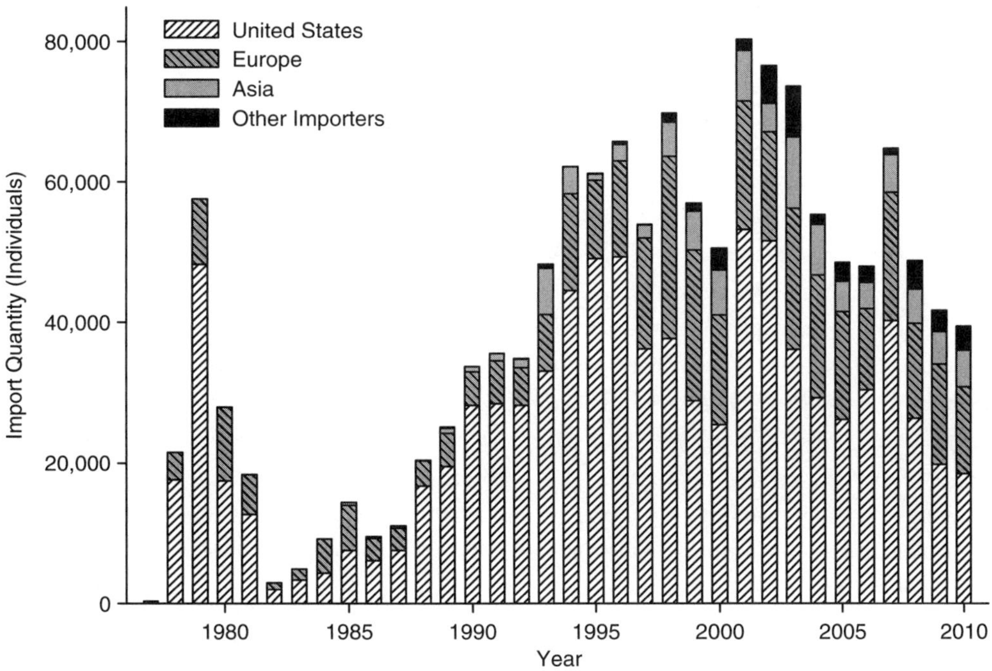

FIGURE 10.3. Annual chameleon imports from 1977 to 2010 broken up into major importing regions or nations. Each bar represents the total import quantity for a year. Each bar is subdivided to depict the proportion of the overall export quantity imported by the United States (the top overall importer of chameleons), Europe, Asia, and all other importing nations.

Wild Caught versus Captive Bred

Of all chameleon species in the trade, relatively few are reliably produced in captivity in any significant quantity and those that are imported as wild-caught individuals frequently have high mortality rates during transit (Abate, 2001; C. Anderson, personal observation). Because many wild-caught species are exported in such large quantities every year, many of these have become species with which exporters fill shipments in order to increase the overall shipment size and value (C. Anderson, personal observation). They are sent in large quantities at a very low price per individual. This low price results in a minimal mark up by the exporter, and a thin profit margin means the animals are not worth the effort to care for as well as the more valuable animals are. The importer then gets large shipments of these species in an already compromised condition. In order to move their inventory of these animals, the importers sell them as quickly as possible at a very small mark-up (Abate, 2001; C. Anderson, personal observation). The result is that these species end up at pet shops in a severely compromised condition. They are then sold retail as quickly as possible for low amounts (as low as $15 or less for *Chamaeleo senegalensis*; C. Anderson, personal observation) and often die in a matter of weeks to months. This trend is further bolstered by the export of large numbers of adult animals, which tend to be more difficult to acclimate to captivity than juveniles.

Unfortunately, because these animals are available for retail sale in such large numbers and at such low prices, there is limited means for anyone interested in these species to make a profit breeding that species in captivity (C. Anderson, personal observation). Even people who are genuinely interested in breeding any of these species in captivity thus tend to give up their efforts quickly because the market is flooded with underpriced alternatives and coming anywhere close to covering their costs is an unobtainable goal.

The entire dynamic of the trade changes, however, in species with annual importations of more modest numbers of individuals. In these cases, the market is no longer flooded and anyone working to propagate that species in captivity is able to sell their offspring at prices that more accurately reflect the expense of producing these animals in captivity. The wild-caught imports that are still available then become valuable animals to infuse into captive breeding groups, with breeders being willing to pay more for animals that are healthier and have a better chance of acclimating successfully and adding to their breeding efforts. This causes the importers to value the condition of these animals, meaning they take better care of them so they can mark them up more. As a result, they demand higher-quality specimens of these species from their exporters, who are then inclined to take better care of these animals because they too can ask more per individual and able to export fewer individuals.

For instance, while *F. pardalis* is the sixth most exported chameleon species, the average number of specimens exported every year dropped by 75% from the period 1991–2001 to the period 2002–2010, resulting in a considerable reduction in the number of imported specimens available in the trade in recent years. The increased exportation rate of *F. pardalis* in the earlier period resulted in great availability of wild-caught imports on the market for relatively low prices—prices that were often lower than the price of captive-bred animals. Following the reduction in annual export quotas for this species, however, interest in breeding this species in captivity increased, causing wild-caught imports to be sought out by breeders to infuse new bloodlines into their breeding groups. At present, these wild-caught imports are frequently more expensive than the captive-bred animals produced later. The result has been that the quality of animals exported has increased and the proportion of individuals that survive once in captivity has increased as well.

Overall, CITES trade data from all exporting countries (as opposed to data limited to exporting countries with native chameleons) indicates that a total of 136,294 captive-bred chameleons were imported between 1985 and 2010 (CITES, 2012a). Of that number, 95% were imported between 1996 and 2010 and 85% in the last 10 years of this time period, indicating a significant increase in the supply of captive-bred species. Sixty-six percent of captive-bred imports were *Chamaeleo calyptratus*. *Furcifer pardalis* makes up the next largest proportion of captive-bred imports, at approximately 11,800 individuals. Slovakia is the largest exporter of captive-bred chameleons, with 26% of all exports (approximately 35,700 individuals), followed by the Ukraine (approximately 21,500 individuals) and the Czech Republic (approximately 18,500 individuals). The United States is the single largest importer of captive-bred chameleons, with 35% of all imports (approximately 47,500 individuals), followed by Japan (approximately 23,000 individuals), and Canada (approximately 10,000 individuals). It should be noted, however, that feral populations of *F. oustaleti*, *C. calyptratus* and *Trioceros jacksonii xantholophus* in the United States is also known to offset a considerable amount of the domestic demand for these species (C. Anderson, personal observation).

Sustainable Trade

The basic tenet of CITES is that international trade in species should be benign to wild populations. The Antsingy Leaf Chameleon (*Brookesia perarmata*) is the only chameleon in Appendix I, but all other CITES-listed chameleons are in Appendix II, which has different mechanisms for managing and monitoring trade. The Significant Trade Review allows CITES to review trade in Appendix-II species in order to identify species for which trade appears to be unsustainable and to recommend remedial actions. In cases in which unsustainable trade is suspected, detailed recommendations aimed at specific Parties are made, with failure to comply likely to lead to a suspension of imports of the species in question. Parties are expected, when called upon by CITES, to demonstrate that collection and export is nondetrimental to wild populations. This is challenging for many Parties, especially those lacking sufficient resources. Carpenter et al. (2005) described the impact on livelihoods, markets and the composition of exports following the 1994 suspension of all Malagasy *Calumma* and all but four *Furcifer* species. As a more up-to-date example, in July 2012 CITES suspended imports of *Trioceros feae* because Equatorial Guinea failed to comply with recommendations made through the Significant Trade Review process.

The wider debate on the impact of trade in chameleons in general is hampered by a lack of information. While Parties are rightly permitted to export species regardless of their IUCN Red List status, taxa that are considered a conservation priority (e.g., Critically Endangered and Endangered) quickly become the focus of attention, especially when export levels appear "high." This can eventually lead to the involvement of the CITES Standing Committee, which may enforce trade suspensions in cases in which Parties are unable to provide convincing evidence that exports of threatened taxa are nondetrimental to wild populations. When considering the likely impact of collection on a species, two important considerations need to be taken into account: (1) the proportion of the species' range over which collection can legally occur and (2) the level and intensity of collection. Put simply, collection outside of protected areas in scattered geographical localities might be preferred for many species. When species are restricted in range, like *Kinyongia fischeri,* even modest collection levels may be harmful. It is noteworthy that this species and *K. tavetana* entered the CITES Significant Trade Process in 2012 because of concerns regarding the potential impact of trade on wild populations. Both species are among the top ten most exported chameleons (Table 10.5).

Annual export quotas are set by individual countries to limit the number of animals exported to sustainable levels. Eleven countries, all from Africa, have used quotas to manage their chameleon exports in the last 10 years (Table 10.8). All countries set export quotas for wild-caught chameleons and three set export quotas for ranched chameleons.

Illegal Trade

A complicating factor of limiting legal trade, unfortunately, is a surge in illegal trade if demand significantly exceeds legal trade limits. These illegal markets are obvious in the extensive captive trade of South African *Bradypodion* species and of Malagasy *Calumma* and *Furcifer* species (excluding *F. pardalis, F. lateralis, F. oustaleti,* and *F. verrucosus*) worldwide, but particularly in Europe (Anderson, personal observation). The nature of this illegal trade takes various forms but includes the open trading of species that have never been legally exported under CITES (e.g., *Bradypodion setaroi*; CITES, 2012a; C. Anderson, personal observation), the importation of species with forged CITES documents that were not issued by the CITES Management Authority of the exporting country (i.e., Cameroon; CITES, 2012b), the mass issuance of CITES documents and subsequent exportation of species from countries where they do not naturally occur (e.g., *Trioceros pfefferi, T. quadricornis,* and *T. wiedersheimi* export from Equatorial Guinea; CITES, 2012a), or the availability of species that have been banned from export for more than 15 years (Carpenter et al., 2004, 2005) and never successfully propagated in captivity in sustainable numbers (e.g., *C. gallus* and *F. antimena*; C. Anderson, personal observation).

One reason many of these species are traded in the open market is the inability of authorities to verify whether or not specimens that are currently in captivity are of legal origin. Because any captive-bred descendants of CITES-listed specimens that were acquired and imported legally are themselves legal, enforcement issues generally require authorities

TABLE 10.8 Summary of Summed Annual Export Quotas Used by 11 African Countries between 2003 and 2012

Country	Wild	Ranched/F1	Total Export Quota
Benin	3,600	104,600	108,200
Cameroon	9,000	0	9,000
Chad	4,000	0	4,000
DRC	106,000	0	106,000
Ethiopia	100,200	0	100,200
Ghana	21,000	0	21,000
Madagascar	84,800	0	84,800
Mozambique	42,000	0	42,000
Niger	39,000	0	39,000
Togo	55,000	85,000	140,000
United Republic of Tanzania	178,000	20,032	198,032

be able to prove that an animal is not a descendant of legally acquired and imported specimens or was not itself legally acquired and imported (Todd, 2011). This is exceedingly difficult because traders can falsely declare specimens as the direct descendants of specimens on old CITES documents from previous legal shipments of that species (Todd, 2011). Smuggled chameleons, or the progeny of smuggled chameleons, can therefore be laundered through and "legal" CITES export documents acquired by almost anyone who has ever acquired legal specimens of that species with little ability from the authorities to prove otherwise unless they are caught in the act of smuggling itself (Todd, 2011). Some countries (e.g., Thailand) have become hotbeds of such laundering operations (Todd, 2011).

10.3 CHAMELEONS AND GLOBAL CHANGE

The world's climate has already warmed by 0.6°C over the past 100 years, and the warming trend is set to continue. Although anthropogenic climate change is often generalized as "global warming," models of future climate suggest far more complex scenarios, including changes in diurnal temperature range, cloud cover, and precipitation and their consequent interactions (Adler et al., 2008; Zhou et al., 2009). Of particular note is an increase in the expected frequency of extreme weather events, such as heat waves, droughts, floods, storms, and cyclones (IPCC, 2011), which may result in disturbance to many ecosystems. For example, Malagasy forests are expected to be negatively affected by an increasing frequency of El Niño events associated with climate change (Ingram and Dawson, 2005).

After habitat destruction, climate change is one of the most serious and widespread threats to biological diversity (IPCC, 2007). Although extant species have undergone significant climatic change in their past, these changes have occurred at a much slower rate, allowing for both extinction and speciation processes (Chapter 7). In addition, many species are under a high degree of threat because of anthropogenic changes to their habitats, so that climate change is acting as an added stressor. In broad terms, climate change has been shown to have effects on the abundance, distribution, and phenology of species, although many studies are suggesting that it is the complex interaction of these changes together with other stressors that can produce devastating impacts on species. For example, climatic warming is hypothesized as acting synergistically with a novel pathogenic fungus to cause extinction and declines of multiple frog species in Central America (Pounds et al., 2006). Fungi and other infectious diseases have been highlighted as potential causes of future biodiversity loss, with climatic changes as a potent cofactor (Fisher et al., 2012).

Although very few studies have addressed the influence of climate change directly on chameleons, there are some general patterns emerging from studies on lizards that can aid in producing an overview. Ectotherms are thought to be particularly vulnerable to thermal climatic changes because of their direct dependence on external heat sources. This is especially true of topical ectotherms, which are intolerant of high temperatures as well as being sensitive to temperature change. In a comparative, macrophysiological analysis, Huey et al. (2009) showed that lowland lizards living in neotropical forests are at a heightened risk from climatic warming. In Mexico, a study showed that increases in air temperature caused lizards to spend more time in the shade to avoid surpassing thermal maxima (Sinervo et al., 2010). This resulted in less foraging time, with the greatest effect observed during their spring breeding season. In another study on temperate lizards, Massot et al. (2008) showed that increasing spring temperatures were coupled with a decrease in juvenile dispersal. Thus, species with low dispersal ability and geographic barriers to movement are likely to be at an increased risk of extinction. The relationships between ambient temperature, which affect lizard fitness through performance (critical thermal maximum and the thermal optimum; Chapter 3), are crucial in our understanding of future environmental changes. Clusella-Trullas et al. (2011) suggested that as temperature variation and precipitation regimes are likely to have been influential in the evolution of lizard performance, predicting future scenarios will rely on these factors in addition to average temperature. The effects of climate on the predators and (mainly) invertebrate prey items of chameleons are as yet unknown. There is a potential to alter timing of both activity period and breeding success, but as yet there are no studies that have attempted to quantify such effects.

For chameleons that breed year round (Chapter 5), changes in climate are unlikely to impact on phenology as they are likely to be able to take advantage of any favorable climatic periods. However, some chameleons are known to have annual life-history traits that are linked to particular climatic events, such as the onset of a rainy season or the change in nest temperatures associated with hatching times (Karsten et al., 2008, Box 5.1). In future scenarios in which extreme weather events fall out of these seasons, or weather patterns generally become unpredictable, this could result in the loss of recruitment for an entire season. It is also possible that changes in ambient temperature during embryogenesis have unexpected effects.

Many studies have used species distribution models to forecast future distributions under different climatic scenarios. Such studies are strongly dependent on the predictive models used, and these are constantly being improved. Existing studies are useful, however, in that they provide a range of scenarios showing differential responses of a range of species. Houniet et al. (2009) reported a range of responses for *Bradypodion* chameleons from projecting distribution models into IPCC climate scenarios for 2080. While most species were predicted to contract their range, one species (*B. transvaalense*) displayed a shift southward, together with range contraction. In Spain, Moreno-Rueda et al. (2011) predicted a northward shift for *Chamaeleo chamaeleon*. Predictions of shifting ranges are then dependent on the dispersal ability of a particular species, coupled with the availability of appropriate habitat to move to and appropriate corridors to disperse through.

Montane species are particularly vulnerable (cf. Janzen, 1967) because of the upslope distribution displacements that occur when cold-adapted species are forced upward in the face of rising temperatures. These typically montane species may eventually run out of sufficiently cool climate to move to and may be forced into extinction. Although the most frequently cited case for this phenomenon is the extinction of a Central American frog, the same study also showed population crashes in arboreal anoline lizards (Pounds et al., 1999). Raxworthy et al. (2008) made a study of changes in minimum altitude of herpetofauna on Tsaratanana Mountain in central Madagascar. Their study included six chameleons, one of which (*Calumma malthe*) was found to display upslope displacement (increasing both minimum and maximum altitudes) over a 10-year period (between 1993 and 2003). Clearly, there are many other examples of chameleons that could be victims of upslope displacement, in the mountains of East and West Africa, as well as in Madagascar, where there are 35 species of *Calumma*, which are endemic to zones within 600 m of mountain summits (Raxworthy and Nussbaum, 2006).

Chameleons are not typical lizards (Chapter 1), and we may not be able to generalize on the effects of climatic change from studies on other lizards groups. For example, if climate change is expected to play a key role in lizard performance, should we expect slow-moving chameleons to be affected? On the other hand, their slow nature may make them less able to disperse and move to areas of more appropriate habitat. Chameleons appear to be unaffected by temperature-dependent sex determination (Andrews, 2005), which affects the majority of reptile species, but they do have unique embryogenic systems that may put them at increased risk (Chapter 5). In order to understand how future climate scenarios are likely to affect this unique family of lizards, researchers will have to understand more of how their ecology and physiology are influenced by climatic variables. It is also important to recognize that some species may benefit from newly disturbed habitats because of extreme weather events, while others have increased risk as they are already under the influence of extreme climatic constraints or narrow environmental niches. A preliminary step may be to identify species that could have an increased risk and for which research could concentrate.

The effects of climate change are difficult to predict, but over the long term, there is little doubt that human-induced climate change will impact our environment. In contrast, habitat loss is the greatest immediate and acute threat to biodiversity globally (Myers et al., 2000; Butchart et al., 2010). Many chameleons have restricted distributions (Chapter 7), and some are endemic to a single forest (e.g., *Bradypodion ngomeense, Brookesia micra, Kinyongia magomberae*) or locality (Table 10.3). In addition, chameleons are habitat-specific and adapted to certain types of vegetation (Chapters 5 and 7). These factors make it unlikely that chameleons can cope well with habitat destruction, alteration, and fragmentation. Although there are a few species that appear to succeed in altered landscapes (e.g., *K. boehmei* in subsistence shambas, *Bradypodion pumilum* and *Bradypodion damaranum* in urban gardens, and *Furcifer lateralis* in disturbed environments) the vegetation structure in transformed settings is probably similar in some way to that of their natural environment, allowing them to endure. These examples should be considered the exception, rather than the rule.

It is not surprising that areas that retain the greatest diversity in Africa are also those under the greatest pressure for habitat alteration (Myers et al., 2000; Burgess et al., 2007; Fjeldså and Burgess, 2008; Hall et al., 2009; Menegon et al., 2008, 2011; Williams, 2012). Indeed, population density and growth are positively correlated with species richness and endemism in top global biodiversity hotspots (Cincotta et al., 2000; Balmford et al., 2001; McKee et al., 2004; Jha and Bawa, 2006). In Africa, areas that have high numbers of chameleon species are in the same regions where human influence on the landscape is pronounced (Fig. 10.4). Three areas of Africa are notable for chameleon diversity—Madagascar, the Eastern Arc Mountains, and the Guinea–Congolian highland forests (Chapter 7). Not surprisingly, population growth in the Eastern Arc region has increased by 24% in the past decade and human population density has reached more than 80 people/km^2. Although the density is lower in Madagascar (43 per km^2), the growth is faster, with the population having increased by 35% in the past decade. Population in the Guinea–Congolian hotspot, which encompasses the highland forests, is growing at twice the global average and now stands at more than 165 per km^2 (Williams, 2012). Population growth globally is straining our limited resources and advancing the degradation of the environment worldwide. Yet these hotspots are reservoirs of our biodiversity, and they are bearing the brunt of direct effects on habitat and ultimately species, such as chameleons.

10.4 THE WAY FORWARD

Conservation of the world's chameleons will rely heavily on our ability to generate sufficient data on biological/environmental requirements, threats, and their effects on populations over the next 20 years. Although provisional, current information suggests that chameleons may face a higher level of threats than reptiles in general, in part because of ongoing trade, but also because of restricted distributions in dwindling forested habitat of the tropics. Although we are poorly informed about how current populations of chameleons can be conserved, we are completely ignorant of what the effects of climate change might bring to these unusual arboreal lizards. In order to change this, we need physiological studies that can help inform us, as well as better models incorporating these data and more, surpassing the current species distributions models with mechanistic models (cf. Kearney and Porter, 2009) for the effects of climate change on chameleons.

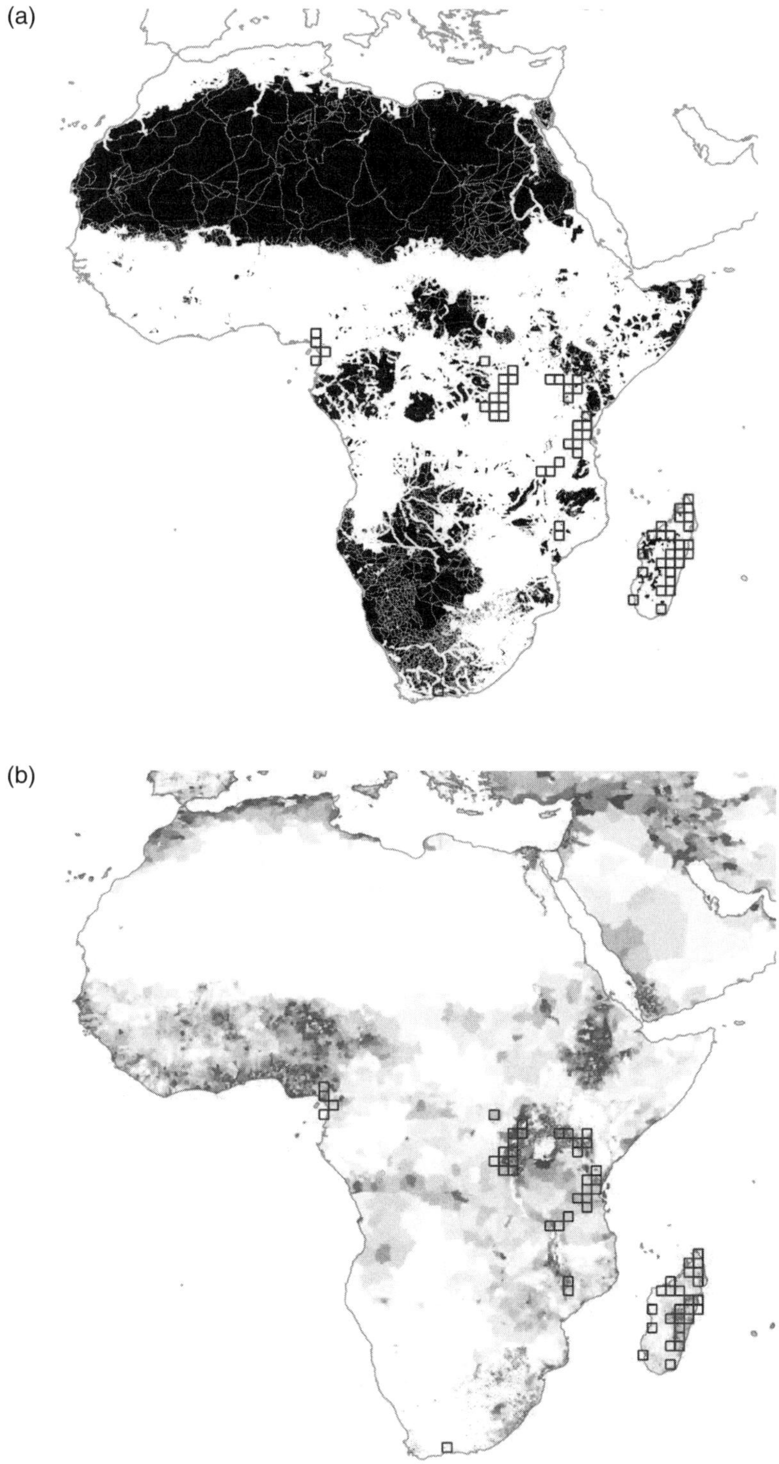

FIGURE 10.4. (a) Human influence in Africa is shown in shades from light (more impact) to dark (less impact), and (b) human population density in light (low density) to dark (high density). Both maps are superimposed with those 1-degree grid cells where high numbers of chameleon species occur. Particularly noteworthy is that in areas with high numbers of species, human influence and population density are also high. Human influence and population density maps obtained from Socioeconomic Data and Applications Center (SEDAC), http://sedac.ciesin.columbia.edu/wildareas/.

While it is hoped that IUCN Red List assessments will be made for all chameleon species in the near future, there is much more work to be undertaken in terms of gathering population/behavioral ecological data on the most threatened species. Chameleon conservation would also benefit from work focused on how global change (climate change and habitat alternation) affects chameleon populations. Furthermore, studies that focus on gathering data pertaining to chameleon population trends for species under trade would ensure confidence in non-detriment findings" (which provide justification that collection and trade does not negatively impact wild populations) and improve CITES regulations for chameleons.

ACKNOWLEDGMENTS

We are grateful to the reviewers who helped to improve earlier versions of the chapter. Richard Jenkins was supported by a Darwin Initiative grant (no. 17010) to the University of Kent and is grateful to Madagasikara Voakajy, the Department of Animal Biology, University of Antananarivo and the government of Madagascar for supporting chameleon research and conservation. The IUCN Red List of Threatened Species is compiled from input from a great many volunteers who have shared their knowledge of chameleons and we would like to take this opportunity to thank them all.

APPENDIX

List of 196 Described Chameleon Species as of 2012, with the Broad Region in Which They Occur

Species	Region
Archaius tigris (Kuhl, 1820)	Seychelles
Bradypodion atromontanum Branch, Tolley, and Tilbury, 2006	Southern Africa
Bradypodion caeruleogula Raw and Brothers, 2008	Southern Africa
Bradypodion caffer (Boettger, 1889)	Southern Africa
Bradypodion damaranum (Boulenger, 1887)	Southern Africa
Bradypodion dracomontanum Raw, 1976	Southern Africa
Bradypodion gutturale (Smith, 1849)	Southern Africa
Bradypodion kentanicum (Hewitt, 1935)	Southern Africa
Bradypodion melanocephalum (Gray, 1865)	Southern Africa
Bradypodion nemorale Raw, 1978	Southern Africa
Bradypodion ngomeense Tilbury and Tolley, 2009	Southern Africa
Bradypodion occidentale (Hewitt, 1935)	Southern Africa
Bradypodion pumilum (Gmelin, 1789)	Southern Africa
Bradypodion setaroi Raw, 1976	Southern Africa
Bradypodion taeniabronchum (Smith, 1831)	Southern Africa
Bradypodion thamnobates Raw, 1976	Southern Africa
Bradypodion transvaalense (Fitzsimons, 1930)	Southern Africa
Bradypodion ventrale (Gray, 1845)	Southern Africa
Brookesia ambreensis Raxworthy and Nussbaum, 1995	Madagascar
Brookesia antakarana Raxworthy and Nussbaum, 1995	Madagascar
Brookesia bekolosy Raxworthy and Nussbaum, 1995	Madagascar
Brookesia betschi Brygoo, Blanc, and Domergue, 1974	Madagascar
Brookesia bonsi Ramanantsoa, 1980	Madagascar
Brookesia brygooi Raxworthy and Nussbaum, 1995	Madagascar
Brookesia brunoi Crottini, Miralles, Glaw, Harris, Lima, and Vences, 2012	Madagascar
Brookesia confidens Glaw, Köhler, Townsend, and Vences, 2012	Madagascar
Brookesia decaryi Angel, 1939	Madagascar
Brookesia dentata Mocquard, 1900	Madagascar
Brookesia desperata Glaw, Köhler, Townsend, and Vences, 2012	Madagascar

(Continued)

Species	Region
Brookesia ebenaui (Boettger, 1880)	Madagascar
Brookesia exarmata Schimmenti and Jesu, 1996	Madagascar
Brookesia griveaudi Brygoo, Blanc, and Domergue, 1974	Madagascar
Brookesia karchei Brygoo, Blanc, and Domergue, 1970	Madagascar
Brookesia lambertoni Brygoo and Domergue, 1970	Madagascar
Brookesia lineata Raxworthy and Nussbaum, 1995	Madagascar
Brookesia lolontany Raxworthy and Nussbaum, 1995	Madagascar
Brookesia micra , 2012	Madagascar
Brookesia minima Boettger, 1893	Madagascar
Brookesia nasus Boulenger, 1887	Madagascar
Brookesia perarmata (Angel, 1933)	Madagascar
Brookesia peyrierasi Brygoo and Domergue, 1974	Madagascar
Brookesia ramanantsoai Brygoo and Domergue, 1975	Madagascar
Brookesia stumpffi Boettger, 1894	Madagascar
Brookesia superciliaris (Kuhl, 1820)	Madagascar
Brookesia therezieni Brygoo and Domergue, 1970	Madagascar
Brookesia thieli Brygoo and Domergue, 1969	Madagascar
Brookesia tristis Glaw, Köhler, Townsend, and Vences, 2012	Madagascar
Brookesia tuberculata Mocquard, 1894	Madagascar
Brookesia vadoni Brygoo and Domergue, 1968	Madagascar
Brookesia valerieae Raxworthy, 1991	Madagascar
Calumma amber Raxworthy and Nussbaum, 2006	Madagascar
Calumma ambreense (Ramanantsoa, 1974)	Madagascar
Calumma andringitraense (Brygoo, Blanc, and Domergue, 1972)	Madagascar
Calumma boettgeri (Boulenger, 1888)	Madagascar
Calumma brevicorne (Günther, 1879)	Madagascar
Calumma capuroni (Brygoo, Blanc, and Domergue, 1972)	Madagascar
Calumma crypticum Raxworthy and Nussbaum, 2006	Madagascar
Calumma cucullatum (Gray, 1831)	Madagascar
Calumma fallax (Mocquard, 1900)	Madagascar
Calumma furcifer (Vaillant and Grandidier, 1880)	Madagascar
Calumma gallus (Günther, 1877)	Madagascar
Calumma gastrotaenia (Boulenger, 1888)	Madagascar
Calumma glawi Böhme, 1997	Madagascar
Calumma globifer (Günther, 1879)	Madagascar
Calumma guibei (Hillenius, 1959)	Madagascar
Calumma guillaumeti (Brygoo, Blanc, and Domergue, 1974)	Madagascar
Calumma hafahafa Raxworthy and Nussbaum, 2006	Madagascar
Calumma hilleniusi (Brygoo, Blanc, and Domergue, 1973)	Madagascar
Calumma jejy Raxworthy and Nussbaum, 2006	Madagascar
Calumma linota (Müller, 1924)	Madagascar
Calumma malthe (Günther, 1879)	Madagascar
Calumma marojezense (Brygoo, Blanc, and Domergue, 1970)	Madagascar
Calumma nasutum (Duméril and Bibron, 1836)	Madagascar
Calumma oshaughnessyi (Günther, 1881)	Madagascar
Calumma parsonii (Cuvier, 1824)	Madagascar
Calumma peltierorum Raxworthy and Nussbaum, 2006	Madagascar
Calumma peyrierasi (Brygoo, Blanc, and Domergue, 1974)	Madagascar

Species	Region
Calumma tarzan Gehring, Pabijan, Ratsoavina, Köhler, Vences, and Glaw, 2010	Madagascar
Calumma tsaratananense (Brygoo and Domergue, 1967)	Madagascar
Calumma tsycorne Raxworthy and Nussbaum, 2006	Madagascar
Calumma vatosoa Andreone, Mattioli, Jesu, and Randrianirina, 2001	Madagascar
Calumma vencesi Andreone, Mattioli, Jesu, and Randrianirina, 2001	Madagascar
Calumma vohibola Gehring, Ratsoavina, Vences, and Glaw, 2011	Madagascar
Chamaeleo africanus Laurenti, 1768	West-central Africa, North Africa
Chamaeleo anchietae Bocage, 1872	West-central Africa
Chamaeleo arabicus (Matschie, 1893)	Arabia
Chamaeleo calcaricarens Böhme, 1985	North Africa
Chamaeleo calyptratus Duméril & Duméril, 1851	Arabia
Chamaeleo chamaeleon (Linnaeus, 1758)	Europe, North Africa, Arabia
Chamaeleo dilepis Leach, 1819	Pan Africa
Chamaeleo gracilis Hallowell, 1842	East Africa, West-central Africa
Chamaeleo laevigatus (Gray, 1863)	East Africa
Chamaeleo monachus (Gray, 1865)	Socotra Island
Chamaeleo namaquensis Smith, 1831	Southern Africa
Chamaeleo necasi Ullenbruch, Krause, Böhme, 2007	West-central Africa
Chamaeleo senegalensis Daudin, 1802	West-central Africa
Chamaeleo zeylanicus Laurenti, 1768	Asia
Furcifer angeli (Brygoo and Domergue, 1968)	Madagascar
Furcifer antimena (Grandidier, 1872)	Madagascar
Furcifer balteatus (Duméril and Bibron, 1851)	Madagascar
Furcifer belalandaensis (Brygoo and Domergue, 1970)	Madagascar
Furcifer bifidus (Brongniart, 1800)	Madagascar
Furcifer campani (Grandidier, 1872)	Madagascar
Furcifer cephalolepis (Günther, 1880)	Comoros
Furcifer labordi (Grandidier, 1872)	Madagascar
Furcifer lateralis (Gray, 1831)	Madagascar
Furcifer major (Brygoo, 1971)	Madagascar
Furcifer minor (Günther, 1879)	Madagascar
Furcifer nicosiai Jesu, Mattioli, and Schimmenti, 1999	Madagascar
Furcifer oustaleti (Mocquard, 1894)	Madagascar
Furcifer pardalis (Cuvier, 1829)	Madagascar
Furcifer petteri (Brygoo and Domergue, 1966)	Madagascar
Furcifer polleni (Peters, 1874)	Comoros
Furcifer rhinoceratus (Boettger, 1893)	Madagascar
Furcifer timoni Glaw, Köhler, and Vences, 2009	Madagascar
Furcifer tuzetae (Brygoo, Bourgat, and Domergue, 1972)	Madagascar
Furcifer verrucosus (Cuvier, 1829)	Madagascar
Furcifer viridis Florio, Ingram, Rakotondravony, Louis, and Raxworthy, 2012	Madagascar

(Continued)

Species	Region
Furcifer willsii (Günther, 1890)	Madagascar
Kinyongia adolfifriderici (Sternfeld, 1912)	East Africa
Kinyongia asheorum Necas, Sindaco, Korený, Kopecná, Malonza, and Modrý, 2009	East Africa
Kinyongia boehmei (Lutzmann and Necas, 2002)	East Africa
Kinyongia carpenteri (Parker, 1929)	East Africa
Kinyongia excubitor (Barbour, 1911)	East Africa
Kinyongia fischeri (Reichenow, 1887)	East Africa
Kinyongia gyrolepis Greenbaum, Tolley, Joma, and Kusamba, 2012	East Africa
Kinyongia magomberae Menegon, Tolley, Jones, Rovero, Marshall, and Tilbury, 2009	East Africa
Kinyongia matschiei (Werner, 1895)	East Africa
Kinyongia multituberculata (Nieden, 1913)	East Africa
Kinyongia oxyrhina (Klaver and Böhme, 1988)	East Africa
Kinyongia tavetana (Steindachner, 1891)	East Africa
Kinyongia tenuis (Matschie, 1892)	East Africa
Kinyongia uluguruensis (Loveridge, 1957)	East Africa
Kinyongia uthmoelleri (Müller, 1938)	East Africa
Kinyongia vanheygeni Necas, 2009	East Africa
Kinyongia vosseleri (Nieden, 1913)	East Africa
Kinyongia xenorhina (Boulenger, 1901)	East Africa
Nadzikambia baylissi Branch and Tolley, 2010	East Africa
Nadzikambia mlanjensis (Broadley, 1965)	East Africa
Rhampholeon acuminatus Mariaux and Tilbury, 2006	East Africa
Rhampholeon beraduccii Mariaux and Tilbury, 2006	East Africa
Rhampholeon boulengeri Steindachner, 1911	East Africa
Rhampholeon chapmanorum Tilbury, 1992	East Africa
Rhampholeon gorongosae Broadley, 1971	Southern Africa
Rhampholeon marshalli Boulenger, 1906	Southern Africa
Rhampholeon moyeri Menegon, Salvidio, and Tilbury, 2002	East Africa
Rhampholeon nchisiensis (Loveridge, 1953)	East Africa
Rhampholeon platyceps Günther, 1893	East Africa
Rhampholeon spectrum (Buchholz, 1874)	West-central Africa
Rhampholeon spinosus (Matschie, 1892)	East Africa
Rhampholeon temporalis (Matschie, 1892)	East Africa
Rhampholeon uluguruensis Tilbury and Emmrich, 1996	East Africa
Rhampholeon viridis Mariaux and Tilbury, 2006	East Africa
Rieppeleon brachyurus (Günther, 1893)	East Africa
Rieppeleon brevicaudatus (Matschie, 1892)	East Africa
Rieppeleon kerstenii (Peters, 1868)	East Africa, North Africa
Trioceros affinis (Rüppel, 1845)	North Africa
Trioceros balebicornutus (Tilbury, 1998)	North Africa
Trioceros bitaeniatus (Fischer, 1884)	East Africa
Trioceros camerunensis (Müller, 1909)	West-central Africa
Trioceros chapini (De Witte, 1964)	West-central Africa
Trioceros conirostratus (Tilbury, 1998)	East Africa

Species	Region
Trioceros cristatus (Stutchbury, 1837)	West-central Africa
Trioceros deremensis (Matschie, 1892)	East Africa
Trioceros ellioti (Günther, 1895)	East Africa
Trioceros feae (Boulenger, 1906)	West-central Africa
Trioceros fuelleborni (Tornier, 1900)	East Africa
Trioceros goetzei (Tornier, 1899)	East Africa
Trioceros hanangensis Krause & Böhme, 2010	East Africa
Trioceros harennae (Largen, 1995)	North Africa
Trioceros hoehnelii (Steindachner, 1891)	East Africa
Trioceros incornutus (Loveridge, 1932)	East Africa
Trioceros ituriensis (Schmidt, 1919)	East Africa, Central Africa
Trioceros jacksonii (Boulenger, 1896)	East Africa
Trioceros johnstoni (Boulenger, 1901)	East Africa, Central Africa
Trioceros kinangopensis Stipala, Lutzmann, Malonza, Wilkinson, Godley, Nyamache, and Evans, 2012	East Africa
Trioceros kinetensis (Schmidt, 1943)	East Africa
Trioceros laterispinis (Loveridge, 1932)	East Africa
Trioceros marsabitensis (Tilbury, 1991)	East Africa
Trioceros melleri (Gray, 1865)	East Africa
Trioceros montium (Buchholz, 1874)	West-central Africa
Trioceros narraioca (Necas, Modry, and Slapeta, 2003)	East Africa
Trioceros ntunte (Necas, Modry, and Slapeta, 2005)	East Africa
Trioceros nyirit Stipala, Lutzmann, Malonza, Wilkinson, Godley, Nyamache, and Evans, 2011	East Africa
Trioceros oweni (Gray, 1831)	West-central Africa
Trioceros perreti (Klaver and Böhme, 1992)	West-central Africa
Trioceros pfefferi (Tornier, 1900)	West-central Africa
Trioceros quadricornis (Tornier, 1899)	West-central Africa
Trioceros rudis (Boulenger, 1906)	East Africa
Trioceros schoutedeni (Laurent, 1952)	East Africa
Trioceros schubotzi (Sternfeld, 1912)	East Africa
Trioceros serratus (Mertens, 1922)	West-central Africa
Trioceros sternfeldi (Rand, 1963)	East Africa
Trioceros tempeli (Tornier, 1900)	East Africa
Trioceros werneri (tornier, 1899)	East Africa
Trioceros wiedersheimi (Nieden, 1910)	West-central Africa

SOURCE: Glaw and Vences, 2007; Tolley and Burger, 2007; Tilbury, 2010; Uetz, 2012.

ABBREVIATIONS

asl	above sea level
cf.	compare
cm	centimeters
e.g.	for example
i.e.	that is
km	kilometers
m	meters
mm	millimeters
Mya	million years ago
Myr	million years
Ri.	Rieppeleon
Rh.	Rhampholeon
sp.	species (singular)
spp.	species (plural)

REFERENCES

Abate, A. 1998. Reports from the field: Parson's chameleon. *Chameleon Information Network* 29:17–25.

Abate, A. 2001. The fate of wild-caught chameleons exported for the pet trade. *Chameleon Information Network* 41:15.

Abu-Ghalyun, Y. 1990. Histochemical and ultrastructural features of the biceps brachii of the African chameleon (*Chamaeleo senegalensis*). *Acta Zoologica* 71:189–192.

Abu-Ghalyun, Y., L. Greenwald, T.E. Hetherington, and A.S. Gaunt. 1988. The physiological basis of slow locomotion in chameleons. *Journal of Experimental Zoology* 245:225–231.

Adams, G.K., R.M. Andrews, and L.M. Noble. 2010. Eggs under pressure: components of water potential of chameleon eggs during incubation. *Physiological and Biochemical Zoology* 83:207–214.

Adams, W.E. 1953. The carotid arch in lizards with particular reference to the origin of the internal carotid artery. *Journal of Morphology* 92:115–155.

Adams, W.E. 1957. The carotid bifurcation in *Chamaeleo. Anatomical Record* 128:651–663.

Adler, R. F., G. Gu, J.-J. Wang, G. J. Huffman, S. Curtis, and D. Bolvin. 2008. Relationships between global precipitation and surface temperature on interannual and longer timescales (1979–2006). *Journal of Geophysical Research* 113:D22104.

Aerts, P., R. Van Damme, B. Vanhooydonck, A. Zaaf, and A. Herrel. 2000. Lizard locomotion: how morphology meets ecology. *Netherlands Journal of Zoology* 50:261–277.

Agnarsson, I., and M. Kuntner. 2012. The generation of a biodiversity hotspot: biogeography and phylogeography of the Western Indian Ocean Islands, pp. 33–82. In K. Anamthawat-Jonsson, Ed., *Current Topics in Phylogenetics and Phylogeography of Terrestrial and Aquatic Systems*. Rijeka, Croatia: InTech.

Akani, G.C., O.K. Ogbalu, and L. Luiselli. 2001. Life-history and ecological distribution of chameleons (Reptilia, Chamaeleonidae) from the rain forests of Nigeria: conservation implications. *Animal Biodiversity and Conservation* 24:1–15.

Ali, J.R., and M. Huber, 2010. Mammalian biodiversity on Madagascar controlled by ocean currents. *Nature* 463:653–680.

Ali, J.R., and D.W. Krause. 2011. Late Cretaceous bioconnections between Indo-Madagascar and Antarctica: refutation of the Gunnerus Ridge causeway hypothesis. *Journal of Biogeography* 38:1855–1872.

Ali, S.M. 1948. Studies on the anatomy of the tail in Sauria and Rhynchocephalia. II. *Chamaeleo zeylanicus* Laurenti. *Proceedings of the Indian Academy of Science* 28B:151–165.

Alifanov, V.R. 1989. New priscagamids (Lacertilia) from the Upper Cretaceous of Mongolia and their systematic position among Iguania. *Paleontological Journal* 23(4):68–80. (Translated from Russian: *Paleontologicheskii Zhurnal* 23(4):73–87.)

Alifanov, V.R. 1991. A revision of *Tinosaurus asiaticus* Gilmor [sic] (Agamidae). *Paleontological Journal* 25(3):148–154. (Translated from Russian: *Paleontologicheskii Zhurnal* 25(3):115–119.)

Alifanov, V.R. 1993. Some peculiarities of the Late Cretaceous and Palaeogene lizard faunas of the Mongolian People's Republic. *Kaupia* 3:9–13.

Alifanov, V.R. 1996. Lizards of the families Priscagamidae and Hoplocercidae (Sauria, Iguania): phylogenetic position and new representatives from the Late Cretaceous of Mongolia. *Paleontological Journal* 30(4):466–483. (Translated from Russian: *Paleontologicheskii Zhurnal* 30(4):100–118.)

Alifanov, V.R. 2000. The fossil record of Cretaceous lizards from Mongolia, pp. 368–389. In M.J. Benton, M.A. Shishkin, D.M. Unwin, and E.N. Kurochkin, Eds., *The Age of Dinosaurs in Russia and Mongolia*. Cambridge, United Kingdom: Cambridge University Press.

Alifanov, V.R. 2004. *Parauromastyx gilmorei* gen. et sp. nov. (Isodontosauridae, Iguania), a new lizard from the Upper Cretaceous of Mongolia. *Paleontological Journal* 38(2):206–210. (Translated from Russian: *Paleontologicheskii Zhurnal* 38(2):87–92.)

Alifanov, V.R. 2009. New acrodont lizards (Lacertilia) from the Middle Eocene of Southern Mongolia. *Paleontological Journal* 43(6):675–685. (Translated from Russian: *Paleontologicheskii Zhurnal* 43(6):68–77.)

Altevogt, R. 1977. *Chamaeleo jacksonii* (Chamaeleonidae)—Beutefang. *Publikationen zu Wissenschaftlichen Filmen. Sektion Biologie* 10(49):3–12 [in German with English summary].

Altevogt, R., and R. Altevogt. 1954. Studien zur Kinematik der Chamaleonenzunge. *Zeitschrift für vergleichende Physiologie* 36:66–77 [in German].

Anderson, C.V., and S.M. Deban. 2010. Ballistic tongue projection in chameleons maintains high performance at low temperature. *Proceedings of the National Academy of Sciences of the United States of America* 107:5495–5499.

Anderson, C.V., and S.M. Deban. 2012. Thermal effects on motor control and *in vitro* muscle dynamics of the ballistic tongue apparatus in chameleons. *Journal of Experimental Biology* 215:4345–4357.

Anderson, C.V., Sheridan, T. and S.M. Deban. 2012. Scaling of the ballistic tongue apparatus in chameleons. *Journal of Morphology* 273(11):1214–1226.

Andreone, F. 2004. Crossroads of herpetological diversity: Survey work for an integrated conservation of amphibians and reptiles in northern Madagascar. *Italian Journal of Zoology* 71:229–235.

Andreone, F., Andriamazava, A., Anjeriniaina, M., Glaw, F., Jenkins, R.K.B., Rabibisoa, N., Rakotomalala, D., Randrianantoandro, J.C., Randrianiriana, J., Randrianizahana , H., Raselimanana, A., Ratsoavina, F., Raxworthy, C.J., and Robsomanitrandrasana, E. 2011a. *Brookesia bonsi*. In: IUCN 2012, IUCN Red List of Threatened Species, Version 2012.1. Accessed at www.iucnredlist.org on July 31, 2012.

Andreone, F., Andriamazava, A., Anjeriniaina, M., Glaw, F., Jenkins, R.K.B., Rabibisoa, N., Rakotomalala, D., Randrianantoandro, J.C., Randrianiriana, J., Randrianizahana , H., Raselimanana, A., Ratsoavina, F., Raxworthy, C.J., and Robsomanitrandrasana, E. 2011b. *Calumma tarzan*. In: IUCN 2012, IUCN Red List of Threatened Species, Version 2012.1. Accessed at www.iucnredlist.org on July 31, 2012.

Andreone, F., Andriamazava, A., Anjeriniaina, M., Glaw, F., Jenkins, R.K.B., Rabibisoa, N., Rakotomalala, D., Randrianantoandro, J.C., Randrianiriana, J., Randrianizahana, H., Raselimanana, A., Ratsoavina, F., Raxworthy, C.J., and Robsomanitrandrasana, E. 2011c. *Calumma hafahafa*. In: IUCN 2012, IUCN Red List of Threatened Species, Version 2012.1. Accessed at www.iucnredlist.org on July 31, 2012.

Andreone, F., Andriamazava, A., Anjeriniaina, M., Glaw, F., Jenkins, R.K.B., Rabibisoa, N., Rakotomalala, D., Randrianantoandro, J.C., Randrianiriana, J., Randrianizahana, H., Raselimanana, A., Ratsoavina, F., Raxworthy, C.J., and Robsomanitrandrasana, E. 2011d. *Furcifer belalandaensis*. In: IUCN 2012, IUCN Red List of Threatened Species, Version 2012.1. Accessed at www.iucnredlist.org on July 31, 2012.

Andreone, F., Glaw, F., Mattioli, F., Jesu, R., Schimmenti, G., Randrianirina, J.E., and M. Vences. 2009. The peculiar herpetofauna of some Tsaratanana rainforests and its affinities with Manongarivo and other massifs and forests of northern Madagascar. *Italian Journal of Zoology* 76:92–110.

Andreone, F., F. Glaw, R. A. Nussbaum, C. J. Raxworthy, M. Vences, and J. E. Randrianirina. 2003. The amphibians and reptiles of Nosy Be (NW Madagascar) and nearby islands: a case study of diversity and conservation of an insular fauna. *Journal of Natural History* 37(17):2119–2149.

Andreone, F., F.M. Guarino, and J.E. Randrianirina. 2005. Life history traits, age profile, and conservation of the Panther Chameleon, *Furcifer pardalis* (Cuvier 1829), at Nosy Be, NW Madagascar. *Tropical Zoology* 18:209–225.

Andreone, F., F. Mattioli, R. Jesu, and J.E. Randrianirina. 2001. Two new chameleons of the genus *Calumma* from north-east Madagascar, with observations on hemipenial morphology in the *Calumma furcifer* group (Reptilia, Squamata). *Herpetological Journal* 11:53–68.

Andrews, R.M. 1971. Structural habitat and time budget of a tropical *Anolis* lizard. *Ecology* 52:262–270.

Andrews, R.M. 2005. Incubation temperature and sex ratio of the veiled chameleon (*Chamaeleo calyptratus*). *Journal of Herpetology* 39:515–518.

Andrews, R.M. 2007. Effects of temperature on embryonic development of the veiled chameleon, *Chamaeleo calyptratus*. *Comparative Biochemistry and Physiology A—Physiology* 148:698–706.

Andrews, R.M. 2008a. Effects of incubation temperature on growth and performance of the veiled chameleon (*Chamaeleo calyptratus*). *Journal of Experimental Zoology* 309A:435–446.

Andrews, R.M. 2008b. Lizards in the slow lane: Thermal biology of chameleons. *Journal of Thermal Biology* 33:57–61.

Andrews, R.M., C. Diaz-Paniagua, A. Marco, and A. Portheault. 2008. Developmental arrest during embryonic development of the common chameleon (*Chamaeleo chamaeleon*) in Spain. *Physiological and Biochemical Zoology* 81:336–344.

Andrews, R.M., and S. Donoghue. 2004. Effects of temperature and moisture on embryonic diapause of the veiled chameleon (*Chamaeleo calyptratus*). *Journal of Experimental Zoology* 301A:629–635.

Andrews, R.M., and K.B. Karsten. 2010. Evolutionary innovations of squamate reproductive and developmental biology in the family Chamaeleonidae. *Biological Journal of the Linnean Society* 100:656–668.

Andrews, R.M., and F.H. Pough. 1985. Metabolism of squamate reptiles: allometries and ecological relationships. *Physiological Zoology* 58:214–231.

Andriatsimietry, R., S.M. Goodman, E. Razafimahatratra, J.W.E. Jeglinski, M. Marquard, and J.U. Ganzhorn. 2009. Seasonal variation in the diet of *Galidictis grandidieri* Wozencraft, 1986 (Carnivora: Eupleridae) in a sub-arid zone of extreme south-western Madagascar. *Journal of Zoology* 279:410–415.

Angel, F. 1933. Sur un genre Malgasche nouveau, de la famille des Chamaeleontidés. *Bulletin du Musée D'Histoire Naturelle Paris* 5:443–446.
Angel, F. 1942. Les lézards de Madagascar. *Mémoires de l'Académie Malgache* 36:1–193
Aouraghe, H., J. Agustí, B. Ouchaou, S. Bailon, J.M. Lopez-Garcia, H. Haddoumi, K.E. Hammouti, A. Oujaa, and B. Bougariane. 2010. The Holocene vertebrate fauna from Guenfouda site, Eastern Morocco. *Historical Biology* 22(1–3):320–326.
Archer, M., D.A. Arena, M. Bassarova, R.M.D. Beck, K. Black, W.E. Boles, P. Brewer, B.N. Cooke, K. Crosby, A. Gillespie, H. Godthelp, S.J. Hand, B.P. Kear, J. Louys, A. Morrell, J. Muirhead, K.K. Roberts, J.D. Scanlon, K.J. Travouillon, and S. Wroe. 2006. Current status of species-level representation in faunas from selected fossil localities in the Riversleigh World Heritage Area, northwestern Queensland. *Alcheringa* Special Issue 1:1–17.
Aristotle (350 BC) Of the chameleon. Book 2, part 11. *Historia Animalium*. Oxford, United Kingdom: Clarendon Press.
Askew, G.N., and R.L. Marsh. 2001. The mechanical power output of the pectoralis muscle of blue-breasted quail (*Coturnix chinensis*): the *in vivo* length cycle and its implications for muscle performance. *Journal of Experimental Biology* 204(21):3587–3600.
Atsatt, R. 1953. Storage of sperm in the female chameleon *Microsaura pumila pumila*. *Copeia* 1953:59.
Augé, M. 1990. La faune de Lézards et d'Amphisbaenes de l'Éocène inférieur de Condé-en-Brie (France). *Bulletin du Muséum national d'Histoire naturelle, Paris*, 4e série, section C, 12:111–141 [in French].
Augé, M. 2005. Evolution des lézards du Paléogène en Europe. *Mémoires du Muséum National d'Histoire Naturelle* 192:1–369 [in French].
Augé, M., and J.C. Rage. 2006. Herpetofaunas from the Upper Paleocene and Lower Eocene of Morocco. *Annales de Paléontologie* 92:235–253.
Augé, M., and R. Smith. 1997. The Agamidae (Reptilia, Squamata) from the Paleogene of Western Europe. *Belgian Journal of Zoology* 127(2):123–138 [in French with English abstract].
Averianov, A., and I. Danilov. 1996. Agamid lizards (Reptilia, Sauria, Agamidae) from the Early Eocene of Kyrgyzstan. *Neues Jahrbuch für Geologie und Paläontologie-Monatshefte* 12:739–750.
Averianov, A.O. 2000. A new species of *Tinosaurus* from the Palaeocene of Kazakhstan (Squamata: Agamidae). *Zoosystematica Rossica* 9(2):459–460.
Averianov, A.O., A.V. Lopatin, P.P. Skutschas, N.V. Martynovich, S.V. Leshchinskiy, A.S. Rezvyi, S.A. Krasnolutskii, and A.V. Fayngertz. 2005. Discovery of Middle Jurassic mammals from Siberia. *Acta Palaeontologica Polonica* 50(4):789–797.
Axelrod, D.I., and P.H. Raven. 1978. Late Cretaceous and Tertiary vegetation history of Africa, pp. 77–130. In M.J.A. Werger, Ed., *Biogeography and Ecology of Southern Africa*. The Hague, The Netherlands: Junk.
Ayala-Guerrero, F., and G. Mexicano. 2008. Sleep and wakefulness in the green iguanid lizard (*Iguana iguana*). *Comparative Biochemistry and Physiology A—Physiology* 151:305–312.
Bagnara, J.T., and M.E. Hadley. 1973. *Chromatophores and Colour Change: The Comparative Physiology of Animal Pigmentation*. Englewood Cliffs, NJ: Prentice-Hall.
Balmford, A., Moore, J.L., Brooks, T., Burgess, N., Hansen, L.A., Williams, P., and C. Rahbek. 2001. Conservation conflicts across Africa. *Science* 291:2616–2619.
Bandyopadhyay, S., D.D. Gillette, S. Ray, and D.P. Sengupta. 2010. Osteology of *Barapasaurus tagorei* (Dinosauria: Sauropoda) from the Early Jurassic of India. *Palaeontology* 53:533–569.
Barej M.F., I. Ineich, V. Gvoždík, N. Lhermitte-Vallarino, N.L. Gonwouo, M. LeBreton, U. Bott, and A. Schmitz. 2010. Insights into chameleons of the genus *Trioceros* (Squamata: Chamaeleonidae) in Cameroon, with the resurrection of *Chamaeleon serratus* Mertens, 1922. *Bonn Zoological Bulletin* 57(2):211–229.

Barnett, K.E., R.B. Cocroft, and L.J. Fleishman. 1999. Possible communication by substrate vibration in a chameleon. *Copeia* 1999:225–228.
Bauer, A.M. 1997. Peritoneal pigmentation and generic allocation in the Chamaeleonidae. *African Journal of Herpetology* 46(2):117–122.
Beddard, F.E. 1904. Contribution to the anatomy of the Lacertilia. (3) On some points in the vascular system of *Chamaeleon* and other lizards. *Proceedings of the Zoological Society of London* 1904(2):6–22.
Beddard, F.E. 1907. Contributions to the knowledge of the systematic arrangement and anatomy of certain genera and species of Squamata. *Proceedings of the Zoological Society of London* 1907:35–45.
Bell, D.A. 1989. Functional anatomy of the chameleon tongue. *Zoologische Jahrbücher. Abteilung für Anatomie und Ontogenie der Tiere* 119:313–336.
Bell, D.A. 1990. Kinematics of prey capture in the chameleon. *Zoologische Jahrbücher. Abteilung für allgemeine Zoologie und Physiologie der Tiere* 94:247–260.
Bennett, A.F. 1985. Temperature and muscle. *Journal of Experimental Biology* 115:333–344.
Bennett, A.F. 2004. Thermoregulation in African chameleons, pp. 234–241. In S. Morris and A. Vosloo, Eds., *Animals and Environments: Proceedings of the Third International Conference of Comparative Physiology and Biochemistry, International Congress Series, Vol 1275*. Amsterdam, The Netherlands: Elsevier.
Bennett, A.F., and W.R. Dawson. 1976. Metabolism, pp. 127–223. In C. Gans and W.R. Dawson, Eds., *Biology of the Reptilia, Volume 5*. London: Academic Press.
Bennett, G. 1875. Notes on the *Chlamydosaurus* or frilled lizard of Queensland and the discovery of a fossil species. *Papers and Proceedings of the Royal Society of Tasmania* 1875:56–58.
Bennis, M., M. El Hassni, J-P. Rio, D. Lecren, J. Repérant, and R. Ward. 2001. A quantitative ultrastructural study of the optic nerve of the chameleon. *Brain Behavior and Evolution* 58:49–60.
Bennis, M., J. Repérant, J-P. Rio, and R. Ward. 1994. An experimental re-evaluation of the primary visual system of the European chameleon, *Chamelaeo chameleon. Brain Behavior and Evolution* 43:173–188.
Bennis, M., J. Repérant, R. Ward, and M. Wasowicz. 1996. Topography of the NADPH-Diaphorase system in the chameleon brain. *Journal of Brain Research* 2:281–288.
Bennis, M., C. Versaux-Botteri, J. Repérant, and J.A. Armengol. 2005. Calbindin, calretinin and parvalbumin immunoreactivity in the retina of the chameleon (*Chamaeleo chameleon*). *Brain Behavior and Evolution* 65:177–187.
Berger, P.J., and G. Burnstock. 1979. Autonomic nervous system, pp. 1–57. In R.G. Northcutt and P. Ulinski, Eds., *Biology of the Reptilia: Neurology*. London: Academic Press.
Bergeson, D. J. 1998. Patterns of suspensory feeding in *Alouatta palliata, Ateles geoffroyi*, and *Cebus capucinus*, pp. 45–60. In E. Strasser, J. Fleagle, A. Rosenberger and H. McHenry, Eds., *Primate Locomotion: Recent Advances*. New York: Plenum Press.
Bergmann, P.J., and D.J. Irschick. 2011. Vertebral evolution and the diversification of Squamate reptiles. *Evolution* 66(4):1044–1058.
Bergmann, P.J., S. Lessard, and A.P. Russell. 2003. Tail growth in *Chamaeleo dilepis* (Sauria: Chamaeleonidae): functional implications of segmental patterns. *Journal of Zoology, London* 261:417–425.
Bergquist, H. 1952. Studies on the cerebral tube in vertebrates: the neuromeres. *Acta Zoologica Stockholm* 33:117–187.
Bickel, R., and J.B. Losos. 2002. Patterns of morphological variation and correlates of habitat use in chameleons. *Biological Journal of the Linnean Society* 76(1):91–103.

Birkhead, T.R., and A.P. Møller. 1993. Sexual selection and the temporal separation of reproductive events: sperm storage data from reptiles, birds and mammals. *Biological Journal of the Linnean Society* 50:295–311.

Blackburn, D.G. 1999. Are viviparity and egg-guarding evolutionarily labile in squamates? *Herpetologica* 55:556–573.

Blackburn, D.G. 2006. Squamate reptiles as model organisms for the evolution of viviparity. *Herpetological Monographs* 20:131–146.

Blanc, C.P. 1972. Les reptiles de Madagascar et des îles voisines, pp. 501–614. In R. Battistini, and G. Vindard, Eds., *Biogeography and ecology in Madagascar.*The Hague, The Netherlands: Junk [in French].

Blanco, M.A., and P.W. Sherman. 2005. Maximum longevities of chemically protected and non-protected fishes, reptiles, and amphibians support evolutionary hypotheses of aging. *Mechanisms of Ageing and Development* 126:794–803.

Blasco, M. 1997a. *Chamaeleo chamaeleon*, pp. 158–159. In J.-P., Gasc, A. Cabela, J. Crnobrnja Isailovic, D. Dolmen, K. Grossenbacher, P. Haffner, J. Lescure, H. Martens, J.P. Martínez-Rica, H. Maurin, M.E. Oliveira, T.S. Sofianidou, M. Veith, and A. Zuiderwijk, Eds., *Atlas of Amphibians and Reptiles in Europe.* Paris, France: Societas Europaea Herpetologica and Muséum National d'Histoire Naturelle.

Blasco, M. 1997b. *Chamaelo chamaeleon* (Linnaeus, 1758) Camaleón común, Camaleão, pp. 190–192. In J.M. Pleguezuelos, Ed., *Distribución y Biogeografía de los anfibios y reptiles en España y Portugal.* Granada, Spain: Editorial Universidad de Granada [in Spanish].

Blob, R.W., and A.A. Biewener. 1999. *In vivo* locomotor strain in the hindlimb bones of *Alligator mississippiensis* and *Iguana iguana*: implications for the evolution of limb bone safety factor and non-sprawling limb posture. *Journal of Experimental Biology* 202:1023–1046.

Bockman, D.E. 1970. The thymus, pp 111–133. In C. Gans and T.S. Parsons, Eds., *Biology of the Reptilia. Volume 3. Morphology C.* New York: Academic Press.

Böhm, M., Collen, B., Baillie, J.E.M., Chanson, J., Cox, N., Hammerson, G., Hoffmann, M., Livingstone, S.R., Ram, M., Rhodin, A.G.J., Stuart, S.N. et al. 2013. The conservation status of the world's reptiles. *Biological Conservation* 157:372–385.

Böhme, M. 2003. The Miocene Climatic Optimum: evidence from ectothermic vertebrates of Central Europe. *Palaeogeography, Palaeoclimatology, Palaeoecology* 195:389–401.

Böhme, M. 2010. Ectothermic vertebrates (Actinopterygii, Allocaudata, Urodela, Anura, Crocodylia, Squamata) from the Miocene of Sandelzhausen (Germany, Bavaria) and their implications for environment reconstruction and palaeoclimate. *Paläontologische Zeitschrift* 84:3–41.

Böhme, W., and C.J.J. Klaver. 1980. The systematic status of *Chamaeleo kinetensis* Schmidt, 1943, from the Imatong mountains, Sudan, with comments on lung and hemipenal morphology within the *Chamaeleo bitaeniatus* group. *Amphibia-Reptilia* 1:3–17.

Boistel, R., A. Herrel, G. Daghfous, P.A. Libourel, E. Boller, P. Taffoureau, and V. Bels. 2010. Assisted walking in Malagasy dwarf chameleons. *Biology Letters* 6(6):740–743.

Bolliger, T. 1992. Kleinsäugerstratigraphie der miozänen Hörnilschüttung (Ostschweiz). *Dokumenta naturae* 75:1–297 [in German].

Bonetti, A. 1998. New life from Roman relics. *BBC Wildlife* 1998 16(7):10–16.

Bonine, K.E., and T. Garland Jr. 1999. Sprint performance of phrynosomatid lizards, measured on a high-speed treadmill, correlates with hindlimb length. *Journal of Zoology, London* 248:255–265.

Bons, J., and N. Bons. 1960. Notes sur la reproduction et le développement de *Chamaeleo chamaeleon* (L.). *Bulletin de la Société des Sciences Naturelles et Physiques du Maroc* 40:323–335.

Born, G. 1879. Die Nasenhöhlen und der Thränennassengang der amnioten Wirbelthiere. *Morphologisches Jahrbuch* 5:62–140 [in German].

Borsuk-Białynicka, M. 1991. Questions and controversies about saurian phylogeny, a Mongolian perspective, pp. 9–10. In Z. Kielan-Jaworowska, N. Heintz, and H.A. Nacerem, Eds., *5th Symposium on Mesozoic Terrestrial Ecosystems and Biota (Extended Abstracts)*. Contributions of the Palaeontological Museum, University of Oslo 364.

Borsuk-Białynicka, M., and S.M. Moody. 1984. Priscagaminae, a new subfamily of the Agamidae (Sauria) from the Late Cretaceous of the Gobi Desert. *Acta Palaeontologica Polonica* 29(1–2):51–81.

Bosworth, W., P. Huchon, and K. McClay. 2005. The Red Sea and Gulf of Aden Basins. *Journal of African Earth Sciences* 43:334–378.

Bourgat, R. 1968. Etude des variations annuelles de la population de *Chamaeleo pardalis* Cuvier de l'Ile de la Reunion. *Vie Milieu* 19:227–231.

Bourgat, R.M. 1973. Cytogénétique des caméléons de Madagascar. Incidences taxonomiques, biogéographiques et phylogénétiques. *Bulletin de la Société Zoologique de France* 98(1):81–90.

Bourgat, R.M., and C.A. Domergue. 1971. Notes sur le *Chamaeleo tigris* Kuhl 1820 des Seychelles. *Annales de l'Université de Madagascar, Série Sciences de la Nature et Methématiques* 8:235–244.

Bowmaker, J.K., E.R. Loew, and M. Ott. 2005. The cone photoreceptors and visual pigments of chameleons. *Journal of Comparative Physiology A* 191:925–932.

Brady, L.D., and R.A. Griffiths. 1999. Status assessment of chameleons in Madagascar. Gland, Switzerland, and Cambridge, United Kingdom: IUCN Species Survival Commission.

Brady, L.D., and R.A. Griffiths. 2003. Chameleon population density estimates, pp. 970–972. In S. Goodman and J. Benstead, Eds., *The Natural History of Madagascar*. Chicago: University of Chicago Press.

Brady, L. D., K. Huston, R.K.B. Jenkins, J.L.D. Kauffmann, J. Rabearivony, G. Raveloson, and M. Rowcliffe. 1996. UEA Madagascar Expedition'93. Final Report. Unpublished Report, University of East Anglia: Norwich.

Brain, C.K. 1961. *Chamaeleo dilepis*—a study on its biology and behavior. *Journal of the Herpetological Association of Rhodesia* 15:15–20.

Bramble, D.M., and D.B. Wake. 1985. Feeding mechanisms of lower tetrapods, pp 230–261. In M. Hildebrand, D.M. Bramble, K.F. Liem, and D.B. Wake, Eds., *Functional Vertebrate Morphology*. Cambridge, United Kingdom: Cambridge University Press.

Branch, W.R. 1998. *Field Guide to the Snakes and Other Reptiles of Southern Africa*. Cape Town, South Africa: Struik.

Branch, W.R., and J. Bayliss. 2009. A new species of *Atheris* (Serpentes: Viperidae) from northern Mozambique. *Zootaxa* 2113:41–54.

Branch, W.R., and K.A. Tolley. 2010. A new species of chameleon (Sauria: Chamaeleonidae: *Nadzikambia*) from Mount Mabu, central Mozambique. *African Journal of Herpetology* 59:157–172.

Briggs, JC. 2003. The biogeographic and tectonic history of India. *Journal of Biogeography* 30:381–388.

Bringsøe, H. 2007. An observation of *Calumma tigris* (Squamata: Chamaeleonidae) feeding on White-footed ants, *Technomyrmex albipes* complex, in the Seychelles. *Herpetological Bulletin* 102:15–17.

Brink, J.M. 1957. Vergelijkend karyologisch onderzoek aan het genus *Chamaeleon*. *Genen en phaenen* 2:35–40.

Broadley, D.G. 1965. A new chameleon from Malawi. *Arnoldia* 31:1–3.

Broadley, D.G. 1966. Studies on the ecology and ethology of African lizards. *Journal of the Herpetological Association of Africa* 2:6–16.

Broadley, D.G. 1973. Predation on birds by reptiles and amphibians in south-eastern Africa. *Honeyguide* 76:19–21
Broadley, D.G. 1983. *FitzSimons' Snakes of Southern Africa* (rev. ed.). Johannesburg, South Africa: Delta Books.
Broadley, D.G., and D.K. Blake. 1979. A field study of *Rhampholeon marshalli marshalli* on Vumba Mountain, Rhodesia (Sauria: Chamaeleonidae). *Arnoldia* 8:1–6.
Brock, G.T. 1941. The skull of the chameleon, *Lophosaura ventralis* (Gray); some developmental stages. *Proceedings of the Zoological Society of London B* 110(3–4):219–241.
Brooks, T.M., R.A. Mittermeier, C.G. Mittermeier, G.A.B. da Fonseca, A.B. Rylands, W.R. Konstant, P. Flick, J. Pilgrim, S. Oldfield, G. Magin, and C. Hilton-Taylor. 2002. Habitat loss and extinction in the hotspots of biodiversity. *Conservation Biology* 16:909–923.
Broschinski, A. 2000. The lizards from the Guimarota mine, pp. 59–68 in T. Martin, and B. Krebs, Eds., *Guimarota: A Jurassic Ecosystem*. Munich: Verlag Dr. Friedrich Pfeil.
Brücke, E. 1852a. Über die Zunge der Chamäleonen. *Sitzungsberichte der Mathematisch-Naturwissenschaftlichen Classe der Kaiserlichen Akademie der Wissenschaften* 8:65–70 [in German].
Brücke, E. 1852b. Untersuchungen be idem Farbwechsel des afrikanischen Chameleons. *Denkschrift der Kaiserlichen Akademie der Wissenschaften in Wien* 4:179–210.
Bruner, H.L. 1907. On the cephalic veins and sinuses of reptiles, with description of a mechanism for raising the venous blood-pressure in the head. *American Journal of Anatomy* 7:1–117.
Brygoo, E.R. 1971. Reptiles Sauriens Chamaeleonidae. Genre *Chamaeleo. Faune de Madagascar* 33:1–318.
Brygoo, E.R. 1978. Reptiles Sauriens Chamaeleonidae. Genre *Brookesia* et complement pour le genre *Chamaeleo. Faune de Madagascar* 47:1–173.
Burgess, N.D., Balmford, A., Cordeiro, N.J., Fjeldså, J., Küper, W., Rahbek, C., Sanderson, E.W., Scharlemann, J.R.P.W., Sommer, J.H., and P.H. Williams. 2007. Correlations among species distributions, human density and human infrastructure across the high biodiversity tropical mountains of Africa. *Biological Conservation* 134:164–177.
Burmeister, E.-G., 1989. Eine Walzenspinne (Solifugae, Galeodidae) als Nahrung des Gemeinen Chamäleons (*Chamaeleo chamaeleon* Linnaeus, 1758). *Herpetofauna* 11:32–34.
Burrage, B.R. 1973. Comparative ecology and behaviour of *Chamaeleo pumilis pumilis* (Gmelin) and *C. namaquensis* A. Smith (Sauria: Chamaeleonidae). *Annals of the South African Museum* 61:1–158.
Bustard, H.R. 1966. Observations on the life history and behavior of *Chamaeleo bitaeniatus* Fischer. *Herpetologica* 22:13–23.
Bustard, H.R. 1967. The comparative behavior of chameleons: fight behavior in *Chamaeleo gracilis* Hallowell. *Herpetologica* 23:44–50.
Butchart, S.H.M., Walpole, M., Collen, B., Van Strien, A., Scharlemann, J.R.P.W., Almond, R.E.A., Baillie, J.E.M., Bomhard, B., Brown, C., Bruno, J., Carpenter, K.E., Carr, G.V.M., Chanson, J., Chenery, A.M., Csirke, J., Davidson, N.C., Dentener, F., Foster, M., Galli, A., Galloway, J.N., Genovesi, P., Gregory, R.D., Hockings, M., Kapos, V., Lamarque, J.-F., Leverington, F., Loh, J., McGeoch, M.A., McRae, L., Minasyan, A., Morcillo, M.H.N., Oldfield, T.E.E., Pauly, D., Quader, S., Revenga, C., Sauer, J.R., Skolnik, B., Spear, D., Stanwell-Smith, D., Stuart, S.N., Symes, A., Tierney, M., Tyrrell, T.D., Vié, J.-C., and R. Watson. 2010. Global biodiversity: indicators of recent declines. *Science* 328:1164–1168.

Butler, M.A. 2005. Foraging mode of the chameleon, *Bradypodion pumilum*: a challenge to the sit-and-wait versus active forager paradigm? *Biological Journal of the Linnean Society* 84:797–808.
Camargo, C.R., M.A. Visconti, and A.M.L. Castrucci. 1999. Physiological color change in the bullfrog, *Rana catesbeiana*. *Journal of Experimental Zoology* 283:160–169.
Camp, C.L. 1923. Classification of the lizards. *Bulletin of the American Museum of Natural History* 48:289–481.
Canella, M.F. 1963. Note di fisiologia dei cromatofori dei vertebrati pecilotermi, particolarmente dei lacertili. *Monitore Zoologico Italiano* 71:430–480.
Canham, M.T. 1999. The identification of specialized scale surface structures and scale arrangements of the ventral portion of a prehensile tail, used for increased rip in the *Chamaeleo* genus. *Chameleon information Network* 33:5–8.
Carothers, J. H. 1986. An experimental confirmation of morphological adaptation: toe fringes in the sand-dwelling lizard *Uma scoparia*. *Evolution* 40(4):871–874.
Carpenter, A.I., and O. Robson. 2005. A review of the endemic chameleon genus *Brookesia* from Madagascar, and the rationale for its listing on CITES Appendix II. *Oryx* 39:375–380.
Carpenter, A.I., Robson, O., Rowcliffe, J.M., and A.R. Watkinson. 2005. The impacts of international and national governance changes on a traded resource: a case study of Madagascar and its chameleon trade. *Biological Conservation* 123:279–287.
Carpenter, A.I., Rowcliffe, J.M., and A.R. Watkinson. 2004. The dynamics of the global trade in chameleons. *Biological Conservation* 120:291–301.
Carpenter, G.C. 1977. Variation and evolution of stereotyped behavior in reptiles, pp. 335–403. In C. Gans and D.W. Tinkle, Eds., *Biology of Reptiles*. London: Academic Press.
Cartmill, M. 1985. Climbing, pp. 73–88. In M. Hildebrand, D. M. Bramble, K. F. Liem and D. B. Wake, Eds., *Functional Vertebrate Morphology*. Cambridge, United Kingdom: Belknap Press.
Case, E.C. 1909. The dorsal spines of *Chameleo cristatus*, Stuch. *Science (Weekly)* 29(755):979.
Čerňanský, A. 2010. A revision of chamaeleonids from the Lower Miocene of the Czech Republic with description of a new species of *Chamaeleo* (Squamata, Chamaeleonidae). *Geobios* 43:605–613.
Čerňanský, A. 2011. A revision of the chameleon species *Chamaeleo pfeili* Schleich (Squamata; Chamaeleonidae) with description of a new material of chamaeleonids from the Miocene deposits of southern Germany. *Bulletin of Geosciences* 86(2):275–282.
Cheke, A.S. 1987. An ecological history of the Mascarene Islands, with particular reference to extinctions and introductions of land vertebrates, pp. 5–89. In A.W. Diamond, Ed., *Studies of Mascarene Island Birds*. Cambridge, United Kingdom: Cambridge University Press.
Cheke, A.S., and J. Hume. 2008. *Lost Land of the Dodo*. London: Poyser.
Chevret, P., and G. Dobigny. 2005. Systematics and evolution of the subfamily Gerbillinae (Mammalia, Rodentia, Muridae). *Molecular Phylogenetics and Evolution* 35:674–688.
Chkhikvadze, V.M. 1985. Preliminary results of the study of Tertiary amphibians and squamate reptiles of the Zaisan Basin. *Voprosy Gerpetologii – Shestaya Vsesoyuznaya 7 Gerpetologicheskaya Konferentsiya, Tashkent, 18–20 sentyabrya 1985, Avtoreferaty dokladov*, 234–235 [in Russian].
Chorowicz, J. 2005. The East African rift system. *Journal of African Earth Sciences* 43:379–410.
Cincotta, R., Wisnewski, J., and R. Engelman. 2000. Human population in the biodiversity hotspots. *Nature* 404:990–992.
CITES. 2012a. CITES trade statistics derived from the CITES Trade Database, Cambridge, United Kingdom: UNEP World Conservation Monitoring Centre. Accessed June 13, 2012.

CITES. 2012b. Notification to the Parties No. 2012/021. Accessed April 11, 2012.

Clothier, J., and J.N. Lythgoe. 1987. Light-induced color changes by the iridophores of the neon tetra, *Paracheirodon innesi*. *Journal of Cell Science* 88:663–668.

Clusella-Trullas, S., Blackburn, T.M., and S.L. Chown. 2011. Climatic predictors of temperature performance curve parameters in ectotherms imply complex responses to climate change. *The American Naturalist* 177:738–751.

Cole, N. 2009. *A Field Guide to the Reptiles and Amphibians of Mauritius*. Vacoas, Mauritius: Mauritian Wildlife Foundation.

Conrad, J.L. 2008. Phylogeny and systematics of Squamata (Reptilia) based on morphology. *Bulletin of the American Museum of Natural History* 310:1–182.

Conrad, J.L., and M.A. Norell. 2007. A complete Late Cretaceous iguanian (Squamata, Reptilia) from the Gobi and identification of a new Iguanian Clade. *American Museum Novitates* 3587:1–47.

Cooper, W.E., and L.J. Vitt. 2002. Distribution, extent, and evolution of plant consumption by lizards. *Journal of Zoology* 257:487–517.

Cooper, W.E., and N. Greenberg. 1992. Reptilian coloration and behavior, pp. 298–422. In C. Gans and D. Crews, Eds., *Biology of the Reptilia*. Chicago: Chicago University Press.

Cope, E.D. 1892. The osteology of the Lacertilia. *Proceedings of the American Philosophical Society* 30:185–219.

Couvreur, T.L.P., Chatrou, L.W., Sosef, M.S.M., and J.E. Richardson. 2008. Molecular phylogenetics reveal multiple tertiary vicariance origins of the African rain forest trees. *BMC Biology* 6:54.

Couvreur, T.L.P., Forest, F., and W.J. Baker. 2011. Origin and global diversification patterns of tropical rain forests: inferences from a complete genus-level phylogeny of palms. *BMC Biology* 9:44.

Covacevich, J., P. Couper, R.E. Molnar, G. Witten, and W. Young, 1990. Miocene dragons from Riversleigh: new data on the history of the family Agamidae (Reptilia: Squamata) in Australia. *Memoirs of the Queensland Museum* 29:339–360.

Crespo, E. G., and M.E. Oliveira. 1989. Atlas da Distribucao dos Anfibios e Répteis de Portugal Continental. Servicio Nacional de Parques Reservas e Conservacao da Naturaleza, Lisboa [in Portuguese].

Crottini, A., D.J. Harris, I.A. Irisarri, A. Lima, S. Rasamison, and G.M. Rosa. 2010. Confirming Domergue: *Ithycyphus oursi* Domergue, 1986 predation upon *Furcifer oustaleti* (Mocquard, 1894). *Herpetology Notes* 3:127–131.

Cuadrado, M. 1998a. The influence of female size on the extent and intensity of mate guarding by males in *Chamaeleo chamaeleon*. *Journal of Zoology* 246:351–358.

Cuadrado, M. 1998b. The use of yellow spot colors as a sexual receptivity signal in females of *Chamaeleo chamaeleon*. *Herpetologica* 54:395–402.

Cuadrado, M. 2000. Body colors indicate the reproductive status of female Common chameleons: experimental evidence for the inter-sex communication function. *Ethology* 106:79–91.

Cuadrado, M. 2001. Mate guarding and social mating system in male common chameleons (*Chamaeleo chamaeleon*). *Journal of Zoology* 255:425–435.

Cuadrado, M., and J. Loman. 1997. Mating behaviour in a chameleon (*Chamaeleo chamaeleon*) population in southern Spain—effects of male and female size, pp. 81–88 in W. Böhme, W. Bischoff and T. Ziegler, Eds., *Herpetologica Bonnensis*. Bonn, Germany: Societas Europaea Herpetologica: Bonn.

Cuadrado, M., and Loman, J. 1999. The effects of age and size on reproductive timing in female *Chamaeleo chamaeleon*. *Journal of Herpetology* 33:6–11.

Cuadrado, M., J. Martin, and P. Lopez. 2001. Camouflage and escape decisions in the common chameleon, *Chamaeleo chamaeleon*. *Biological Journal of the Linnean Society* 72:547–554.

Cuvier, G. 1805. Lecons d'Anatomie Comparée, Tome III. Paris: Recueillies et Publiés par L. Duvernoy [in French].
Daniels, S.R., and J. Bayliss. 2012. Neglected refugia of biodiversity: mountainous regions in Mozambique and Malawi yield two novel freshwater crab species (Potamonautidae: Potamonautes). *Zoological Journal of the Linnean Society* 164:498–509.
Dart, R.A. 1934. The dual structure of the neopallium: its history and significance. *Journal of Anatomy* 69:3–19.
daSilva, J.M., and K.A. Tolley. 2013. Ecomorphological variation and sexual dimorphism in a recent radiation of dwarf chameleons (*Bradypodion*). *Biological Journal of the Linnean Society* 109(1): 113–130.
Datta, P.M., and S. Ray. 2006. Earliest lizard from the Late Triassic (Carnian) of India. *Journal of Vertebrate Paleontology* 26(4):795–800.
Davenport, T.R.B., W.T. Stanley, E.J. Sargis, D.W. De Luca, N.E. Mpunga, S.J. Machaga, and L.E. Olson. 2006. A new genus of African monkey, *Rungwecebus:* morphology, ecology, and molecular phylogenetics. Science 312:1378–1381.
D'Cruze, N.C., and J.A. Sabel. 2005. *Ptychadena mascareniensis* (Mascarene ridged frog): predation on an endemic malagasy chameleon. *Herpetological Bulletin* 93:26–27.
de Groot, J.H., and J.L. van Leeuwen. 2004. Evidence for an elastic projection mechanism in the chameleon tongue. *Proceedings of the Royal Society B* 271(1540):761–770.
De Quieroz, K. 1995. Phylogenetic approaches to classification and nomenclature, and the history of taxonomy (an alternative interpretation). *Herpetological Review* 26(2):79–81.
de Stefano, G. 1903. I sauri del Quercy appartenentialla collenzione Rossignol. *Atti della Societa Italiana di Scienze Naturalie del Museo Civico di Storia Naturale di Milano* 42:382–418 [in Italian].
Delfino, M., T. Kotsakis, M. Arca, C. Tuveri, G. Pitruzzella, and L. Rook. 2008. Agamid lizards from the Plio-Pleistocene of Sardinia (Italy) and an overview of the European fossil record of the family. *Geodiversitas* 30(3):641–656.
Dewevre, L.S. 1895. Le méchanime de la projection de la langue chez le caméléon. *Journal de l'anatomie et de la physiologie normales et pathologiques de l'homme et des animaux* 31:343–360 [in French].
Diaz-Paniagua, C. 2007. Effect of cold temperature on the length of incubation of *Chamaeleo chamaeleon. Amphibia-Reptilia* 28:387–392.
Diaz-Paniagua, C., and M. Cuadrado. 2003. Influence of incubation conditions on hatching success, embryo development and hatchling phenotype of common chameleon (*Chamaeleo chamaeleon*) eggs. *Amphibia-Reptilia* 24:429–440.
Díaz-Paniagua, C., M. Cuadrado, M.C. Blázquez, and J.A. Mateo. 2002. Reproduction of *Chamaeleo chamaeleon* under contrasting environmental conditions. *Herpetological Journal* 12:99–104.
Dierenfeld, E.S., E.B. Norkus, K. Caroll, and G.W. Ferguson. 2002. Carotenoids, vitamin A and vitamin E concentrations during egg development in panther chameleons (*Furcifer pardalis*). *Zoo Biology* 21:295–303.
Dimaki, M., A.K. Hundsdörfer, and U. Fritz. 2008. Eastern Mediterranean chameleons (*Chamaeleo chamaeleon, Ch. africanus*) are distinct. *Amphibia-Reptilia* 29:535–540.
Dimaki, M., E.D. Valakos, and A. Legakis. 2000. Variation in body temperatures of the African Chameleon *Chamaeleo africanus* Laurenti, 1768 and the Common Chameleon *Chamaeleo chamaeleon* (Linnaeus, 1758). *Belgian Journal of Zoology* 130:87–91.
Dong, Z.M. 1965. A new species of *Tinosaurus* from Lushih, Honan. *Vertebrata PalAsiatica* 9(1):79–83 [in Chinese with English summary].

Døving, K.B., and D. Trotier. 1998. Structure and function of the vomeronasal organ. *Journal of Experimental Biology* 201(21):2913–2925.
Drake, R.E., J.A. Van Couvering, M.H. Pickford, G.H. Curtis, and J.A. Harris. 1988. New chronology for the Early Miocene mammalian faunas of Kisingiri, Western Kenya. *Journal of the Geological Society, London* 145:479–491.
Duke-Elder, S. 1957. System of ophthalmology. Vol. I. The eye in evolution. London: Kimpton.
Dunson, W.A. 1976. Salt glands in reptiles, pp. 413–445. In C. Gans and W.R. Dawson, Eds., *Biology of the Reptilia. Volume 5. Physiology A.* New York: Academic Press.
Duvernoy, L.G. 1836. Sur les mouvements de la langue du chameleon. *Comptes Rendus Hebdomadaires des Séances de l'Académie des Sciences, Paris* 2:349–351 [in French].
Edinger, T. 1955. The size of parietal foramen and organ in reptiles. A rectification. *Bulletin of the Museum of Comparative Zoology at Harvard College* 114:1–34.
Edgar, J.I. 1979. Fatbody and liver cycles in two tropical lizards *Chamaeleo hohneli* and *Chamaeleo jacksoni* (Reptilia, Lacertilia, Chamaeleonidae). *Journal of Herpetology* 13(1):113–117.
El Hassni, M., S. Ba M'Hamed, J. Repérant, and M. Bennis. 1997. Quantitative and topographical study of retinal ganglion cells in the chameleon (*Chamaeleo chameleon*). *Brain Research Bulletin* 44:621–625.
Emmett, D.A. 2004. Altitudinal distribution of the Short-Tailed Pygmy Chameleon (*Rhampholeon brevicaudatus*) and the Usambara Pitted Pygmy Chameleon (*R. temporalis*) in Tanzania. *African Herp News* 37:12–13.
Engelbrecht, D. van Z. 1951. Contributions to the cranial morphology of the chamaeleon *Microsaura pumila* Daudin. *Annale van die Uniwersiteit van Stellenbosch.* 27(1):3–31.
Estes, R. 1983a. *Sauria Terrestria, Amphisbaenia (Handbuch der Paläoherpetologie).* Stuttgart, Germany: Gustav Fischer Verlag.
Estes, R. 1983b. The fossil record and the early distribution of lizards, pp. 365–398. In A.G.J. Rhodin, and K. Miyata, Eds., *Advances in Herpetology and Evolutionary Biology : Essays in Honor of E. E. Williams.* Cambridge, MA: Museum of Comparative Zoology, Harvard University.
Estes, R., K. de Queiroz, and J. Gauthier. 1988. Phylogenetic relationships within Squamata, pp. 119–281. In R. Estes, and G. Pregill, Eds., *Phylogenetic Relationships of the Lizard Families.* Stanford, CA: Stanford University Press.
Etheridge, R. 1967. Lizard caudal vertebrae. *Copeia* 1967(4):699–721.
Evans, S.E. 1998. Crown group lizards from the Middle Jurassic of Britain. *Palaeontographica, Abt. A* 250:123–154.
Evans, S.E. 2003. At the feet of the dinosaurs: the origin, evolution and early diversification of squamate reptiles (Lepidosauria: Diapsida). *Biological Reviews* 78:513–551.
Evans, S.E., and M.E.H. Jones. 2010. The origin, early history and diversification of lepidosauromorph reptiles, pp. 27–44. In S. Bandyopadhyay, Ed., *New Aspects of Mesozoic Biodiversity.* Lecture Notes in Earth Sciences 132. Berlin: Springer Verlag.
Evans, S.E., G.V.R. Prasad, and B.K. Manhas. 2001. Rhynchocephalians (Diapsida: Lepidosauria) from the Jurassic Kota Formation of India. *Zoological Journal of the Linnean Society* 133:309–334.
Evans, S.E., G.V.R. Prasad, and B.K. Manhas. 2002. An acrodont iguanian from the Mesozoic Kota Formation of India. *Journal of Vertebrate Paleontology* 22:299–312.
Farrell, A.P., A.K. Gamperl, and E.T. Francis. 1998. Comparative Aspects of Heart Morphology, pp. 375–424. In C. Gans and A.S. Gaunt, Eds., *Biology of the Reptilia. Volume 19. Morphology G.* Ithaca, NY: Society for the Study of Amphibians and Reptiles.
Fejfar, O., and H.H. Schleich. 1994. Ein Chamäleonfund aus dem unteren Orleanium des Braunkohlen-Tagebaus Merkur-Nord (Nordböhmen). *Courier Forschungsinstitut Senckenberg* 173:167–173 [in German].

Ferguson, G.W., W.H. Gehrmann, T.C. Chen, E.S. Dierenfeld, and M.F. Holick. 2002. Effects of artificial ultraviolet light exposure on reproductive success of the female panther chameleon (*Furcifer pardalis*) in captivity. *Zoo Biology* 21:525–537.

Ferguson, G.W., W.H. Gehrmann, K.B. Karsten, S.H. Hammack, Michele McRae, T.C. Chen, N.P. Lung, and M.F. Holick. 2003. Do panther chameleons bask to regulate endogenous vitamin D3 production. *Physiological and Biochemical Zoology* 76:52–59.

Ferguson, G.W., W.H. Gehrmann, K.B. Karsten, A.J. Landwer, E.N. Carman, T.C. Chen, and M.F. Holick. 2005. Ultraviolet exposure and vitamin D synthesis in a sun-dwelling and shade-dwelling species of *Anolis*: Are there adaptations for lower ultraviolet B and dietary vitamin D3 availability in the shade? *Physiological and Biochemical Zoology* 78:193–200.

Ferguson, G.W., J.B. Murphy, J.B. Ramanamanjato, and A.P. Raselimanana. 2004.*The Panther Chameleon. Color Variation, Natural History, Conservation, and Captive Management.* Malabar, FL: Grieger Publishing.

Filhol, H. 1877. Recherches sur les Phosphorites du Quercy. Pt. II. *Annales Sciences Géologiques* 8:1–338.

Fischer, M.S., Krause, C., and K.E. Lilje. 2010. Evolution of chameleon locomotion, or how to become arboreal as a reptile. *Zoology* 113(2):67–74.

Fisher, M.C., Henk, D.A., Briggs, C.J., Brownstein, J.S., Madoff, L.C., McCraw, S.L., and S.J. Gurr. 2012 Emerging fungal threats to animal, plant and ecosystem health. *Nature* 484:186–194.

Fitch, H.S. 1981. Sexual size differences in reptiles. *University of Kansas Museum of Natural History Miscellaneous Publication* 70:1–72.

Fitzinger, L. 1843. *Systema Reptilium, fasciculus primus, Amblyglossae.* Braumüller & Siedel: Wien.

Fitzsimons, V.F. 1943. Chamaeleonidae: the lizards of South Africa. *Transvaal Museum Memoirs* 1:151–174.

Fjeldså, J., and N.B. Burgess. 2008. The coincidence of biodiversity patterns and human settlement in Africa. *African Journal of Ecology* 46:33–42.

Fjeldså, J., and J.C. Lovett. 1997. Geographical patterns of old and young species in African forest biota: the significance of specific montane areas as evolutionary centres. *Biodiversity and Conservation* 6:322–346.

Flanders, M. 1985. Visually guided head movement in the African chameleon. *Vision Research* 25:935–942.

Fleishman, L.J. 1985. Cryptic movement in the vine snake *Oxybelis aeneus*. *Copeia* 1985:242–245.

Florio, A.M., C.M. Ingram, H.A. Rakotondravony, E.E. Louis Jr., and C.J. Raxworthy. 2012. Detecting cryptic diversity in the widespread and morphologically conservative carpet chameleon (*Furciferlateralis*) of Madagascar. *Journal of Evolutionary Biology* 25:1399–1414.

Forister, M.L., A.C. McCall, N.J. Sanders, J.A. Fordyce, J.H. Thorne, J. O'Brien, D.P. Waetjen, and A.M. Shapiro. 2010. Compounded effects of climate change and habitat alteration shift patterns of butterfly diversity. *Proceedings of the National Academy of Sciences of the United States of America* 107:2088–2092.

Foster, K.L., and T.E. Higham. 2012. How forelimb and hindlimb function changes with incline and perch diameter in the green anole (*Anolis carolinensis*). *Journal of Experimental Biology* 215(13):2288–2300.

Fournier, M., N. Chamot-Rooke, C. Petit, P. Huchon, A. Al-Kathiri, L. Audin, M.-O. Beslier, E. d'Acremont, O. Fabbri, J.-M. Fleury, K. Khanbari, C. Lepvrier, S. Leroy, B. Maillot and S. Merkouriev. 2010. Arabia-Somalia plate kinematics, evolution of the Aden-Owen-Carlsberg triple junction, and opening of the Gulf of Aden. *Journal of Geophysical Research* 115:B04102.

Fox, D.L. 1976. *Animal Biochromes and Structural Colours: Physical, Chemical, Distributional and Physiological Features of Coloured Bodies in the Animal World.* Berkeley: University of California Press.

Fox, H. 1977. The urogenital system of reptiles, pp. 1–157. In C. Gans and T.S. Parsons, Eds., *Biology of the Reptilia. Volume 6. Morphology E.* New York: Academic Press.

Frank, G.H. 1951. Contributions to the cranial morphology of *Rhampholeon platyceps* Günther. *Annale van die Uniwersiteit van Stellenbosch* 27(2):33–67.

Friis, I., S. Demissew, and P. van Breugel. 2010. Atlas of the potential vegetation of Ethiopia. Copenhagen: Royal Danish Academy of Science and Letters.

Frost, D.R., and R. Etheridge. 1989. A phylogenetic analysis and taxonomy of the iguanian lizards (Reptilia: Squamata). *University of Kansas Museum of Natural History Miscellaneous Publications* 81:1–65.

Frost, D. R., R. Etheridge, D. Janies, and T.A. Titus. 2001. Total evidence, sequence alignment, evolution of polychrotid lizards, and a reclassification of the iguania (Squamata: Iguania). *American Museum Novitates* 3343:1–38.

Furbringer, M. 1900. Zur vergleichenden Anatomie des Brustschulterapparates und der Schultermuskeln IV. *Jenaische Zeitschrift für Medizin und Naturwissenschaft* 34:215–718 [in German].

Gabe, M. 1970. The adrenal, pp. 263–318. In C. Gans and T.S. Parsons, Eds., *Biology of the Reptilia. Volume 3. Morphology C.* New York: Academic Press.

Gabe, M., and M. Martoja. 1961. Contribution a l'histologie de la glande surrénale des Squamata (Reptiles). *Archive d'Anatomie Microscopique et de Morphologie Experimentale* 50:1–34 [in French].

Gamble, T., A.M. Bauer, E. Greenbaum, and T.R. Jackman. 2008. Evidence for Gondwanan vicariance in an ancient clade of gecko lizards. *Journal of Biogeography* 35:88–104.

Gans, C. 1967. The chameleon. *Natural History* 76:52–59.

Gao, K., and D. Dashzeveg. 1999. New lizards from the Middle Eocene Mergen Formation of the Mongolian Gobi Desert. *Paläontologische Zeitschrift* 73:327–335.

Gao, K., and M. Norell. 2000. Taxonomic composition and systematics of Late Cretaceous lizard assemblages from Ukhaa Tolgod and adjacent localities, Mongolian Gobi desert. *Bulletin of the American Museum of Natural History* 249:1–118.

Garber, P.A., and J.A. Rehg. 1999. The ecological role of the prehensile tail in white-faced capuchins (*Cebus capucinus*). *American Journal of Physical Anthropology* 110:325–339.

García, G., and M. Vences. 2002. *Furcifer oustaleti* (Oustalet's chameleon). diet. *Herpetological Review* 33:134–135.

Garland, T. Jr., and J. B. Losos. 1994. Ecological morphology of locomotor performance in squamate reptiles, pp. 240–302. In P.C. Wainwright and S.M. Reilly, Eds., *Ecological Morphology: Integrative Organismal Biology.* Chicago: University of Chicago Press.

Gasc, J.-P. 1963. Adaptation a la marche arboricole chez le cameleon. *Archive d'Anatomie, d'Histologie et d'Embryologie Normales et Expérimentales* 46:81–115 [in Italian].

Gasc, J.-P. 1981. Axial Musculature, pp. 355–435. In C. Gans and T.S. Parsons, Eds., *Biology of the Reptilia. Volume 11. Morphology F.* New York: Academic Press.

Gaubert, P., and P. Cordeiro-Estrela. 2006. Phylogenetic systematics and tempo of evolution of the Viverrinae (Mammalia, Carnivora, Viverridae) within feliformians: implications for faunal exchanges between Asia and Africa. *Molecular Phylogenetics and Evolution* 41:266–278.

Gauthier, J.A., M. Kearney, J.A. Maisano, O. Rieppel, and D.B. Behlke. 2012. Assembling the squamate tree of life: perspectives from the phenotype and the fossil record. *Bulletin of the Peabody Museum of Natural History* 53:3–308.

GEF (Global Environmental Facility). 2002. Project Brief: Conservation and Management of the Eastern Arc Mountain Forests, Tanzania. Global Environmental Faciliity: Arusha, Tanzania.

Gehring, P.-S., and N. Lutzmann. 2011. Anmerkungen zum Zungentest-Verhalten bei Chamäleons. *Elaphe* 19(2):12–15 [in German].

Gehring, P.-S., N. Lutzmann, S. Furrer, and R. Sossinka. 2008. Habitat preferences and activity patterns of *Furcifer pardalis* (Cuvier, 1829) in the Masoala Rain Forest Hall of the Zurich Zoo. *Salamandra* 44:129–140

Gehring, P.-S., M. Pabijan, F.M. Ratsoavina, J. Köhler, M. Vences, and F. Glaw. 2010. A Tarzan yell for conservation: a new chameleon, *Calumma tarzan* sp. n., proposed as a flagship species for the creation of new nature reserves in Madagascar. *Salamandra* 46:167–179.

Gehring, P.S., F.M. Ratsoavina, M. Vences, and F. Glaw. 2011. *Calumma vohibola*, a new chameleon species (Squamata: Chamaeleonidae) from the littoral forests of eastern Madagascar. *African Journal of Herpetology* 60(2):130–154.

Gehring, P.-S., K.A. Tolley, F.S. Eckhardt, T.M. Townsend, T. Ziegler, F. Ratsoavina, F. Glaw, and M. Vences. 2012. Hiding deep in the trees: discovery of divergent mitochondrial lineages in Malagasy chameleons of the *Calumma nasutum* group. *Ecology and Evolution* 2:1468–1479.

Germershausen, G. 1913. Anatomische Untersuchungen über den Kehlkopf der Chamaeleonen. *Sitzungsberichte der Gesellschaft naturforschender Freunde zu Berlin* 1913:462–535 [in German].

Gheerbrandt, E., and J.C. Rage. 2006. Palaeobiogeography of Africa: how distinct from Gondwana and Laurasia. *Palaeogeography, Palaeoclimatology, Palaeoecology* 241:224–246.

Gilmore, C.W. 1943. Fossil lizards of Mongolia. *Bulletin of the American Museum of Natural History* 81(4):361–384.

Girdler, R.W., and P. Styles. 1978 Seafloor spreading in the western Gulf of Aden. *Nature* 271(5646):615–617.

Girons, H.S. 1970. The pituitary gland, pp. 135–199. In C. Gans and T.S. Parsons, Eds., *Biology of the Reptilia. Volume 3. Morphology C.* New York: Academic Press.

Glaw, F., J. Köhler, T.M. Townsend, and M. Vences. 2012. Rivaling the world's smallest reptiles: discovery of miniaturized and microendemic new species of leaf chameleons (*Brookesia*) from northern Madagascar. *PLoS ONE* 7:e31314.

Glaw, F., J. Köhler, and M. Vences. 2009. A distinctive new species of chameleon of the genus *Furcifer* (Squamata: Chamaeleonidae) from the Montagne d'Ambre rainforest of northern Madagascar. *Zootaxa* 2269:32–42.

Glaw, F., and M. Vences. 2007. *A Field Guide to the Amphibians and Reptiles of Madagascar*, 3rd ed. Köln, Germany: Vences and Glaw.

Glaw, F., M. Vences, T. Ziegler, W. Böhme, and J. Köhler. 1999. Specific distinctness and biogeography of the dwarf chameleons *Brookesia minima*, *B. peyrierasi* and *B. tuberculata* (Reptilia: Chamaeleonidae): evidence from hemipenal and external morphology. *Journal Zoology London* 247:225–238.

Gnanamuthu, C.P. 1930. The anatomy and mechanism of the tongue of *Chamaeleon carcaratus* (Merrem). *Proceedings of the Zoological Society of London* 31:467–486.

Gnanamuthu, C.P. 1937. Comparative study of the hyoid and tongue of some typical genera of reptiles. *Proceedings of the Zoological Society of London B* 107(1):1–63.

Goldby, F., and H.J. Gamble. 1957. The reptilian cerebral hemispheres. *Biological Reviews of the Cambridge Philosophical Society* 32:383–420.

Gonwouo, L.N., M. LeBreton, C. Wild, L. Chiro, P. Ngassam, and M.N. Tchamba. 2006. Geographic and ecological distribution of the endemic montane chameleons along the Cameroon mountain range. *Salamandra* 42:213–230.

Goodman, B.A., Miles, D.B., and L. Schwarzkopf. 2008. Life on the rocks: habitat use drives morphological and performance evolution in lizards. *Ecology* 89:3462–3471.
Goodman, S.M., and J.P. Benstead. 2003. *The Natural History of Madagasar.* Chicago: University of Chicago Press.
Goodman, S.M., and J.P. Benstead. 2005. Updated estimates of biotic diversity and endemism for Madagascar. *Oryx* 39:73–77.
Gordon, D.H., W. D. Haacke, and N.H.G. Jacobsen. 1987. Chromosomal studies of relationships in Gekkonidae, Chamaeleonidae and Scincidae in South Africa (abstract in Proceedings of the first HAA conference, Stellenbosch). *Journal of the Herpetological Association of Africa* 36:77.
Gray, J.E. 1865. Revision of the genera and species of Chamaeleonidae with the description of some new species. *Proceedings of the Zoological Society of London*1864:465–479.
Greenbaum, E., K.A. Tolley, A. Joma, and C. Kusamba. 2012. A new species of chameleon (Sauria: Chamaeleonidae: *Kinyongia*), from the Northern Albertine Rift, Central Africa. *Herpetologica* 68(1):60–75.
Griffiths, C.J. 1993. The geological evolution of East Africa, pp. 9–21. In J.C. Lovett and S.K. Wasser, Eds., *Biogeography and Ecology of the Rain Forests of Eastern Africa.* Cambridge, United Kingdom: Cambridge University Press.
Gugg, W. 1939. Der Skleralring der plagiotremen Reptilien. *Zoologische Jahrbücher. Abteilung für Anatomie und Ontogenie der Tiere* 65:339–416 [in German].
Gundy, G.C,. and G.Z. Wurst. 1976. The occurrence of parietal eyes in recent Lacertilia (Reptilia). *Journal of Herpetology* 10:113–121.
Guppy, M., and W. Davison. 1982. The hare and the tortoise: metabolic strategies in cardiac and skeletal muscles of the skink and the chameleon. *Journal of Experimental Zoology* 220:289–295.
Haagner, G.V., and W.R. Branch. 1993. Notes on predation on some Cape dwarf chameleons. *The Chameleon* 1:9–10.
Haas, G. 1937. The structure of the nasal cavity in *Chamaeleo chamaeleon* (Linnaeus). *Journal of Morphology* 61(3):433–451.
Haas, G. 1947. Jacobsons organ in the chameleon. *Journal of Morphology* 81(2):195–207.
Haas, G. 1952. The fauna of layer B of the Abu Usba Cave. *Israel Exploration Journal* 2:35–47.
Haas, G. 1973. Muscles of the Jaws and Associated Structures in the Rhynchocephalia and Squamata, pp. 285–490. In C. Gans and T.S. Parsons, Eds., *Biology of the Reptilia. Volume 4. Morphology D.* New York: Academic Press.
Hagey, T.J., J.B. Losos, and L.J. Harmon. 2010. Cruise foraging of invasive chameleon (*Chamaeleo jacksonii xantholophus*) in Hawai'i. *Breviora* 519:1–7.
Haines, R.W. 1952. The shoulder joint of lizards and the primitive reptilian shoulder mechanism. *Journal of Anatomy* 86:412–422.
Haker, H., H. Misslich, M. Ott, M.A. Frens, V. Henn, K. Hess, and P.S. Sandor. 2003. Three-dimensional vestibular eye and head reflexes of the chameleon: characteristics of gain and phase and effects of eye position on orientation of ocular rotation axes during stimulation in yaw direction. *Journal of Comparative Physiology A* 189: 509–517.
Hale, M.E. 1996. Functional morphology of ventral tail bending and prehensile abilities of the seahorse, *Hippocampus kuda. Journal of Morphology* 227:51–65.
Hall, J., Burgess, N.D., Lovett, J., Mbilinyi, B., and R.E. Gereau. 2009. Conservation implications of deforestation across an elevational gradient in the Eastern Arc Mountains, Tanzania. *Biological Conservation* 142:2510–2521.
Hallermann, J. 1994. Zur morphologie der ethmoedalregion der Iguania (Squamata); eine vergleichend-anatomische Untersuchung. *Bonner Zoologische Monographien* 35:1–133 [in German with English summary].

Halpern, M. 1992. Nasal chemical senses in reptiles: Structure and function. Pp 424–532 in C. Gans and D. Crews, Eds., *Biology of the Reptilia, Volume 18, Physiology E*. Chicago: University of Chicago Press.

Harkness, L. 1977. Chameleons use accommodation cues to judge distance. *Nature* 267(5609):346–349.

Hart, N.S. 2001. The visual ecology of avian photoreceptors. *Progress in Retinal and Eye Research* 20:675–703.

Hawlitschek, O., B. Brückmann, J. Berger, K. Green, and F. Glaw. 2011. Integrating field surveys and remote sensing data to study distribution, habitat use, and conservation status of the herpetofauna of the Comoro Islands. *Zookeys* 144:21–79.

Hazard, L.C. 2004. Sodium and potassium secretion by Iguana salt glands, pp. 84–93. In A.C. Alberts, R.L. Carter, W.K. Hayes and E.P. Martins, Eds. *Iguanas: Biology and Conservation*. Berkeley: University of California Press.

Heads, M. 2005. Dating nodes on molecular phylogenies: a criteque of molecular biogeography. *Cladistics 21:62–78.*

Hébert, H., C. Deplus, P. Huchon, K. Khanbari and L. Audin. 2001. Lithospheric structure of a nascent spreading ridge inferred from gravity data: the western Gulf of Aden *Journal of Geophysical Research* 106:B11.

Hebrard, J.J. 1980. Habitats and sleeping perches of three species of chameleon in Kenya. *American Zoology* 20:842.

Hebrard, J.J., and T. Madsen. 1984. Dry season intersexual habitat partitioning by flap-necked chameleons (*Chamaeleo dilepis*) in Kenya. *Biotropica* 16:69–72.

Hebrard, J.L., S.M. Reilly, and M. Guppy. 1982. Thermal ecology of *Chameleo hoehnelii* and *Mabuya varia* in the Aberdare mountains: constraints of heterothermy in an alpine habitat. *Journal of the East African Natural History Society* 176:1–6.

Hecht, M., and R. Hoffstetter. 1962. Note préliminaire sur les amphibiens et les squamates du Landenien supérieur et du Tongrien de Belgique. *Bulletin de l'Institut Royal des Sciences Naturelles de Belgique* 39:1–30 [in French].

Hedges, B.S., and N. Vidal. 2009. Lizards, snakes, and amphisbaenians (Squamata), pp. 383–389. In B.S. Hedges and S. Kumar, Eds., *The Timetree of Life*. New York: Oxford University Press.

Herrel, A. 2007. Herbivory and foraging mode in lizards, pp. 209-236 In S.M. Reilly, L.D. McBrayer and D.B. Miles, Eds., *Lizard Ecology: The evolutionary consequences of foraging mode*. Cambridge: Cambridge University Press.

Herrel, A., S.M. Deban, V. Schaerlaeken, J.-P. Timmermans, and D. Adriaens. 2009. Are morphological specializations of the hyolingual system in chameleons and salamanders tuned to demands on performance? *Physiological and Biochemical Zoology* 82(1):29–39.

Herrel, A., R.S. James, and R. Van Damme. 2007a. Fight versus flight: Physiological basis for temperature-dependent behavioral shifts in lizards. *Journal of Experimental Biology* 210 (10):1762–1767.

Herrel, A., G.J. Measey, B. Vanhooydonck, and K.A. Tolley. 2011. Functional consequences of morphological differentiation between populations of the Cape Dwarf Chameleon (*Bradypodion pumilum*). *Biological Journal of the Linnean Society* 104:692–700.

Herrel, A., G.J. Measey, B. Vanhooydonck, and K.A. Tolley. 2012. Got it clipped? The effect of tail clipping on tail gripping performance in chameleons. *Journal of Herpetology* 46 (1):91–93.

Herrel, A., J.J. Meyers, P. Aerts, and K.C. Nishikawa. 2000. The mechanics of prey prehension in chameleons. *Journal of Experimental Biology* 203(21):3255–3263.

Herrel, A., J.J. Meyers, P. Aerts, and K.C. Nishikawa. 2001a. Functional implications of supercontracting muscle in the chameleon tongue retractors. *Journal of Experimental Biology* 204 (21):3621–3627.

Herrel, A., J.J. Meyers, K.C. Nishikawa, and F. De Vree. 2001b. Morphology and histochemistry of the hyolingual apparatus in chameleons. *Journal of Morphology* 249(2):154–170.

Herrel, A., J.J. Meyers, J.-P. Timmermans, and K.C. Nishikawa. 2002. Supercontracting muscle: producing tension over extreme muscle lengths. *Journal of Experimental Biology* 205: 2167–2173.

Herrel, A., V. Schaerlaeken, J.J. Meyers, K.A. Metzger, and C.F. Ross. 2007b. The evolution of cranial design and performance in squamates: consequences of skull-bone reduction on feeding behavior. *Integrative and Comparative Biology* 47:107–117.

Herrel, A., K.A. Tolley, G.J. Measey, J.M. daSilva, D.F. Potgieter, R. Biostel, and B. Vanhooydonck. 2013. Slow but tenacious: an analysis of running and gripping performance in chameleons. *Journal of Experimental Biology* 216:1025–1030.

Herrmann, P.A., and H.W. Herrmann. 2005. Egg and clutch characteristics of the mountain chameleon, *Chamaeleo montium*, in southwestern Cameroon. *Journal of Herpetology* 39:154–157.

Higham, T.E., M.S. Davenport, and B.C. Jayne. 2001. Maneuvering in an arboreal habitat: the effects of turning angle on the locomotion of three sympatric ecomorphs of *Anolis* lizards. *Journal of Experimental Biology* 204 (23):4141–4155.

Higham, T.E., and B.C. Jayne. 2004a. *In vivo* muscle activity in the hindlimb of the arboreal lizard, *Chamaeleo calyptratus*: general patterns and effects of incline. *Journal of Experimental Biology* 207(2):249–261.

Higham, T.E., and B.C. Jayne. 2004b. Locomotion of lizards on inclines and perches: hindlimb kinematics of an arboreal specialist and a terrestrial generalist. *Journal of Experimental Biology* 207(2):233–248.

Higham, T.E., and A.P. Russell. 2010. Divergence in locomotor performance, ecology, and morphology between two sympatric sister species of desert-dwelling gecko. *Biological Journal of the Linnean Society* 101:860–869.

Hill, A.V. 1950. The dimensions of animals and their muscular dynamics. *Science Progress* 38:209–230.

Hillenius, D. 1959. The differentiation within the genus *Chamaeleo* Laurenti 1768. *Beaufortia*, 8(89):1–92.

Hillenius, D. 1978a. Notes on chameleons. IV: A new chameleon form the Miocene of Fort Ternan, Kenya (Chamaeleonidae, Reptilia). *Beaufortia* 28:9–15.

Hillenius, D. 1978b. Notes on chameleons. V: The chameleons of north Africa and adjacent countries, *Chamaeleo chamaleon* (Linnaeus) (Sauria, Chamaeleonidae). *Beaufortia* 28:37–55.

Hillenius, D. 1986. The relationship of *Brookesia, Rhampholeon* and *Chamaeleo* (Chamaeleonidae, Reptilia). *Bijdragen tot de Dierkunde* 56(1):29–38.

Hillenius, D. 1988. The skull of *Chamaeleo nasutus* adds more information to the relationship of *Chamaeleo* with *Rhampholeon* and *Brookesia* (Chamaeleonidae, Reptilia). *Bijdragen Tot De Dierkunde* 58(1):7–11.

Hockey, P.A.R., W.R.J. Dean, and P.G. Ryan. 2005. *Roberts—Birds of Southern Africa*, 7th ed. Cape Town, South Africa: Trustees of the John Voelcker Bird Book Fund.

Hódar, J.A., J.M. Pleguezuelos, and J.C. Poveda. 2000. Habitat selection of the common chameleon (*Chamaeleo chamaeleon*) (L.) in an area under development in southern Spain: implications for conservation. *Biological Conservation* 94: 63–68.

Hofer, U., H. Baur, and L.-F. Bersier. 2003. Ecology of three sympatric species of the genus *Chamaeleo* in a tropical upland forest in Cameroon. *Journal of Herpetology* 37(1):203–207.

Hoffmann, M., C. Hilton-Taylor, A. Angulo, M. Böhm, T.M. Brooks, S.H.M. Butchart, K.E. Carpenter, J. Chanson, B. Collen, N.A. Cox, et al. 2010. The impact of conservation on the status of the world's vertebrates. *Science* 330:1503–1509.

Hoffstetter, R. 1967. Coup d'oeil sur les Sauriens (Lacertiliens) des couches de Purbeck (Jurassique supérieur d'Angleterre, Résumé d'un mémoire). *Colloque international du CNRS* 163:349–371 [in French].

Hoffstetter, R., and J.-P.Gasc. 1969. Vertebrae and Ribs of Modern Reptiles. Pp. 201–310 in C. Gans, Ed., *Biology of the Reptilia. Volume 1. Morphology A.* New York: Academic Press.

Hofman, A., L.R. Maxon, and J.W. Arntzen. 1991. Biochemical evidence pertaining to the taxonomic relationships within the family Chamaeleonidae. *Amphibia-Reptilia* 12:245–265.

Hogben, L., and D. Slome. 1931. The pigmentary effector system VI. The dual character of endocrine co-ordination in amphibian color change. *Proceedings of the Royal Society of London, Series B—Biological Sciences* 108:10–53.

Hogben, L.T., and L. Mirvish. 1928. The pigmentary effector system. V. The nervous control of excitement pallor in reptiles. *Journal of Experimental Biology* 5:295–308.

Holmes, RB., A.M. Murray, P. Chatrath, Y.S. Attia, and E.L. Simons. 2010. Agamid lizard (Agamidae: Uromastycinae) from the lower Oligocene of Egypt. *Historical Biology* 22:215–223.

Honda, M., H. Ota, M. Kobayashi, J. Nabhitabhata, H.-S. Yong, S. Sengoku, and T. Hikida. 2000. Phylogenetic relationships of the family Agamidae (Reptilia: Iguania) inferred from mitochondrial DNA sequences. *Zoological Science* 17:527–537.

Hooijer, D.A. 1961. The fossil vertebrates of Ksâr'Akil, a Palaeolithic rock shelter in the Lebanon. *Zoologische Verhandellingen* 49:3–67.

Hopkins, K.P., and K.A. Tolley. 2011. Morphological variation in the Cape Dwarf Chameleon (*Bradypodion pumilum*) as a consequence of spatially explicit habitat structure differences. *Biological Journal of the Linnean Society* 102(4):878–888.

Hou, L. 1974. Paleocene Lizards from Anhui, China. *Vertebrata PalAsiatica* 12(3):193–202.

Hou, L. 1976. New Materials of Palaeocene Lizards of Anhui. *Vertebrata PalAsiatica* 14(1):48–52.

Houniet, D.T., W. Thuiller, and K.A. Tolley. 2009. Potential effects of predicted climate change on the endemic South African Dwarf Chameleons, *Bradypodion. African Journal of Herpetology* 59:28–35.

Houston, J. 1828. On the structure and mechanism of the tongue of the chameleon. *Transactions of the Royal Irish Academy* 15:177–201.

Huey, R.B., and A.F. Bennett. 1987. Phylogenetic studies of coadaptation: Preferred temperatures versus optimal performance temperatures of lizards. *Evolution* 41 (5):10 98–1115.

Huey, R. B., C. A. Deutsch, J. J. Tewksbury, L. J. Vitt, P. E. Hertz, H. J. Álvarez-Pérez, and T. Garland Jr. 2009. Why tropical forest lizards are vulnerable to climate warming. *Proceedings of the Royal Society London, B* 276:1939–1948.

Huey, R.B., and E.R. Pianka. 1981. Ecological consequences of foraging mode. *Ecology* 62:991–999.

Huey, R.B., and R.D. Stevenson. 1979. Integrating thermal physiology and ecology of ectotherms: A discussion of approaches. *American Zoologist* 19:357–366.

Hugall, A.F., R. Foster, M. Hutchinson, and M.S.Y. Lee. 2008. Phylogeny of Australian agamid lizards based on nuclear and mitichondrial genes: implications for morphological evolution and biogeography. *Biological Journal of the Linnean Society* 93:343–358.

Hugall, A.F., and M.S.Y. Lee. 2004. Molecular claims of Gondwanan age for Australian agamid lizards are untenable. *Molecular Biology and Evolution* 21(11):2102–2110.

Humphreys C.W. 1990. Observations on nest excavations, egg laying and the incubation period of Marshall's Dwarf Chameleon *Rhampholeon marshalli* Boulenger 1906. *Zimbabwe Science News* 24(1/3):3–4.
Hunt, D.M., S.E. Wilkie, J.K. Bowmaker, and S. Poopalasundaram. 2001. Vision in the ultraviolet. *Cellular and Molecular Life Sciences* 58:1583–1598.
Hurle, J.M., Garcia-Martinez, V., Ganan, Y., Climent, V. and M. Blasco. 1987. Morphogenesis of the prehensile autopodium in the common chameleon (*Chamaeleo chamaeleo*). *Journal of Morphology* 194 (2):187–194.
Hutchinson, M.N., A. Skinner, and M.S.Y. Lee. 2012. *Tikiguania* and the antiquity of squamate reptiles (lizards and snakes). *Biology Letters* 8 (4):665–669.
Ingram, J.C., and T.P. Dawson. 2005. Climate change impacts and vegetation response on the island of Madagascar. *Philosophical Transactions of the Royal Society A* 363:55–59.
Intergovernmental Panel on Climate Change (IPCC). 2007. *Fourth Assessment Report: Climate Change 2007, The Physical Science Basis*. Cambridge, United Kingdom: Cambridge University Press.
Intergovernmental Panel on Climate Change (IPCC). 2011. IPCC SREX Summary for Policymakers. Accessed at www.ipcc.ch/news_and_events/docs/ipcc34/SREX_FD_SPM_final.pdf on November 21, 2011.
Irschick, D.J., C.C. Austin, K. Petren, R.N. Fisher, J.B. Losos, and O. Ellers. 1996. A comparative analysis of clinging ability among pad-bearing lizards. *Biological Journal of the Linnean Society* 59:21–35.
Irschick, D.J., and J.B. Losos. 1998. A comparative analysis of the ecological significance of maximal locomotor performance in Caribbean *Anolis* lizards. *Evolution* 52:219–226.
Irschick, D.J., T.E. Macrini, S. Koruba, and J. Forman. 2000. Ontogenetic differences in morphology, habitat use, behavior, and sprinting capacity in two West Indian Anolis lizards. *Journal of Herpetology* 34(3):444–451.
Irwin, M.T., P.C. Wright, C. Birkinshaw, B.L. Fisher, C.J. Gardner, J. Glos, S.M. Goodman, P. Loiselle, P. Rabeson, J.-L. Raharison, M.J. Raherilalao, D. Rakotondravony, A. Raselimanana, J. Ratsimbazafy, J.S. Sparks, L. Wilmé, L., and J.U. Ganzhorn. 2010. Patterns of species change in anthropogenically disturbed forests of Madagascar. *Biological Conservation* 143:2351–2362.
IUCN. 2012. IUCN Red List of Threatened Species. Version 2012.1. Accessed at www.iucnredlist.org on June 19, 2012.
Jackson, J.C. 2007. Reproduction in dwarf chameleons (*Bradypodion*) with particular reference to *B. pumilum* occurring in fire-prone fynbos habitat. Ph.D. thesis. University of Stellenbosch, South Africa.
Jackson, J.F. 1973. Distribution and population phenetics of the Florida scrub lizard, Sceoloporus woodi. *Copeia* 1973:746–761.
Jacobs, B.F. 2004. Palaeobotanical studies from tropical Africa: relevance to the evolution of forest, woodland and savannah biomes. *Philosophical Transactions of the Royal Society of London Series B-Biological Sciences* 359:1573–1583.
Janzen, D.H. 1967. Why mountain passes are higher in the tropics? *American Naturalist* 101:233–249.
Jenkins, R.K.B., L.D. Brady, M. Bisoa, J. Rabearivonyc, and R.A. Griffiths. 2003. Forest disturbance and river proximity influence chameleon abundance in Madagascar. *Biological Conservation* 109:407–415.
Jenkins, R.K.B., L.D. Brady, K. Huston, J.L.D. Kauffmann, J. Rabearivony, G. Raveloson, and M. Rowcliffe. 1999. The population status of chameleons within Ranomafana National Park, Madagascar. *Oryx* 33:38–47.

Jenkins, R.K.B., J. Rabearivony, and H. Rakotomanana. 2009. Predation on chameleons in Madagascar: a review. *African Journal of Herpetology* 58:131–136.
Jha, S., and K.S. Bawa. 2006. Population growth, human development, and deforestation in biodiversity hotspots. *Conservation Biology* 20:906–912.
Johnson, M.K., and A.P. Russell. 2009. Configuration of the setal fields of *Rhoptropus* (Gekkota: Gekkonidae): functional, evolutionary, ecological and phylogenetic implications of observed pattern. *Journal of Anatomy* 214:937–955.
Jollie, M. 1962. *Chordate Morphology*. New York: Reinhold Publishing.
Joshi, M., and B.S. Kotlia. 2010. First Report of the Late Pleistocene fossil lizards from Narmada Basin, Central India. *Earth Science India* 3(1):1–8.
Källén, B. 1951a. Contributions to the knowledge of the medial wall of the reptilian forebrain. *Acta Anatomy* 13:90–100.
Källén, B. 1951b. On the ontogeny of the reptilian forebrain. Nuclear structures and ventricular sulci. *Journal of Comparative Neurology* 95:307–347.
Kaloloha, A., C. Misandeau, and P.-S. Gehring. 2011. Notes on the diversity and natural history of the snake fauna of Ambodiriana—Manompana, a protected rainforest site in north-eastern Madagascar. *Herpetology Notes* 4:397–402.
Karsten, K.B., L.N. Andriamandimbiarisoa, S.F. Fox, and C.J. Raxworthy. 2008. A unique life history among tetrapods: An annual chameleon living mostly as an egg. *Proceedings of the National Academy of Sciences of the United States of America* 105:8980–8984.
Karsten, K.B., L.N. Andriamandimbiarisoa, S.F. Fox, and C.J. Raxworthy. 2009b. Population densities and conservation assessments for three species of chameleons in the Toliara region of southwestern Madagascar. *Amphibia-Reptilia* 30:341–350.
Karsten, K.B., L.N. Andriamandimbiarisoa, S.F. Fox, and C.J. Raxworthy. 2009c. Social behavior of two species of chameleons in Madagascar: insights into sexual selection. *Herpetologica* 65:54–69.
Karsten, K.B., G.W. Ferguson, T.C. Chen, and M.F. Holick. 2009a. Panther chameleons, *Furcifer pardalis*, behaviorally regulate optimal exposure to UV on dietary vitamin D3 status. *Physiological and Biochemical Zoology* 82:218–225.
Kashyap, H.V. 1960. Morphology of the reptilian heart. *Bulletin of the Zoological Society of India, Nagpur* 3:23–34.
Kassarov, L. 2003. Are birds the primary selective force leading to evolution of mimicry and aposematism in butterflies? An opposing point of view. *Behaviour* 140:433–451.
Kathariner, L. 1894. Anatomie und Mechanismus der Zunge der Vermilinguier. *Jenaische Zeitschrift für Medizin und Naturwissenschaft* 29:247–270 [in German].
Kauffmann, J.L.D., L.D. Brady, and R.K.B. Jenkins. 1997. Behavioural observations of the chameleon *Calumma oshaughnessyi oshaughnessyi* in Madagascar. *Herpetological Journal* 7:77–80.
Kearney, M., and W. Porter. 2009. Mechanistic niche modelling: combining physiological and spatial data to predict species' ranges. *Ecology Letters* 12:334–350.
Kelso, E.C., and P.A. Verrell. 2002. Do male veiled chameleons, *Chamaeleo calyptratus*, adjust their courtship displays in response to female reproductive status? *Ethology* 108:495–512.
Keren-Rotem, T., A. Bouskila, and E. Geffen. 2006. Ontogenetic habitat shift and risk of cannibalism in the common chameleon (*Chamaeleo chamaeleon*). *Behavioral Ecology and Sociobiology* 59:723–731.
Kirmse, W., R. Kirmse, and E. Milev. 1994. Visuomotor operation in transition from object fixation to prey shooting in chameleons. *Biological Cybernetics* 71:209–214.

Klaver, C. 1979. A review of *Brookesia* systematics with special reference to lung morphology. *Bonner Zoologische Beiträge* 30:163–175.

Klaver, C., and W. Böhme. 1986. Phylogeny and classification of the Chamaeleonidae (Sauria) with special reference to hemipenis morphology. *Bonner Zoologische Monographien* 22:1–64.

Klaver, C., and W. Böhme. 1992. The species of the *Chamaeleo cristatus* group from Cameroon and adjacent countries, West Africa. *Bonn Zoological Bulletin* 43:433–476.

Klaver, C.J.J. 1973. Lung anatomy: aid in chameleon-taxonomy. *Beaufortia* 20(269):155–177.

Klaver, C.J.J. 1977. Comparative lung-morphology in the genus *Chamaeleo* Laurenti, 1768 (Sauria: Chamaeleonidae) with a discussion of taxonomic and zoogeographic implications. *Beaufortia* 25(327):167–199.

Klaver, C.J.J. 1979. A review of *Brookesia* systematics with special reference to lung morphology. *Bonner Zoologishe Beiträge Heft* 1–2(30):163–175.

Klaver, C.J.J. 1981. Lung morphology in the Chamaeleonidae (Sauria) and its bearing upon phylogeny, systematics and zoogeography. *Zeitschrift fuer Zoologische Systematik und Evolutionsforschung* 19:36–58.

Klaver, C.J.J., and W. Böhme. 1997. Chamaeleonidae. *Das Tierreich* 112, I-XV:1–85.

Knoll, A., F. Glaw, and J. Köhler. 2009. The Malagasy snake *Pseudoxyrhopus ambreensis* preys upon chameleon eggs by shell slitting. *Herpetology Notes* 2:161–162.

Koreny, L. 2006. *Phylogeny of East-African chameleons.* MSc thesis, Faculty of Biological Sciences, University of South Bohemia, Ceske Budejovice.

Kosuch, J., M. Vences, and W. Böhme. 1999. Mitochondrial DNA sequence data support the allocation of Greek mainland chameleons to *Chamaeleo africanus. Amphibia-Reptilia* 20:440–443.

Kraus, F., A. Medeiros, D. Preston, C.S. Jarnevich, and G.H. Rodda. 2012. Diet and conservation implications of an invasive chameleon, *Chamaeleo jacksonii* (Squamata: Chamaeleonidae) in Hawaii. *Biological Invasions* 14:579–593.

Krause, C., and M.S. Fischer. 2013. Biodynamics of climbing: effects of substrate orientation on the locomotion of a highly arboreal lizard (*Chamaeleo calyptratus*). *Journal of Experimental Biology* 216(18)1448–1457.

Krause, D.W., S.E. Evans, and K. Gao. 2003. First definitive record of a Mesozoic lizard from Madagascar. *Journal of Vertebrate Paleontology* 23(4):842–856.

Krause, D.W., R.R. Rogers, C.A. Forster, J.H. Hartman, J.H. Buckley, and S.D. Sampson. 1999. The Late Cretaceous vertebrate fauna of Madagascar: implications for Gondwanan paleobiogeography. *GSA Today* 9:1–7.

Kumazawa, Y. 2007. Mitochondrial genomes from major lizard families suggest their phylogenetic relationships and ancient radiations. *Gene* 388:19–26.

Laffan, S.W., E. Lubarsky, and D.F. Rosauer. 2010. Biodiverse, a tool for the spatial analysis of biological and related diversity. *Ecography* 33:643–647 (version 0.14).

Lakjer, T. 1926. Studien über die Trigeminus-versorgte Kaumuskulatur der Sauropsiden. Copenhagen: C.A. Reitzel [in German].

Land, M.F. 1995. Fast-focus telephoto eye. *Nature* 373:658–659.

Largen, M.J., and S. Spawls. 2010. The amphibians of Ethiopia and Eritrea. Frankfurt am Main, Germany: Edition Chimaira.

Le Berre, F. 1995. *The new chameleon handbook.* Barron's: Hong Kong, China.

Le Gall, B., P. Nonnotte, J. Rolet, M. Benoit, H. Guillou, M. Mousseau-Nonnotte, J. Albaric, and J. Deverchère. 2008. Rift propagation at craton margin: distribution of faulting and volcanism in the North Tanzanian divergence (East Africa) during Neogene times. *Tectonophysics* 448:1–19.

Leakey, L.S.B. 1965. *Olduvai Gorge 1951–1961. Vol.1. A preliminary report on the geology and fauna.* Cambridge, United Kingdom: Cambridge University Press.
Leblanc, E. 1924. Les muscles orbitaires des reptiles. Étude des muscles chez *Chameleo vulgaris. Comptes Rendus de l'Académie des Sciences Paris* 179:996–998 [in French].
Leblanc, E. 1925. Les muscles orbitaires des reptiles. Étude des muscles chez *Chamaeleo vulgaris. Bulletin de la Société d'Histoire Naturelle d'Afrique du Nord* 16 :49–61 [in French].
Lecuru, S. 1968a. Etude des variations morphologiques du sternum, des clavicules et de l'interclavicule des lacertiliens. *Annales des Sciences Naturelles: Zoologie et Biologie Animale. Série 12* 10:511–544 [in French].
Lecuru, S. 1968b. Remarques sur le scapulo-coracoide des lacertiliens. *Annales des Sciences Naturelles: Zoologie et Biologie Animale. Série 12* 10:475–510 [in French].
Lee, D.-C., A.N. Halliday, J.G. Fitton, and G. Poli. 1994. Isotopic variations with distance and time in the volcanic islands of the Cameroon line: evidence for a mantle plume origin. *Earth and Planetary Science Letters* 123:119–138.
Leidy, J. 1872. Remarks on fossils from Wyoming. *Proceedings of the Natural Academy of Sciences of Philadelphia* 1872:122.
Leidy, J. 1873. Contributions to the extinct vertebrate fauna of western territories. *Report of the United States Geological Survey of the Territories* 1:14–358.
Lever, C. 2003. *Naturalized Reptiles and Amphibians of the World.* New York: Oxford University Press.
Li, J. 1991a. Fossil reptiles from Hetaoyuan Formation, Xichuan, Henan. *Vertebrata PalAsiatica* 29(3):190–203.
Li, J. 1991b. Fossil reptiles from Zhaili Member, Hedi Formation, Yuanqu, Shanxi. *Vertebrata PalAsiatica* 29(4):276–285.
Li, P.P., K. Gao, L.-H. Hou, and X. Xu. 2007. A gliding lizard from the Early Cretaceous of China. *Proceedings of the National Academy of Sciences of the United States of America* 104(13):5507–5509.
Lin, E.J.I., and C.E. Nelson. 1981. Comparative reproductive biology of two sympatric tropical lizards, *Chamaeleo jacksonii* Boulenger and *Chamaeleo hoehnelii* Steindachner (Sauria: Chamaeleonidae). *Amphibia-Reptilia* 3/4:287–311.
Lin, J. 1980. Desiccation tolerance and thermal maxima in the lizards. *Chamaeleo jacksoni* and *C. hohneli. Copeia* 1980:363–366.
Lin, J., and C.E. Nelson. 1980. Comparative reproductive biology of two sympatric tropical lizards *Chamaeleo jacksonii* Boulenger and *Chamaeleo hoehnelii* Steindachner (Sauria: Chamaeleonidae). *Amphibia-Reptilia* 1:287–311.
Linder, H.P., H.M. de Klerk, J. Born, N.D. Burgess, J. Fjeldså, and C. Rahbek. 2012. The partitioning of Africa: statistically defined biogeographical regions in sub-Saharan Africa. *Journal of Biogeography* 39:1189–1205.
Linder, H.P., J. Lovett, J.M. Mutke, W. Barthlott, N. Jürgens, T. Rebelo, and W. Küper. 2005. A numerical re-evaluation of the sub-Saharan phytochoria of mainland Africa. *Biologiske Skrifter* 55:229–252.
Lloyd, C.N.V. 1974. Feeding behaviour in the green mamba, *Dendroaspis angusticpes* (A. Smith). *Journal of the Herpetological Association of Africa* 12:1–12.
Loader, S.P., D.J. Gower, K.M. Howell, N. Doggart, M.O. Rödel, B.T. Clarke, R.O. de Sá, B.L. Cohen, and M. Wilkinson. 2004. Phylogenetic relationships of African Microhylid frogs inferred from DNA sequences of mitochondrial 12S and 16S ribosomal rRNA genes. *Organisms Diversity and Evolution* 4:227–235.
Losos, J.B. 1990. The evolution of form and function: morphology and locomotor performance in West Indian *Anolis* lizards. *Evolution* 44(5):1189–1203.

Losos, J.B., and D.L. Mahler. 2011. Adaptive radiation: the interaction of ecological opportunity, adaptation, and speciation, pp. 381–420. In M.A. Bell, D.J. Futuyma, W.F. Eanes and J.S. Levinton, Eds., *Evolution Since Darwin: The First 150 Years*. Sunderland, MA: Sinauer Associates.

Losos, J.B., and B. Sinervo. 1989. The effects of morphology and perch diameter on sprint performance of *Anolis* lizards. *Journal of Experimental Biology* 145:23–30.

Losos, J.B., B.M. Walton, and A.F. Bennett. 1993. Trade-offs between sprinting and clinging ability in Kenyan chameleons. *Functional Ecology* 7:281–286.

Loveridge, A. 1923. Notes on East African snakes, collected 1918–1923. *Proceedings of the Zoological Society of London* 1923:871–897.

Loveridge, A. 1953. Zoological results of a fifth expedition to East Africa III. Reptiles from Nyasaland and Tete. *Bulletin of the Museum of Comparative Zoology* 110:143–322.

Loveridge A. 1957. Checklist of the reptiles and amphibians of East Africa (Uganda, Kenya, Tanganyika, Zanzibar). *Bulletin of the Museum of Comparative Zoology (Harvard)* 117(2):153–362.

Lovett, J.C. 1993. Climatic history and forest distribution in eastern Africa. Pp. 23–29 in J.C. Lovett and S. Wasser, Eds., *Biogeography and ecology of the rain forests of Eastern Africa*. Cambridge, United Kingdom: Cambridge University Press.

Lovett J.C. and S.K. Wasser. 1993. *Biogeography and ecology of the rain forests of eastern Africa*. Cambridge University Press: Cambridge.

Lowin, A.J. 2012. Chameleon species composition and density estimates of three unprotected dry deciduous forests between Montagne d'Ambre Parc National and Ankarana Réserve Spéciale in northern Madagascar. *Herpetology Notes* 5:107–113.

Lubosch, W. 1932. Bemerkungen über die Zungenmuskulatur des Chamäleons. *Morphologisches Jahrbuch* 71:158–170 [in German].

Lubosch, W. 1933. Untersuchunden über die Visceralmuskulatur der Sauropsiden. *Gegenbaurs. Morphologisches Jahrbuch* 72:584–666 [in German].

Luiselli, L. 2006. Nonrandom co-occurrence patterns of rainforest chameleons. *African Journal of Ecology* 45:336–346.

Luiselli, L., F.M. Angelici, and G.C. Akani. 2000. Reproductive ecology and diet of the Afro-tropical tree snake *Rhamnophis aethiopissa* (Colubridae). *Herpetological Natural History* 7:163–171.

Luiselli, L., G.C. Akani, and F.M. Angelici. 2001. Diet and foraging behaviour of three ecologically little-known African forest snakes: *Meizodon coronatus, Dipsadoboa duchesnei* and *Hapsidophrys lineatus. Folia Zoologica* 50:151–158.

Luiselli, L., and L. Rugiero. 1996. *Chamaeleo chamaeleon*. Diet. *Herpetological Review* 27:78–79.

Luppa, H. 1977. Histology of the digestive tract, pp. 225–313. In C. Gans and T.S. Parsons, Eds., *Biology of the Reptilia. Volume 6. Morphology E.* New York: Academic Press.

Lutz, G.J., and L.C. Rome. 1996. Muscle function during jumping in frogs, II. Mechanical properties of muscle: implications for system design. *American Journal of Physiology* 271(2 Pt 1):C571-C578.

Lutzmann, N. 2000. Phytophagie bei Chamäleons. *Draco* 1:82.

Lutzmann, N. 2004. Females carrying males in chameleon courtship. *Reptilia (GB)* 35:34–36.

Lynn, W.G. 1970. The thyroid, pp. 201–234. In C. Gans and T.S. Parsons, Eds., *Biology of the Reptilia. Volume 3. Morphology C.* New York: Academic Press.

Lynn, W.G., and G.A. Walsh. 1957. The morphology of the thyroid gland in the Lacertilia. *Herpetologica* 13(3):157–162.

Macey, J.R., Kuehl, J.V., Larson, A., Robinson, M.D., Ugurtas, I.H., Ananjeva, N.B., Rahman, H., Javed, H.I., Osman, R.M., Doumma, A. and T.J. Papenfuss. 2008. Socotra Island the forgotten fragment of Gondwana: unmasking chameleon lizard history with complete mitochondrial genomic data. *Molecular Phylogenetics and Evolution* 49:1015–8.

Macey, J.R., A. Larson, N.B. Ananjeva, Z. Fang, and T.J. Papenfuss. 1997a. Two novel gene orders and the role of light-strand replication in rearrangement of the vertebrate mitochondrial genome. *Molecular Biology and Evolution* 14:91–104.

Macey, J.R., A. Larson, N.B. Ananjeva, and T.J. Papenfuss. 1997b. Evolutionary shifts in three major structural features of the mitochodrial genome among iguanian lizards. *Journal of Molecular Evolution* 44:660–674.

Macey, J.R., J.A. Schulte II, and A. Larson. 2000a. Evolution and phylogenetic information content of mitochondrial genomic structural features illustrated with acrodont lizards. *Systematic Biology* 49(2):257–277.

Macey, J.R., J.A. Schulte II, J.J. Fong, I. Das, and T. Papenfuss. 2006. The complete mitochondrial genome of an agamid lizards from the Afro-Asian subfamily Agaminae and the phylogenetic position of *Bufoniceps* and *Xenagama*. *Molecular Phylogenetics and Evolution* 39:881–886.

Macey, J.R., J.A. Schulte II, A. Larson, N.B. Ananjeva, Y. Wang, R. Pethiyagoda, N. Rastegar-Pouyani, and T.J. Papenfuss. 2000b. Evaluating trans-Tethys migration: an example using acrodont lizard phylogenetics. *Systematic Biology* 49(2):233–256.

Mackay, J.Y. 1886. The arterial system of the chamaeleon (*Chamaeleo vulgaris*). *Proceedings of the Philosophical Society of Glasgow* 17:353–365.

Macleod, N., P.F. Rawson, P.L. Forey, F.T. Banner, M.K. Boudagher-Fadel, P.R. Bown, J.A. Burnett, P. Chambers, S. Culver, S.E. Evans, C. Jeffery, M.A. Kaminski, A.R. Lord, A.C. Milner, A.R. Milner, N. Morris, E. Owen, B.R. Rosen, A.B. Smith, P.D. Taylor, E. Urquhart, and Y.R. Young. 1997. The Cretaceous-Tertiary biotic transition. *Journal of the Geological Society* 154:265–292.

Malan, M.E. 1945. Contributions to the comparative anatomy of the nasal capsule and the organ of Jacobson of the Lacertilia. *Annale van die Universiteit van Stellenbosch* 24:69–138.

Maley, J. 1996. The African rain forest-main characteristics of changes in vegetation and climate from the Upper Cretaceous to the Quaternary. *Proceedings of the Royal Society of Edinburgh Section B: Biology* 104:31–73.

Mariaux, J., N. Lutzmann, and J. Stipala. 2008. The two-horned chameleons of East Africa. *Zoological Journal of the Linnean Society*152:367–391.

Mariaux, J., and C.R. Tilbury. 2006. The pygmy chameleons of the Eastern Arc Range (Tanzania): evolutionary relationships and the description of three new species of *Rhampholeon* (Sauria: Chamaeleonidae). *Herpetological Journal* 16(3):315–331.

Markwick P.J., and P.J. Valdes. 2004. Palaeo-digital elevation models for use as boundary conditions in coupled ocean-atmosphere GCM experiments: a Maastrichtian (Late Cretaceous) example. *Palaeogeography, Palaeoclimatology, Palaeoecology* 213:37–63.

Marsh, O. 1872. Preliminary description of new Tertiary reptiles. Parts I and II. *American Journal of Science* 4:298–309.

Martin, J. 1992. *Masters of Disguise: A Natural History of Chameleons*. New York: Facts on File.

Massot, M., J. Clobert, and R. Ferriere. 2008. Climate warming, dispersal inhibition and extinction risk. *Global Change Biology* 14:461–469.

Masterson, A.N.B. 1994. Do flap-necked chameleons eat birds? *Honeyguide* 40:186.

Masterson, A.N.B. 1999. Another chameleon basher: the crested barbet. *Honeyguide* 45:142.

Mates, J.W.B. 1978. Eye movements of African chameleons: spontaneous saccade timing. *Science* 199:1087–1088.

Matthee, C.A., C.R. Tilbury, and T. Townsend. 2004. A phylogenetic review of the African leaf chameleons: genus *Rhampholeon* (Chamaeleonidae): the role of vicariance and climate change in speciation. *Proceedings of the Royal Society B* 271:1967–1975.

Matthey, R. 1957. Cytologie comparée et taxonomie des Chamaeleontidae (Reptilia - Lacertilia). *Revue suisse de zoologie* 64:709–732.

Matthey, R., and J.M. van Brink. 1956. Note preliminaire sur la cytologie chromosomique comparée des Caméléons. *Revue suisse de zoologie* 63:241–246.

Matthey, R., and J.M. van Brink. 1960. Nouvelle contribution à la cytology comparée des Chamaeleontidae (Reptilia-Lacertilia). *Bulletin de la Société vaudoise des sciences naturelles* 67:241–246.

Mattingly, W.B., and B.C. Jayne. 2004. Resource use in arboreal habitats: structure affects locomotion of four ecomorphs of *Anolis* lizards. *Ecology* 85 (4):1111–1124.

Maul, L.C., K.T. Smith, R. Barkai, A. Barash, P. Karkanas, R. Shahack-Gross, and A. Gopher. 2011. Microfaunal remains at Middle Pleistocene Qesem Cave, Israel: Preliminary results on small vertebrates, environment and biostratigraphy. *Journal of Human Evolution* 60(4):464–480.

Mayer, A.F. 1835. Analecten für vergleichende Anatomie. Bonn, Germany: Eduard Weber [in German].

McCarthy, T., and B. Rubidge. 2005. The story of earth and life: a southern African perspective on a 4.6-billion-year journey. Cape Town, South Africa: Struik Publishers.

McKee, J.K., P.W. Sciulli, C.D. Fooce, and T.A. Waite. 2004. Forecasting global biodiversity threats associated with human population growth. *Biological Conservation* 115:161–164.

Measey, J. 2008. Das Taita-Zweihornchamäleon - auf der Suche nach Chamäleons in ihrem natürlichen Habitat. *Chamaeleo Mitteilungsblatt* 37:17–24.

Measey, G.J., K. Hopkins, and K.A. Tolley. 2009. Morphology, ornaments and performance in two chameleon ecomorphs: is the casque bigger than the bite? *Zoology* 112:217–226.

Measey, G.J., A.D. Rebelo, A. Herrel, B. Vanhooydonck, and K.A. Tolley. 2011.Diet, morphology and performance in two chameleon morphs: do harder bites equate with harder prey? *Journal of Zoology* 285(4):247–255.

Measey, G.J., and K.A. Tolley. 2011. Sequential fragmentation of Pleistocene forests in an East Africa biodiversity hotspot: chameleons as a model to track forest history. *PLoS ONE* 6:e26606.

Meiri, S. 2008. Evolution and ecology of lizard body sizes. *Global Ecology and Biogeography* 17:724–734.

Meldrum, D.J. 1998. Tail-assisted hind limb suspension as a transitional behavior in the evolution of the platyrrhine prehensile tail, pp 145–156. In E. Strasser, J. Fleagle, A. Rosenberger and H. McHenry, Eds., *Primate Locomotion: Recent Advances.* New York: Plenum Press.

Melville, J., E.G. Ritchie, S.N.J. Chapple, R.E. Glor, and J.A. Schulte II. 2011. Evolutionary origins and diversification of dragon lizards in Australia's tropical savannas. *Molecular Phylogenetics and Evolution* 58(2):257–270.

Melville, J., and R. Swain. 2000. Evolutionary relationships between morphology, performance and habitat openness in the lizard genus *Niveoscincus* (Scincidae: Lygosominae). *Biological Journal of the Linnean Society* 70:667–683.

Menegon, M., C. Bracebridge, N. Owen, and S.P. Loader. 2011. Herpetofauna of montane areas of Tanzania. 4. Amphibians and reptiles of Mahenge Mountains, with comments on biogeography, diversity, and conservation. *Fieldiana Life and Earth Sciences* 4:103–111

Menegon, M., N. Doggart, and N. Owen. 2008. The Nguru Mountains of Tanzania, an outstanding hotspot of herpetofaunal diversity. *Acta Herpetologica* 3:107–127.

Menegon, M. and T. Davenport. 2008. The amphibian fauna of the Eastern Arc Mountains of Kenya and Tanzania. Pp. 63 in Stuart, SN., Hoffmann, M., Chanson, J.S., Cox, N.A., Berridge, R.J., Ramani P., and B.E. Young, Eds., *Threatened Amphibians of the World.* Lynx Edicions: Barcelona, Spain.

Menegon, M., and S. Salvidio. 2005. Amphibian and reptile diversity in the southern Udzungwa Scarp Forest Reserve, South-Eastern Tanzania, pp. 205–212. In B.A. Huber, B.J. Sinclair and K.H. Lampe Eds., *African Biodiversity: Molecules, Organisms, Ecosystems*. Proceedings of the 5th International Symposium on Tropical Biodiversity, Museum Koenig, Bonn. New York: Springer.

Menegon, M., K.A. Tolley, T. Jones, F. Rovero, A.R. Marshall, and C.R. Tilbury. 2009. A new species of chameleon (Sauria: Chamaeleonidae: *Kinyongia*) from the Magombera forest and Udzungwa Mountains National Park, Tanzania. *African Journal of Herpetology* 58(2): 59–70.

Mertens, R. 1966. Chamaeleonidae. *Das Tierreich, Berlin* 83:1–37.

Metcalf, J., N. Bayly, M. Bisoa, and J. Rabearivony. 2005. Edge effect from paths on two chameleon species in Madagascar. *African Journal of Herpetology* 54:99–102.

Metcalfe, I. 1996a. Pre-Cretaceous evolution of SE Asian terrones. Pp. 97–122 in R. Hall, and D.J. Blundell, Eds., *Tectonic Evolution of Southeast Asia*. London: Geological Society. Special Publication 106.

Metcalfe, I. 1996b. Gondwanaland dispersion, Asian accretion and evolution of Eastern Tethys. *Australian Journal of Earth Sciences* 43:605–623.

Methuen, P.A., and J. Hewitt. 1914. A contribution to our knowledge of the anatomy of chamaeleons. *Transactions of the Royal Society of South Africa* 4(2):89–104.

Meyers, J.J., A. Herrel, and K.C. Nishikawa. 2002. Comparative study of the innervation patterns of the hyobranchial musculature in three iguanian lizards: *Sceloporus undulates, Pseudotrapelus sinaitus,* and *Chamaeleo jacksonii. Anatomical Record* 267(2):177–189.

Meyers, J.J., and K.C. Nishikawa. 2000. Comparative study of tongue protrusion in the three iguanian lizards, *Sceloporus undulates, Pseudotrapelus sinaitus* and *Chamaeleo jacksonii. Journal of Experimental Biology* 203 (18):2833–2849.

Meyers, R.A., and B.M. Clarke. 1998. How do flap-necked chameleons move their flaps? *Copeia* 1998(3):759–761.

Miehe, S., and G. Miehe. 1994. *Ericaceous forests and heathlands in the Bale Mountains of South Ethiopia:. Ecology and Man's Impact*. Reinbek, Germany: Warnke.

Mittermeier, R.A., P. Robles Gil, M. Hoffman, J. Pilgrim, T. Brooks, C. Goettsch Mittermeier, J. Lamoreux, and G.A.B. da Fonseca. 2004. *Hotspots Revisited*. Mexico City: CEMEX, Agrupación Sierra Madre, S.C.

Mivart, S.G. 1870. On the myology of *Chamaeleon parsonii. Proceedings of the Scientific Meetings of the Zoological Society of London* 57:850–890.

Monadjem, A., M.C. Schoeman, A. Reside, D.V. Pio, S. Stoffberg, J. Bayliss, F.P.D. Cotterill, M. Curran, M. Kopp, and P.J. Taylor. 2010. A recent inventory of the bats of Mozambique with documentation of seven new species for the country. *Acta Chiropterologica* 12:371–391.

Montuelle, S., G. Daghfous, and V. Bels. 2008. Effect of locomotor approach on feeding kinematics in the green anole (*Anolis carolinensis*). *Journal of Experimental Zoology* 309A(9):563–567.

Moody, S. 1980. The phylogenetic relationships of taxa within the lizard family Agamidae. Ph.D. thesis. University of Michigan.

Moody, S., and Z. Roček. 1980. *Chamaeleo caroliquarti* (Chamaeleonidae, Sauria), a new species from the Lower Miocene of central Europe. *Věstník Ústředního ústavu geologického* 55:85–92.

Mooi, R.D., and A.C. Gill. 2010. Phylogenies without synapomorphies—a crisis in fish systematics: time to show some character. *Zootaxa* 2450:26–40

Morrison, R.L., W.C. Sherbrooke, and S.K. Frostmason. 1996. Temperature-sensitive, physiologically active iridophores in the lizard *Urosaurus ornatus*: an ultrastructural analysis of color change. *Copeia* 1996:804–812.

Moreno-Rueda, G., J.M. Pleguezuelos, M. Pizarro, and A. Montori. 2011. Northward shifts of the distributions of Spanish reptiles in association with climate change. *Conservation Biology* 26:278–283.

Mörs, T. 2002. Biostratigraphy and paleoecology of continental Tertiary vertebrate faunas in the Lower Rhine Embayment (NW-Germany). *Netherlands Journal of Geosciences/Geologie en Mijnbouw* 81:177–183.

Mörs, T., F. von der Hocht, and B. Wutzler, 2000. Die erste Wirbeltierfauna aus der miozänen Braunkohle der Niederrheinischen Bucht (Ville-Schichten,Tagebau Hambach). *Paläontologische Zeitschrift* 74:145–170 [in German].

Müller, R., and T. Hildenhagen. 2009. Untersuchungen zu Subdigital- und Subcaudalstrukturen bei Chamäleons (Sauria: Chamaelaeonidae). *Sauria* 31(3):41–54 [in German with English summary].

Müller, U.K., and S. Kranenbarg. 2004. Power at the tip of the tongue. *Science* 304 (5668):217–218.

Mutungi, G. 1992. Slow locomotion in chameleons: histochemical and ultrastructural characteristics of muscle fibers isolated from iliofibularis muscle of Jackson's chameleon (*Chamaeleo jacksonii*). *Journal of Experimental Zoology* 263:1–7.

Myers, N., R.A. Mittermeier, C.G. Mittermeier, G.A.B. Da Fonseca, and J. Kent. 2000. Biodiversity hotspots for conservation priorities. *Nature* 403:853–858.

Nagy, Z.T., G. Sonet, F. Glaw, and M. Vences. 2012. First large-scale DNA barcoding assessment of reptiles in the biodiversity hotspot of Madagascar, based on newly designed COI primers. *PloS ONE* 7:e34506.

Nečas, P. 2004. *Chameleons: Nature's Hidden Jewels*, 2nd ed. Frankfurt am Main, Germany: Chimaira.

Nečas, P. 2009. Ein neues Chamäleon der Gattung *Kinyongia* Tilbury Tolley & Branch 2006 aus den Poroto-Bergen, Süd-Tansania (Reptilia: Sauria: Chamaeleonidae). *Sauria* 31(2):41–48.

Nečas, P., and W. Schmidt. 2004. *Stump-tailed Chameleons: Miniature Dragons of the Rainforest. The Genera Brookesia and Rhampholeon.* Frankfurt am Main, Germany: Chimaira Buchhandelsgesellschaft mbH.

Nečas, P., R. Sindaco, L. Koreny, J. Kopecna, P.K. Malonza, and D. Modry. 2009. *Kinyongia asheorum* sp. n., a new montane chameleon from the Nyiro Range, northern Kenya (Squamata: Chamaeleonidae). *Zootaxa* 2028:41–50.

Nechaeva, M.V., I.G. Makarenko, E.B. Tsitrin, and N.P. Zhdanova. 2005. Physiological and morphological characteristics of the rhythmic contractions of the amnion in veiled chameleon (*Chameleo calyptratus*) embryogenesis. *Comparative Biochemistry and Physiology A—Physiology* 140: 19–28.

Nelson, G., and P.Y. Ladiges. 2009. Biogeography and the molecular dating game: a futile revival of phentics? *Bulletin de la Societe Geologique de France* 180(1):39–43.

Nessov, L.A. 1988. Late mesozoic amphibians and lizards of Soviet Middle Asia. *Acta Zoologica Cracoviensia* 31:475–486.

Nonnotte, P., H. Guillou, B. Le Gall, M. Benoit, J. Cotten, and S. Scaillet. 2008. New K-Ar age determinations of Kilimanjaro volcano in the North Tanzanian diverging rift, East Africa. *Journal of Volcanology and Geothermal Research* 173:99–112.

Norris, K.S., and W.R. Dawson. 1964. Observations on the water economy and electrolyte excretion of chuckwallas (Lacertilia, *Sauromalus*). *Copeia* 1964:638–646.

Northcutt, R.G. 1978. Forebrain and midbrain organization in lizards and its phylogenetic significance, pp. 11–64. In N. Greenberg and P.D. MacLean, Eds., *Behavior and Neurology of Lizards.* Rockville, MD: National Institute of Mental Health.

Nussbaum, R.A., C.J. Raxworthy, A.P. Raselimanana, and J.-B. Ramanamanjato. 1999. Amphibians and reptiles of the Réserve Naturelle Intégrale d'Andohahela, Madagascar, pp. 155–173. In S.M. Goodman, Ed., *A Floral and Faunal Inventory of the Réserve Naturelle Intégrale d'Andohahela, Madagascar: With Reference to Elevational Variation*. Fieldiana Zoology, new series, 94. Chicago: Field Museum of Natural History.

Ogg, J.G., G. Ogg, and F.M. Gradstein. 2008. *The concise geologic time scale*. Cambridge, United Kingdom: Cambridge University Press.

Ogilvie, P.W. 1966. An anatomical and behavioral investigation of a previously undescribed pouch found in certain species of the genus *Chamaeleo*. PhD thesis, University of Oklahoma.

Okajima, Y., and Y. Kumazawa. 2010. Mitochondrial genomes of acrodont lizards: timing of gene rearrangements and phylogenetic and biogeographic implications. *BMC Evolutionary Biology* 10(141):1–15.

Ord, T.J., and J.A. Stamps. 2009. Species identity cues in animal communication. *American Naturalist* 174:585–593.

Osorio, D., A. Miklosi, and Z. Gonda. 1999. Visual ecology and perception of coloration patterns by domestic chicks. *Evolutionary Ecology* 13:673–689.

Ott, M. 2001. Chameleons have independent eye movements but synchronise both eyes during saccadic prey tracking. *Experimental Brain Research* 139:173–179.

Ott, M., and F. Schaeffel. 1995. A negatively powered lens in the chameleon. *Nature* 373:692–694.

Ott, M., F. Schaeffel, and W. Kirmse. 1998. Binocular vision and accommodation in prey-catching chameleons. *Journal of Comparative Physiology A—Sensory Neural and Behavioural Physiology* 182:319–330.

Parcher, S.R. 1974. Observations on the Natural Histories of Six Malagasy Chamaeleontidae [sic]. *Zeitschrift für Tierzuchtung und Zuchtungsbiologie* 34:500–523.

Parker, H.W. 1942. The lizards of British Somaliland. *Bulletin of the Museum of Comparative Zoology at Harvard College* 91:1–101.

Parker, W.K. 1881. On the structure of the skull in the chameleons. *Transactions of the Zoological Society of London* 11:77–105.

Parsons, T.S. 1970. The nose and Jacobson's organ, pp. 99–191. In C. Gans and T.S. Parsons, Eds. *Biology of the Reptilia. Volume 2. Morphology B*. New York: Academic Press.

Parsons, T.S., and J.E. Cameron. 1977. Internal relief of the digestive tract, pp. 159–223. In C. Gans and T.S. Parsons, Eds., *Biology of the Reptilia. Volume 6. Morphology E*. New York: Academic Press.

Patnaik, R., and H.H. Schleich. 1998. Fossil micro-reptiles from Pliocene Siwalik sediments of India. *Veroffentlichungen aus dem Fuhlrott Museum* 4:295–300.

Patrick, D.A., P. Shirk, J.R. Vonesh, E.B. Harper, and K.M. Howell. 2011. Abundance and roosting ecology of chameleons in the East Usambara Mountains of Tanzania and the potential effects of harvesting. *Herpetological Conservation and Biology* 6:422–431.

Paulo, O.S., I. Pinto, M.W. Bruford,W.C. Jordan, and R.A. Nichols. 2002. The double origin of Iberian peninsular chameleons. *Biological Journal of the Linnean Society* 75:1–7.

Paxton, J.R. 1991. Interaction between laughing doves and chameleon. *Honeyguide* 37:180–181.

Peaker, M., and J.L. Linzell. 1975. *Salt Glands in Birds and Reptiles*. Cambridge, United Kingdom: Cambridge University Press.

Pearson, R.G., and C.J. Raxworthy. 2009. The evolution of local endemism in Madagascar: watershed versus climatic gradient hypotheses evaluated by null biogeographic models. *Evolution* 63:959–967.

Perry, S.F. 1998. Lungs: Comparative Anatomy, Functional Morphology, and Evolution, pp. 1–92. In C. Gans and A.S. Gaunt, Eds., *Biology of the Reptilia. Volume 19. Morphology G*. Ithaca, NY: Society for the Study of Amphibians and Reptiles.

Peterson, J.A. 1973. Adaptation for arboreal locomotion in the shoulder region of lizards. PhD thesis, University of Chicago.

Peterson, J.A. 1984. The locomotion of *Chamaeleo* (Reptilia: Sauria) with particular reference to the forelimb. *Journal of Zoology, London* 202:1–42.

Pettigrew, J.D., S.P. Collin, and M. Ott. 1999. Convergence of specialised behaviour, eye movements and visual optics in the sandlance (Teleostei) and the chameleon (Reptilia). *Current Biology* 9(8):421–424.

Pianka, E.R. 1986. *Ecology and natural history of desert lizards: analyses of the ecological niche and community structure.* Princeton, NJ: Princeton University Press.

Pianka, E.R., and L.J. Vitt. 2003. *Lizards: Windows to the Evolution of Diversity.* Berkeley: University of California Press.

Pickford, M. 1986. Sediment and fossil preservation in the Nyanza Rift system of Kenya. *Geological Society Special Publication* 25:345–362.

Pickford, M. 2001. Africa's smallest ruminant: a new tragulid from the Miocene of Kenya and the biostratigraphy of East African Tragulidae. *Geobios* 34(4):437–447.

Pickford, M., Y. Sawada, R. Tayama, Y. Matsuda, T. Itaya, H. Hyodo, and B. Senut. 2006. Refinement of the age of the Middle Miocene Fort Ternan Beds, Western Kenya, and its implications for Old World biochronology. *Comptes Rendus Geoscience* 338:545–555.

Pitman, C.R.S. 1958. Snake and lizard predation of birds. *Bulletin of the British Ornithology Club* 78:120–124.

Pleguezuelos, J.M., J.C. Poveda, R. Monterrubio, and D. Ontiveros. 1999. Feeding habits of the common chameleon, *Chamaeleo chamaeleon* in the southeastern Iberian Peninsula. *Israel Journal of Zoology* 45:267–276.

Plumptre, A.J., T.R.B. Davenport, M. Behangana, R. Kityo, G. Eilu, P. Ssegawa, C. Ewango, D. Meirte, C. Kahindo, M. Herremans, J.K. Peterhans, J.D. Pilgrim, M. Wilson, M. Languy, and D. Moyer. 2007. The biodiversity of the Albertine Rift. *Biological Conservation* 134:178–194.

Poglayen-Neuwall, I. 1954. Die Kiefermuskulatur der Eidechsen und ihre Innervation. *Zeitschrift für Wissenschaftliche Zoologie* 158:79–132 [in German].

Pook, C., and C. Wild. 1997. The phylogeny of the *Chamaeleo (Trioceros) cristatus* species group from Cameroon inferred from direct sequencing of the mitochondrial 12S ribosomal RNA gene: Evolutionary and paleobiogeographic implications, pp. 297–306. In W. Böhme, W. Bischoff and T. Ziegler, Eds., *Herpetologia Bonnensis.* Bonn, Germany: Societas Europaea Herpetologica.

Potgieter, D. 2012. *Investigating the presence of ecomorphological forms in Bradypodion damaranum using molecular and morphometric techniques.* M.Sc. thesis. Stellenbosch University, Stellenbosch.

Pounds, J.A., M.R. Bustamante, L.A. Coloma, J.A. Consuegra, M.P. Fogden, P.N. Foster, E. La Marca, et al. 2006. Widespread amphibian extinctions from epidemic disease driven by global warming. *Nature* 439:161–167.

Pounds, J.A., M.L.P. Fogden, and J.H. Campbell. 1999. Biological response to climate change on a tropical mountain. *Nature* 398:611–615.

Poynton, J., and R. Boycott. 1996. Species turnover between Afromontane and eastern African lowland faunas: patterns shown by amphibians. *Journal of Biogeography* 23:669–680.

Poynton, J.C., S.P. Loader, E. Sherratt, and B.T. Clarke. 2006. Amphibian diversity in East African biodiversity hotspots: altitudinal and latitudinal patterns. *Biodiversity and Conservation* 16:1103–1118.

Prasad, G.V.R., and S. Bajpai. 2008. Agamid lizards from the Early Eocene of Western India: Oldest Cenozoic lizards from South Asia. *Palaeontologia Electronica* 11(1):1–19.

Prasad, J. 1954. The temporal region in the skull of *Chamaeleon zeylanicus* Laurenti. *Current Science* 23:235–236.

Prieto, J., M. Böhme, H. Maurer, K. Heissig, and H. Abdul Aziz. 2009. Biostratigraphy and sedimentology of the Fluviatile Untere Serie (Early and Middle Miocene) in the central part of the North Alpine Foreland Basin: implications for palaeonvironment and climate. *International Journal of Earth Sciences (Geologische Rundschau)* 98:1767–1791.

Prothero, D., and R. Estes. 1980. Late Jurassic lizards from Como Bluff, Wyoming and their palaeobiogeographic significance. *Nature* 286:484–486.

Quay, W.B. 1979. The parietal eye-pineal complex, pp. 245–406. In C. Gans, R.G. Northcutt and P. Ulinski, Eds., *Biology of the Reptilia. Volume 9. Neurology A.* New York: Academic Press.

Rabearivony, J. 1999. Conservation and status of assessment of *Brookesia,* the dwarf chameleons of Madagascar. M.Sc. thesis, University of Kent, United Kingdom.

Rabearivony, J. 2012. Etude bio-écologique et conservation des caméléons dans les habitats écotoniques des rivières malgaches. Thèse de Doctorat. Facultés des Sciences, Université d'Antananarivo.

Rabearivony, J., L.D. Brady, R.K. Jenkins, and O.R. Ravoahangimalala. 2007. Habitat use and abundance of a low-altitude chameleon assemblage in eastern Madagascar. *Herpetological Journal* 17:247–254.

Rage, J.C. 1972. Les amphibiens et les reptiles du du Würmien II de la grotte de l'Hortus. *Études Quaternaires* 1:297–298 [in French].

Rage, J.C. 1987. Lower vertebrates from the early-Middle Eocene Kuldana Formation of Kohat (Pakistan): Squamata. *Contributions from the Museum of Paleontology University of Michigan* 27:187–193.

Rage, J.C., and M. Augé. 1993. Squamates from the Cainozoic of the western part of Europe: a review. *Revue de Paléobiologie* special volume 7:199–216.

Raholdina, A.M.F. 2012. Etude écologique et analyse structural de la population de *Furcifer campani* (Grandidier, 1872) dans le massif de l'Ankaratra. Mémoire de DEA, Facultés des Sciences, Université d'Antananarivo.

Rana, R.S. 2005. Lizard fauna from the Intertrappean (Late Cretaceous-Early Palaeocene) beds of Peninsular India. *Gondwana Geological Magazine Nagpur* 8:123–132.

Randrianantoandro, J.C., R.R. Andriatsimanarilafy, H. Rakotovololonalimanana, E.F. Hantalalaina, D. Rakotondravony, O.R. Ramilijaona, J. Ratsimbazafy, G.F. Razafindrakoto, and R.K.B. Jenkins. 2009. Population assessments of chameleons from two montane sites in Madagascar. *Herpetological Conservation and Biology* 5:23–31.

Randrianantoandro, J.C., R. Randrianavelona, R.R. Andriatsimanarilafy, E.F. Hantalalaina, D. Rakotondravony, and R.K.B. Jenkins. 2007a. Roost site characteristics of sympatric dwarf chameleons (genus *Brookesia*) from western Madagascar. *Amphibia-Reptilia* 28:577–581.

Randrianantoandro, J.C., R. Randrianavelona, R.R. Andriatsimanarilafy, E.F. Hantalalaina, D. Rakotondravony, M. Randrianasolo, H.L. Ravelomanantsoa, and R.K.B. Jenkins. 2007b. Identifying important areas fro the conservation of dwarf chameleons (*Brookesia* spp.) in Tsingy de Bemaraha National Park, western Madagascar. *Oryx* 42:578–583.

Randrianantoandro, J.C., B. Razafimahatratra, M. Soazandry, J. Ratsimbazafy, and R.K.B. Jenkins. 2010. Habitat use by chameleons in a deciduous forest in western Madagascar. *Amphibia-Reptilia* 31:27–35.

Raselimanana, A.P. 2008. Herpétofaune des forêts sèches malgaches. *Malagasy Nature* 1:46–75.

Raselimanana, A.P., and D. Rakotomalala. 2003. Chamaeleonidae, chamaeleons, pp. 960–969. In S.M. Goodman and J.P. Benstead, Eds., *The Natural History of Madagascar.* Chicago: University of Chicago Press.

Raselimanana, A. P., C.J. Raxworthy, and R.A. Nussbaum. 2000. Herpetofaunal species diversity and elevational distribution within the Parc National de Marojejy, Madagascar, pp. 157–174. In S. M. Goodman, *A Floral and Faunal Inventory of the Parc National de Marojejy, Madagascar: With Reference to Elevational Variation*. Fieldiana: Zoology, new series, 97. Chicago: Field Museum of Natural History.

Rathke, H. 1857. Untersuchungen über die Aortenwurzeln und die von ihnen ausgehenden Arterien der Saurier. *Denkschriften/Akademie der Wissenschaften in Wien, Mathematisch-Naturwissenschaftliche Klasse* 13:51–142 [in German].

Raw, L.R.G. 1976. A survey of the dwarf chameleons of Natal, South Africa, with descriptions of three new species (Sauria: Chamaeleonidae). *Durban Museum Novitates* 11(7):139–161.

Raxworthy, C.J. 1988. Reptiles, rainforest and conservation in Madagascar. *Biological Conservation* 43:181–211.

Raxworthy, C.J. 1991. Field observations on some dwarf chameleons (*Brookesia* spp.) from rainforest areas of Madagascar, with the description of a new species. *Journal of Zoology, London* 224:11–25.

Raxworthy, C.J., M.R.J. Forstner, and R.A. Nussbaum. 2002. Chameleon radiation by oceanic dispersal. *Nature* 415:784–787.

Raxworthy, C.J., and R.A. Nussbaum. 1995. Systematics, speciation and biogeography of the dwarf chameleons (*Brookesia*: Reptilia, Squamata, Chamaeleonitidae) of northern Madagascar. *Journal of Zoology, London* 235:525–558.

Raxworthy, C.J., and R.A. Nussbaum. 1996. Montane amphibian and reptile communities in Madagascar. *Conservation Biology* 10:750–756.

Raxworthy, C.J., and R.A. Nussbaum. 2006. Six new species of occipital-lobed *Calumma* chameleons (Squamata: Chamaeleonidae) from montane regions of Madagascar, with a new description and revision of *Calumma brevicorne*. *Copeia* 2006(4):711–734.

Raxworthy, C. J., R.G. Pearson, N. Rabibisoa, A.M. Rakotondrazafy, J.-B. Ramanamanjato, A.P. Raselimanana, S. Wu, R.A. Nussbaum, and D.A. Stone. 2008. Extinction vulnerability of tropical montane endemism from warming and upslope displacement: a preliminary appraisal for the highest massif in Madagascar. *Global Change Biology* 14:1703–1720.

Razafimahatratra, B., A. Mori, and M. Hasegawa. 2008. Sleeping site pattern and sleeping behavior of *Brookesia decaryi* (Chamaeleonidae) in Ampijoroa dry forest, northwestern Madagascar. *Current Herpetology* 27:93–99.

Reaney, L.T., S. Yee, J.B. Losos, and M.J. Whiting. 2012. Ecology of the flap-necked chameleon *Chamaeleo dilepis* in southern Africa. *Breviora* 532:1–18.

Regal, P.J. 1978. Behavioral differences between reptiles and mammals: an analysis of activity and mental capabilities, pp. 183–202. In N. Greenberg and P.D. Maclean, Eds., *Behavior and neurobiology of lizards*. Washington, DC: Department of Health, Education and Welfare.

Reid, J.C. 1986. A list with notes of Lizards of the Calabar area of southern Nigeria, pp 699–704. In Z. Roček, Ed., *Studies in Herpetology*. Prague, Czech Republic: Charles University.

Reilly, S.M. 1982. Ecological notes on *Chamaeleo schubotzi* from Mount Kenya. *Journal of the Herpetological Association of Africa* 18:28–30.

Reisinger, W.J., D.M. Stuart-Fox, and B.F.N. Erasmus. 2006. Habitat associations and conservation status of an endemic forest dwarf chameleon (*Bradypodion* sp.) from South Africa. *Oryx* 40:183–188.

Rewcastle, S.C. 1981. Stance and gait in tetrapods: an evolutionary scenario, pp 239–267. In M.H. Day, Ed., *Vertebrate Locomotion*. London: Academic Press.

Rewcastle, S.C. 1983. Fundamental adaptations in the lacertilian hind limb: a partial analysis of the sprawling limb posture and gait. *Copeia* 1983 (2):476–487.

Reynoso, V.-H., 1998. *Huehuecuetzpalli mixtecus* gen. et sp. nov: a basal squamate (Reptilia) from the Early Cretaceous of Tepexi de Rodríguez, Central México. *Philosophical Transactions of the Royal Society of London B* 353:477–500.

Ribbing, L. 1938. Die Muskeln und Nerven der Extremitaten, pp. 543–682. In L. Bolk, E. Goppert, E. Kallius and W. Lubosch, Eds., *Handbuch der vergleichenden Anatomie der Wirbeltiere.* Berlin: Urban and Schwarzenberg [in German].

Rice, M.J. 1973. Supercontracting striated muscle in a vertebrate. *Nature* 243:238–240.

Richter, B., and M. Fuller. 1996. Palaeomagnetism of the Sibumasu and Indochina blocks: Implications for the extrusion tectonic model, pp. 203–224. In R. Hall, and D. Blundell, Eds., *Tectonic Evolution of Southeast Asia.* London: Geological Society Special Publication 106.

Rieppel, O. 1981. The skull and jaw adductor musculature in chameleons. *Revue Suisse de Zoologie* 88(2):433–445.

Rieppel, O. 1987. The phylogenetic relationships within the Chamaeleonidae, with comments on some aspects of cladistics analysis. *Zoological Journal of the Linnean Society* 89(1):41–62.

Rieppel, O. 1993. Studies on skeleton formation in reptiles. II. *Chamaeleo hoehnelii* (Squamata: Chamaeleoninae), with comments on the homology of carpal and tarsal bones. *Herpetologica* 49(1):66–78.

Rieppel, O., and C. Crumly. 1997. Paedomorphosis and skull structure in Malagasy chamaeleons (Reptilia: Chamaeleoninae). *Journal of Zoology, London* 243(2):351–380.

Rieppel, O., A. Walker, and I. Odhiambo. 1992. A preliminary report on a fossil chamaeleonine (Reptilia: Chamaeleoninae) skull from the Miocene of Kenya. *Journal of Herpetology* 26(1):77–80.

Rigby, B.J., N. Hirai, J.D. Spikes, and H. Eyring. 1959. The mechanical properties of rat tail tendon. *Journal of General Physiology* 43:265–283.

Roček, Z. 1984. Lizards (Reptilia: Sauria) from the Lower Miocene locality Dolnice (Bohemia, Czechoslovakia). *Rozpravy Československé Akademie Věd* 94(1):1–69.

Rocha, S., M.A. Carretero, and D.J. Harris. 2005. Mitochondrial DNA sequence data suggests two independent colonizations of the Comoros archipelago by chameleons of the genus *Furcifer. Belgian Journal of Zoology* 135(1):39–42.

Rodrigues, A.S.L., J.D. Pilgrim, J.F. Lamoreux, M. Hoffmann, and T.M. Brooks. 2006. The value of the IUCN Red List for conservation. *Trends in Ecology and Evolution* 21:71–76.

Romanoff, A.L. 1960. *The avian embryo: structural and functional development.* New York: Macmillan.

Rome, L.C. 1990. Influence of temperature on muscle recruitment and muscle function in vivo. *American Journal of Physiology* 259(2 Pt 2):R210-R222.

Romer, A.S. 1956. Osteology of the Reptiles. Chicago: University of Chicago Press.

Ross, H.H. 1964. Book Review: Principles of numerical taxonomy. *Systematic Zoology* 13:106–108.

Russell, A.P., and A. M. Bauer. 2008. The appendicular locomotor apparatus of *Sphenodon* and normal-limbed squamates, pp. 1–466. In C. Gans, A. S. Gaunt and K. Adler, Eds., *Biology of the Reptilia. Volume 21. Morphology I.* Ithaca, NY: Society for the Study of Amphibians and Reptiles.

Russell, A.P., and V. Bels. 2001. Biomechanics and kinematics of limb-based locomotion in lizards: review, synthesis and prospectus. *Comparative Biochemistry and Physiology A* 131:89–112.

Russell, A.P., and T.E. Higham. 2009. A new angle on clinging in geckos: incline, not substrate, triggers the deployment of the adhesive system. *Proceedings of the Royal Society B* 276(1673):3705–3709.

Russell, A.P., and M.K. Johnson. 2007. Real-world challenges to, and capabilities of, the gekkotan adhesive system: contrasting the rough and the smooth. *Canadian Journal of Zoology* 85:1228–1238.

Sahni, A. 2010. Indian Cretaceous terrestrial vertebrates: cosmopolitanism and endemism in a geodynamic plate tectonic framework, pp. 91–104. In S. Bandyopadhyay Ed., *New Aspects of Mesozoic Biodiversity*. Lecture Notes in Earth Sciences 132. Berlin: Springer Verlag.

Salzmann, U., and P. Hoelzmann. 2005. The Dahomey Gap: an abrupt climatically induced rain forest fragmentation in West Africa during the late Holocene. *The Holocene* 15(2):190–199.

Sándor, P.S., M.A. Frens, and V. Henn. 2001. Chameleon eye position obeys Listing's law. *Vision Research* 41:2245–2251.

Sathe, A.M. 1959. Trunk musculature of *Chamaeleon vulgaris* (Reptilia). *First All-India Congress of Zoology, Jabalpur. Abstracts of Papers* October 24–29, 1959:16.

Schaefer, N. 1971. A few thoughts concerning the life span of chameleons. *Journal of the Herpetological Association of Africa* 8:21–24.

Schleich, H.H. 1983. Die mittelmiozäne Fossil-Lagerstätte Sandelzhausen. 13. *Chamaeleo bavaricus* sp. nov., ein neuer Nachweis aus dem Jungtertiär Süddeutchlands. *Mitteilungen der Beyerischen Staatssamlung für Palaeontologie und Historische Geologie* 23:77–81 [in German].

Schleich, H.H. 1984. Neue Reptilienfunde aus dem Tertiär Deutschlands 2. *Chamaeleo pfeili* sp. nov. von der untermiozänen Fossilfundstelle Rauscheröd/Niederbayern (Reptilia, Sauria, Chamaeleonidae). *Mitteilungen der Bayerischen Staatssamlung für Palaeontologie und Historische Geologie* 24:97–103 [in German].

Schleich, H.H. 1994. Neue Reptilfunde aus dem Tertiär Deutschlands. 15. Neue Funde fossiler Chamäleonen aus dem Neogen Süddeutschlands. *Courier Forschunginstitut Senckenberg* 173:175–195 [in German].

Schleich, H.-H., and W. Kästle. 1979. Hautstrukturen als Kletteranpassungen bei *Chamaeleo* und *Cophotis*. *Salamandra* 15(2):95–100 [in German with English summary].

Schleich, H.-H., and W. Kästle. 1985. Skin structures of Sauria extremities—SEM-studies of four families. *Fortschritte der Zoologie* 30:99–101.

Schmidt, W.J. 1909. Beiträge zur Kenntnis der Parietalorgane der Saurien. *Zeitschrift für Wissenschaftliche Zoologie* 92:359–425 [in German].

Schmidt-Nielsen, K. 1963. Osmotic regulation in higher vertebrates. *Harvey Lectures* 58:53–93.

Schulte II, J.A., J. Melville, and A. Larson, 2003. Molecular phylogenetic evidence for ancient divergence of lizard taxa on either side of Wallace's Line. *Proceedings of the Royal Society of London B: Biological Sciences* 270:597–603.

Schulte, J.A., and F. Moreno-Roark. 2010. Live birth among Iguanian lizards predates Pliocene- Pleistocene glaciations. *Biology Letters* 6:216–218.

Schuster, M. 1984. Zum Beutefangverhalten von *Chamaeleo jacksonii* Boulenger, 1896 (Sauria: Chamaeleonidae). *Salamandra* 20 (1):21–31 [in German with English summary].

Schwartz, J.H., and B. Maresca. 2006. Do molecular clocks run at all? A critique of molecular systematics. *Biological Theory* 1(4):357–371.

Schwenk, K. 1983. Functional morphology and evolution of the chameleon tongue tip. *American Zoologist* 23(4):1028.

Schwenk, K. 1985. Occurrence, distribution and functional significance of taste buds in lizards. *Copeia* 1985(1):91–101.

Schwenk, K. 1995. Of tongues and noses—chemoreception in lizards and snakes. *Trends in Ecology and Evolution* 10:7–12.

Schwenk, K. 2000. Feeding in Lepidosaurs. pp. 175–291 in K. Schwenk, Ed., *Feeding: Form, Function, and Evolution in Tetrapod Vertebrates*. Academic Press: San Diego: USA.

Schwenk, K., and D.A. Bell. 1988. A cryptic intermediate in the evolution of chameleon tongue projection. *Experientia* 44:697–700.

Schwenk, K., and G.S. Throckmorton. 1989. Functional and evolutionary morphology of lingual feeding in squamate reptiles: phylogenetics and kinematics. *Journal of Zoology, London* 219:153–175.
Scotese C. R. 2002. The Paleomap Project. Accessed at www.scotese.com on August 15, 2012.
Secord, R., S.L. Wing, and A. Chew. 2008. Stable isotopes in early Eocene mammals as indicators of forest canopy structure and resource partitioning. *Paleobiology* 34:282–300.
Seiffert, J. 1973. Upper Jurassic lizards from central Portugal. *Memóres Serviços Geológicos de Portugal* (Nova Série 22):1–85.
Senn, D.G., and R.G. Northcutt. 1973. The forebrain and midbrain of some squamates and their bearing on the origin of snakes. *Journal of Morphology* 140:135–152.
Seward, D., D. Grujic, and G. Schreurs. 2004. An insight into the breakup of Gondwana: identifying events through low-temperature thermochronology from the basement rocks of Madagascar. *Tectonics* 23:C3007
Sewertzoff, S.A. 1923. Die Entwickelungsgeschichte der Zunge des *Chamaeleo bilineatus*. *Revue Zoologique Russe* 3:263–283 [in Russian with German translation].
Shanklin, W.M. 1930. The central nervous system of *Chameleon vulgaris*. *Acta Zoologica Stockholm* 11:425–490.
Shanklin, W.M. 1933. The comparative neurology of the nucleus opticus tegmenti with special reference to *Chameleon vulgaris*. *Acta Zoologica Stockholm* 14:163–184.
Shine, R. 1985. The evolution of viviparity in reptiles: an ecological analysis, pp. 605–694. In C. Gans and F. Billett, Eds., *Biology of the Reptilia*. Volume 15. New York: Wiley.
Shine, R., and G.P. Brown. 2008. Adapting to the unpredictable: reproductive biology of vertebrates in the Australian wet-dry tropics. *Philosophical Transactions of the Royal Society B* 363:63–373.
Shine, R., P.S. Harlow, W.R. Branch, and J.K. Webb. 1996. Life on the lowest branch: sexual dimorphism, diet, and reproductive biology of an African twig snake, *Thelotornis capensis* (Serpentes, Colubridae). *Copeia* 1996:290–299.
Shine, R., and M.B. Thompson. 2006. Did embryonic responses to incubation conditions drive the evolution of reproductive modes in squamate reptiles? *Herpetological Monographs* 20:159–171.
Siebenrock, F. 1893. Das Skelet von *Brookesia superciliaris* Kuhl. *Sitzungsberichte der Mathematisch-Naturwissenschaftlichen Classe der Kaiserlichen Akademie der Wissenschaften* 102:71–118 [in German].
Siegel, J.M. 2008. Do all animals sleep? *Trends in Neurosciences* 31:208–213.
Sillman, A.J., J.K. Carver, and E.R. Loew. 1999. The photoreceptors and visual pigments in the retina of a boid snake, the ball python (*Python regius*). *Journal of Experimental Biology* 202:1931–1938.
Sillman, A.J., V.I. Govardovskii, P. Rohlich, J.A. Southard, and E.R. Loew. 1997. The photoreceptors and visual pigments of the garter snake (*Thamnophis sirtalis*): a microspectrophotometric, scanning electron microscopic and immunocytochemical study. *Journal of Comparative Physiology A* 181:89–101.
Sillman, A.J., J.L. Johnson, and E.R. Loew. 2001. Retinal photoreceptors and visual pigments in *Boa constrictor imperator*. *Journal of Experimental Zoology* 290:359–365.
Simonetta, A. 1957. Sulla possibilita che esistano relazioni tra meccanismi cinetici del cranio e morfologia dell'orecchio medio. *Monitore Zoologico Italiano* 65:48–55 [in Italian].
Sinervo, B., and J.B. Losos. 1991. Walking the tight rope: arboreal sprint performance among *Sceloporus occidentalis* lizard populations. *Ecology* 72:1225–1233.
Sinervo, B., F. Mendez-de-la-Cruz, D.B. Miles, B. Heulin, E. Bastiaans, M. Villagran-Santa Cruz, R. Lara-Resendiz, N. Martinez-Mendez, M.L Calderon-Espinosa, R.N. Meza-Lazaro,

H. Gadsden, L.J. Avila, M. Morando, I.J. De la Riva, P.V. Sepulveda, C.F.D. Rocha, N. Ibarguengoytia, C.A. Puntriano, M. Massot, V. Lepetz, T.A. Oksanen, D.G. Chapple, A.M. Bauer, W.R. Branch, J. Clobert, and J.W. Sites Jr. 2010. Erosion of lizard diversity by climate change and altered thermal niches. *Science* 328:894–899.

Singh, L.a.K., L.N. Acharjyo, and H.R. Bustard. 1983. Observations of the reproductive biology of the Indian chameleon *Chamaeleo zeylanicus. Journal of the Bombay Natural History Society* 81:86–92.

Skinner, J.H. 1959. Ontogeny of the breast-shoulder apparatus of the South African lacertilian, *Microsaura pumila pumila* (Daudin). *Annale van die Uniwersiteit van Stellenbosch* 35(1):5–66.

Slaby, O. 1984. Morphogenesis of the nasal apparatus in a member of the genus *Chamaeleon* L. (Morphogenesis of the nasal capsule, the epithelial nasal tube and the organ of Jacobson in Sauropsida. VIII). *Folia Morphologica* 32(3):225–246.

Slatyer, C., D. Rosauer, and F. Lemckert. 2007. An assessment of endemism and species richness patterns in the Australian Anura. *Journal of Biogeography* 34:583–596.

Smith, K.T. 2009. Eocene lizards of the clade *Geiseltaliellus* from Messel and Geiseltal, Germany, and the early radiation of Iguanidae (Squamata: Iguania). *Bulletin of the Peabody Museum of Natural History* 50(2):219–306.

Smith, K.T., S.F.K. Schaal, S. Wei, and C.-T. Li. 2011. Acrodont iguanians (Squamata) from the Middle Eocene of the Huadian Basin of Jilin Province, China, with a critique of the taxon "*Tinosaurus*." *Vertebrata PalAsiatica* 49(1):69–84.

So, K.-K.J., P.C. Wainwright, and A.F. Bennet. 1992. Kinematics of prey processing in *Chamaeleo jacksonii*: conservation of function with morphological specialization. *Journal of Zoology, London* 226:47–64.

Spawls, S. 2000. The chameleons of Ethiopia: an annotated checklist, key and field notes. *Walia* 21:3–13.

Spawls, S., K. Howell, R. Drewes, and J. Ashe. 2004. A Field Guide to the Reptiles of East Africa. London: A & C Black.

Spickler, J.C., S.C. Sillett, S.B., Marks, and H.H. Welsh. 2006. Evidence of a new niche for a North American salamander: *Aneides vagrans* residing in the canopy of old-growth redwood forest. *Herpetological Conservation and Biology* 1:16–26.

Stamps, J.A. 1977. Social behavior and spacing patterns in lizards, pp. 264–334 in C. Gans and D.W. Tinkle, Eds., *Biology of the Reptilia, Volume 7, Ecology and Behavior A*. New York: Academic Press.

Stefanelli, A. 1941. I centri motori dell'occhio e le loro connessioni nel *Chamaeleon vulgaris*, con riferimenti comparative in altri rettili. *Archivio Italiano di Anatomia e di Embriologia* 45:360–412 [in Italian].

Stevens, M., and S. Merilaita. 2009. Animal camouflage: current issues and new perspectives. *Philosophical Transactions of the Royal Society B* 364:423–427.

Stipala, J., N. Lutzmann, P.K. Malonza, L. Borghesio, P. Wilkinson, B. Godley, and M.R. Evans. 2011. A new species of chameleon (Sauria: Chamaeleonidae) from the highlands of northwest Kenya. *Zootaxa* 3002:1–16.

Stipala, J., N. Lutzmann, P.K. Malonza, P. Wilkinson, B. Godley, J. Nyamache, and M.R. Evans. 2012. A new species of chameleon (Squamata: Chamaeleonidae) from the Aberdare Mountains in the central highlands of Kenya. *Zootaxa* 3391:1–22.

Stuart, S., J.S. Chanson, N.A. Cox, B.E. Young, A.S.L. Rodrigues, D.L. Fishman, and R.B. Waller. 2004. Status and trends of amphibian declines and extinctions worldwide. *Science* 306:1783–1786.

Stuart, S.N., and R.J. Adams. 1990. Biodiversity in sub-saharan Africa and its islands: conservation, management, and sustainable use. *Occasional Papers of the IUCN Species Survival Commission No. 6, VI.* Gland, Switzerland: IUCN.

Stuart-Fox, D. 2009. A test of Rensch's rule in dwarf chameleons (*Bradypodion spp.*), a group with female-biased sexual size dimorphism. *Evolutionary Ecology* 23:425–433.

Stuart-Fox, D.M., D. Firth, A. Moussalli, and M.J. Whiting. 2006b. Multiple signals in chameleon contests: designing and analysing animal contests as a tournament. *Animal Behaviour* 71:1263–1271.

Stuart-Fox, D., and A. Moussalli. 2007. Sex-specific ecomorphological variation and the evolution of sexual dimorphism in dwarf chameleons (*Bradypodion spp.*). *Journal of Evolutionary Biology* 20:1073–1081.

Stuart-Fox, D., and A. Moussalli. 2008. Selection for social signalling drives the evolution of chameleon colour change. *PLoS Biology* 6(1):e25.

Stuart-Fox, D., and A. Moussalli. 2009. Camouflage, communication and thermoregulation: lessons from colour changing organisms. *Philosophical Transactions of the Royal Society B* 364:463–470.

Stuart-Fox, D., and A. Moussalli. 2011. Camouflage in color changing animals: trade-offs and constraints, pp. 237–253. In M. Stevens and S. Merilaita, Eds., *Animal Camouflage: Mechanisms and Function.* Cambridge, United Kingdom: Cambridge University Press.

Stuart-Fox, D., A. Moussalli, and M.J. Whiting. 2007. Natural selection on social signals: Signal efficacy and the evolution of chameleon display coloration. *American Naturalist* 170:916–930.

Stuart-Fox, D., A. Moussalli, and M.J. Whiting. 2008. Predator-specific camouflage in chameleons. *Biology Letters* 4:326–329.

Stuart-Fox, D.M., and M.J. Whiting. 2005. Male dwarf chameleons assess risk of courting large, aggressive females. *Biology Letters* 1:231–234.

Stuart-Fox, D., M.J. Whiting, and A. Moussalli. 2006a. Camouflage and colour change: antipredator responses to bird and snake predators across multiple populations in a dwarf chameleon. *Biological Journal of the Linnean Society* 88:437–446.

Takahashi, H. 2008. Fruit feeding behavior of a chameleon *Furcifer oustaleti*: comparison with insect foraging tactics. *Journal of Herpetology* 42:760–763.

Talavera, R., and F. Sanchíz. 1983. Restos pliocénicos de Camaleón común, *Chamaeleo chamaeleon* (L.) de Málaga. *Boletín de la Real Sociedad Española de Historia Natural (Geología)* 81:81–84 [in Spanish].

Tauber, E.S., H.P. Roffwarg, and E.D. Weitzman. 1966. Eye movements and electroencephalogram activity during sleep in diurnal lizards. *Nature* 212:1612–1613.

Tauber, E.S., J.A. Rojas-Ramírez, and R. Hernández-Peón. 1968. Electrophysiological and behavioral correlates of wakefulness and sleep in the lizard *Ctenosaura pectinata. Electroencephalography and Clinical Neurophysiology* 24:424–433.

Thomas, C.D., A. Cameron, R.E. Green, M. Bakkenes, L.J. Beaumont, Y.C. Collingham, B.F.N. Erasmus, M.F. de Siqueira, A. Grainger, L. Hannah, L. Hughes, B. Huntley, A.S. van Jaarsveld, G.F. Midgley, L. Miles, M.A. Ortega-Huerta, A. Townsend Peterson, O.L. Phillips, and S.E. Williams. 2004. Extinction risk from climate change. *Nature* 427:145–148.

Thomas, H., J. Roger, S. Sen, J. Dejax, M. Schuler, Z. Al-Sulaimani, C. Bourdillon de Grassac, G. Breton, F. de Broin, G. Camoin, H. Cappetta, R.P. Carriol, C. Cavelier, C. Chaix, J.Y. Crochet, G. Farjanel, M. Gayet, E. Gheerbrant, A. Lauriat-Rage, D. Noël, M. Pickford, A.F. Poignant, J.C. Rage, J. Roman, J.M. Rouchy, S. Secrétan, B. Sigé, P. Tassy, and

S. Wenz. 1991. Essai de reconstitution des milieux de sédimentation et de vie des primates anthropoïdes de l'Oligocène de Taqah (Dhofar, Sultanat d'Oman). *Bulletin de la Société Géologique de France* 162:713–724 [in French].

Tilbury, C. 2010. *Chameleons of Africa—An Atlas, Including the Chameleons of Europe, the Middle East and Asia*. Frankfurt am Main, Germany: Edition Chimaira.

Tilbury, C.R. 1992. A new dwarf forest chameleon (Sauria: *Rhampholeon* Günther 1874) from Malawi, central Africa. *Tropical Zoology* 5:1–9.

Tilbury, C.R., and K.A. Tolley. 2009a. A re-appraisal of the systematics of the African genus *Chamaeleo* (Reptilia: Chamaeleonidae). *Zootaxa* 2079:57–68.

Tilbury, C.R., and K.A. Tolley. 2009b. A new species of dwarf chameleon (Sauria; Chamaeleonidae, *Bradypodion* Fitzinger) from KwaZulu Natal, South Africa with notes on recent climatic shifts and their influence on speciation in the genus. *Zootaxa* 2226:43–57.

Tilbury, C.R., K.A. Tolley, and W.R. Branch. 2006. A review of the systematics of the genus *Bradypodion* (Sauria: Chamaeleonidae), with the description of two new genera. *Zootaxa* 1363:23–38.

Tinkle, D.W., and J.W. Gibbons. 1977. The distribution and evolution of viviparity in reptiles. *Miscellaneous Publications Museum of Zoology, University of Michigan* 154:1–55.

Todd, M. 2011. Trade in Malagasy Reptiles and Amphibians in Thailand. Petaling Jaya, Selangor, Malaysia: TRAFFIC Southeast Asia.

Toerien, M.J. 1963. The sound-conducting systems of lizards without tympanic membranes. *Evolution* 17(4):540–547.

Tolley, K.A., and M. Burger. 2007. *Chameleons of Southern Africa*. Cape Town, South Africa: Struik.

Tolley, K.A., M. Burger, A.A. Turner, and C.A. Matthee. 2006. Biogeographic patterns and phylogeography of dwarf chameleons (*Bradypodion*) in an African biodiversity hotspot. *Molecular Ecology* 15(3):781–793.

Tolley, K.A., B.M. Chase, and F. Forest. 2008. Speciation and radiations track climate transitions since the Miocene Climatic Optimum: a case study of southern African chameleons. *Journal of Biogeography* 35:1402–1414.

Tolley, K.A., and G.J. Measey. 2007. Chameleons and vineyards in the Western Cape of South Africa: is automated grape harvesting a threat to the Cape Dwarf Chameleon (*Bradypodion pumilum*)? *African Journal of Herpetology* 56:85–89.

Tolley, K.A., R.N.V. Raw, R. Altwegg, and G.J. Measey. 2010. Chameleons on the move: survival and movement of the Cape Dwarf Chameleon, *Bradypodion pumilum*, within a fragmented urban habitat. *African Zoology* 45:99–106.

Tolley, K.A., C.R. Tilbury, W.R. Branch, and C.A. Matthee. 2004. Phylogenetics of the Southern African dwarf chameleons, *Bradypodion* (Squamata: Chamaeleonidae). *Molecular Phylogenetics and Evolution* 30:354–365.

Tolley, K.A., C.R. Tilbury, G.J. Measey, M. Menegon, W.R. Branch, and C.A. Matthee. 2011. Ancient forest fragmentation or recent radiation? Testing refugial speciation models in chameleons within an African biodiversity hotspot. *Journal of Biogeography* 38:1748–1760.

Tolley, K.A., T.M. Townsend, and M. Vences. 2013. Large-scale phylogeny of chameleons suggests African origins and rapid Eocene radiation. *Proceedings of the Royal Society of London Series B—Biological Sciences* 280(1759):20130184

Townsend, T., and A. Larson. 2002. Molecular phylogenetics and mitochondrial genomic evolution in the Chamaeleonidae (Reptilia, Squamata). *Molecular Phylogenetics and Evolution* 23(1):22–36.

Townsend, T.M., A. Larson, E. Louis, and J.R. Macey. 2004. Molecular phylogenetics of Squamata: the position of snakes, amphisbaenians, and dibamids, and the root of the squamate tree. *Systematic Biology* 53:735–757.
Townsend, T.M., D.G. Mulcahy, B.P. Noonan, B.P., J.W. Sites Jr., C.A. Kuczynski, J.J. Wiens, and T.W. Reeder. 2011a. Phylogeny of iguanian lizards inferred from 29 nuclear loci, and a comparison of concatenated and species-tree approaches for an ancient, rapid radiation. *Molecular Phylogenetics and Evolution* 61:363–380.
Townsend, T.M., K.A. Tolley, F. Glaw, W. Böhme, and M. Vences. 2011b. Eastward from Africa: palaeocurrent-mediated chameleon dispersal to the Seychelles islands. *Biology Letters* 7:225–228.
Townsend, T.M., D.R. Vieites, F. Glaw, and M. Vences. 2009. Testing species-level diversification hypotheses in Madagascar: the case of microendemic *Brookesia* leaf chameleons. *Systematic Biology* 58(6):641–656.
Toxopeus, A.G., J.P. Kruijt, and D. Hillenius. 1988. Pair-bonding in chameleons. *Naturwissenschaften* 75:268–269.
Trost, E. 1956. Über die lage des foramen parietals bei rezenten Reptilien und Labyrinthodontia. *Acta Anatomy* 26:318–339 [in German with English summary].
Uetz, P. 2012. The Reptile Database. Accessed at www.reptile-database.org on August 15, 2012.
Ullenbruch, K., P. Krause, and W. Böhme 2007. A new species of the *Chamaeleo dilepis* group (Sauria: Chamaeleonidae) from West Africa. *Tropical Zoology* 20:1–17.
Uller, T., D. Stuart-Fox, and M. Olsson. 2010. Evolution of primary sexual characters in reptiles, pp. 426–453. In A. Córdoba-Aguilar and J.L. Leonard, Eds., *The Evolution of Primary Sexual Characters in Animals*. Oxford, United Kingdom: Oxford University Press.
Underwood, G. 1970. The eye, pp. 1–97. In C. Gans, C. and T.S. Parsons, Eds. *Biology of the Reptilia. Volume 2. Morphology B*. New York: Academic Press.
Upchurch, G.R., B.L. Otto-Bliesner, and C. Scotese. 1998. Vegetation—atmosphere interactions and their role in global warming during the latest Cretaceous. *Philosophical Transactions of the Royal Society of London. Series B: Biological Sciences* 353:97–112.
Upchurch, G.R.J., B.L. Otto-Bliesner, and C.R. Scotese. 1999. Terrestrial vegetation and its effects on climate during the latest Cretaceous. *Geological Society of America Special Papers* 332:407–426.
Van Bocxlaer, I., S.P. Loader, K. Roelants, S.D. Biju, M. Menegon, and F. Bossuyt. 2010. Gradual adaptation toward a range-expansion phenotype initiated the global radiation of toads. *Science* 327:679–682.
van der Meulen, A.J., I. García-Paredes, M.A. Álvarez-Sierra, L.W. van den hoek Ostende, K. Hordijk, A. Oliver, P. López-Guerrero, V. Hernández-Ballarín, and P. Peláez-Campomanes. 2011. Biostratigraphy or biochronology? Lessons from the Early and Middle Miocene small mammal events in Europe. *Geobios* 44:309–321.
van Leeuwen, J.L. 1997. Why the chameleon has spiral-shaped muscle fibres in its tongue. *Philosophical Transactions of the Royal Society of London Series B* 352(1353):573–589.
van Zinderen Bakker, E.M. 1975. The origin and palaeoenvironment of the Namib Desert biome. *Journal of Biogeography* 2:65–73.
Van Heygen, G., and E. Van Heygen. 2004. Eerste waarnemingen in de vrije natuur van het voortplantingsgedrag bij de tijgerkameleon *Calumma tigris* (Kuhl 1820). *TERRA—Antwerpen* 40:49–51.
Vanhooydonck, B., A. Herrel, R. Van Damme, J.J. Meyers, and D.J. Irschick. 2005. The relationship between dewlap size and performance changes with age and sex in a green anole (*Anolis carolinensis*) lizard population. *Behavioral Ecology and Sociobiology* 59(1):157–165.

Vanhooydonck, B., and R. Van Damme. 1999. Evolutionary relationships between body shape and habitat use in lacertid lizards. *Evolutionary Ecology Research* 1:785–805.

Vanhooydonck, B., Van Damme, R. and P. Aerts. 2002. Variation in speed, gait characteristics and microhabitat use in lacertid lizards. *Journal of Experimental Biology* 205:1037–1046.

Vanhooydonck, B., R. Van Damme, A. Herrel, and D.J. Irschick. 2007. A performance based approach to distinguish indices from handicaps in sexual selection studies. *Functional Ecology* 21:645–652.

Vejvalka, J. 1997. Obojživelníci (Amphibia: Caudata, Salientia) a plazi (Reptilia: Lacertilia, Choristodera) miocenní lokality Merkur–sever (Česká republika). M.Sc. Thesis, Charles University, Prague [in Czech].

Vences, M., F. Glaw, and C. Zapp. 1999. Stomach content analyses in Malagasy frogs of the genera *Tomopterna, Aglyptodactylus, Boophis* and *Mantidactylus* (Amphibia: Ranidae). *Herpetozoa* 11:109–116.

Vences, M., J. Kosuch, M.-O. Rödel, S. Lötters, A. Channing, F. Glaw, and W. Böhme. 2004. Phylogeography of *Ptychadena mascareniensis* suggests transoceanic dispersal in a widespread African-Malagasy frog lineage. *Journal of Biogeography* 31:593–601.

Vences, M., D.R. Vieites, F. Glaw, H. Brinkmann, J. Kosuch, M. Veith, and A. Meyer. 2003. Multiple overseas dispersal in amphibians. *Proceedings of the Royal Society of London Series B—Biological Sciences* 270:2435–2442.

Vences, M., K.C. Wollenberg, D.R. Vieites, and D.C. Lees. 2009. Madagascar as a model region of species diversification. *Trends in Ecology and Evolution* 24:456–465.

Vidal, N., and S.B. Hedges. 2005. The phylogeny of squamate reptiles (lizards, snakes, and amphisbaenians) inferred from nine nuclear protein-coding genes. *Comptes Rendus Biologies* 328:1000–1008.

Vidal, N., and S.B. Hedges. 2009. The molecular evolutionary tree of lizards, snakes, and amphisbaenians. *Comptes Rendus Biologies* 332:129–139.

Vinson, J., and J.-M. Vinson. 1969. The saurian fauna of the Mascarene islands. *Mauritius Institute Bulletin* 6:203–320.

Visser, J.G.J. 1972. Ontogeny of the chondrocranium of the chameleon, *Microsaura pumila* (Daudin). *Annale van die Universiteit van Stellenbosch* 47A:1–68.

Vitt, L. J. 2000. Ecological consequences of body size in neonatal and small-bodied lizards in the neotropics. *Herpetological Monographs* 14:388–400.

Von Volker, J.S. 1999. Litho- und biostratigraphische Untersuchungen in der Oberen Süsswassermolasse des Landkreises Biberach a. d. Riß (Oberschwaben) Stuttgarter. *Beiträge zur Naturkunde Serie B (Geologie und Paläontologie)* 276:1–167.

Vrolik, W. 1827. *Natuur - en Ontleedkundige Opmerkingen over den Chameleon*. Amsterdam: Meyer Warnars.

Wager, V.A. 1986. *The Life of the Chameleon*. Durban, South Africa: Wildlife Society.

Wainwright, P.C., and A.F. Bennett. 1992a. The mechanism of tongue projection in chameleons. I. Electromyographic tests of functional hypotheses. *Journal of Experimental Biology* 168:1–21.

Wainwright, P.C., and A.F. Bennett. 1992b. The mechanism of tongue projection in chameleons. II. Role of shape change in a muscular hydrostat. *Journal of Experimental Biology* 168:23–40.

Wainwright, P.C., D.M. Kraklau, and A.F. Bennett. 1991. Kinematics of tongue projection in *Chamaeleo oustaleti*. *Journal of Experimental Biology* 159:109–133.

Wall, G.L. 1942. *The Vertebrate Eye and its Adaptive Radiation*. New York: Hafner.

Wallach, V., W. Wüster, and D.G. Broadley. 2009. In praise of subgenera: taxonomic status of cobras of the genus *Naja* Laurenti (Serpentes: Elapidae). *Zootaxa* 2236:26–36.

Walter, R.C., P.C. Manega, R.L. Hay, R.E. Drake, and G.H. Curtis. 1991. Laser-fusion $^{40}Ar/^{39}Ar$ dating of Bed I, Olduvai Gorge, Tanzania. *Nature* 354:145–149.
Walton, B.M., and A.F. Bennett. 1993. Temperature-dependent color change in Kenyan chameleons. *Physiological Zoology* 66:270–287.
Wang, Y., and J.L. Li. 2008. Squamata, pp. 115–137. In J.L. Li, X.C. Wu, and F. Zhang, Eds., *The Chinese Fossil Reptiles and Their Kin*. Beijing, China: Science Press.
Wells, N.A. 2003. Some hypotheses on the Mesozoic and Cenozoic paleoenvironmental history of Madagascar, pp. 16–34. In S.M. Goodman and J.P. Benstead, Eds., *The Natural History of Madagascar*. Chicago: University of Chicago Press.
Werner, F. 1902a. Einer Monographie der Chamaleonten. *Zoologische Jahrbuecher. Systematik* 15:295–460.
Werner, F. 1902b. Zur Kenntnis des Skeletes von *Rhampholeon spectrum*. *Arbeiten aus dem Zoologishen Instituten der Universität Wien und der Zoologishen Station in Triest* 14:259–290.
Werner, F. 1911. Chamaeleontidae. *Das Tierreich* 27, I-XI:1–52.
Wessels, B.R., and B. Maritz. 2009. *Bitis schneideri* (Namaqua Dwarf Adder). Diet. *Herpetological Review* 40:440.
Wever, E.G. 1968. The ear of the chameleon: *Chamaeleo senegalensis* and *Chamaeleo quilensis*. *Journal of Experimental Zoology* 168(4):423–436.
Wever, E.G. 1969a. The ear of the chameleon: the round window problem. *Journal of Experimental Zoology* 171:1–5.
Wever, E.G. 1969b. The ear of the chameleon: *Chamaeleo höhnelii* and *Chamaeleo jacksoni*. *Journal of Experimental Zoology* 171(3):305–312.
Wever, E.G. 1973. Function of middle ear in lizards: divergent types. *Journal of Experimental Zoology* 184(1):97–125.
Wever, E.G., and Y.L. Werner. 1970. The function of the middle ear in lizards: *Crotaphytus collaris* (Iguanidae). *Journal of Experimental Zoology* 175(3):327–341.
Wheeler, P.E. 1984. An investigation of some aspects of the transition from ectothermic to endothermic metabolism in vertebrates. PhD thesis. University of Durham, North-Carolina.
White, F. 1983. The vegetation of Africa, a descriptive memoir to accompany the UNESCO/AETFAT/UNSO Vegetation Map of Africa (3 Plates, Northwestern Africa, Northeastern Africa, and Southern Africa, 1:5,000,000). Paris: UNESCO.
Wickens, G.E. 1976. *The Flora of Jebel Marra (Sudan Republic) and its Geographical Affinities*. London: Royal Botanic Gardens, Kew.
Wiens, J.J., M.C. Brandley, and T.W. Reeder. 2006. Why does a trait evolve multiple times within a clade? Repeated evolution of snake-like body form in squamate reptiles. *Evolution* 61:123–141.
Wiens, J.J., C.A. Kuczynski, T. Townsend, T.W. Reeder, D.G. Mulcahy, and J.W. Sites, Jr. 2010. Combining phylogenomics and fossils in higher level squamate reptile phylogeny: molecular data change the placement of fossil taxa. *Systematic Biology* 59:674–688.
Wild, C. 1994. Ecology of the Western pygmy chameleon *Rhampholeon spectrum* Buchholz 1874 (Sauria: Chamaeleonidae). *British Herpetological Society Bulletin* 49:29–35.
Wilkinson, M., S.P. Loader, D.J. Gower, J.A. Sheps, and B.L. Cohen. 2003. Phylogenetic relationships of African caecilians (Amphibia: Gymnophiona): insights from mitochondrial rRNA gene sequences. *African Journal of Herpetology* 52:83–92.
Williams, J. 2012. Humans and biodiversity: population and demographic trends in the hotspots. *Population & Environment* Epub before print.

Williams, S.C., and L.D. McBrayer. 2011. Attack-based indices, not movement patterns, reveal intraspecific variation in foraging behavior. Behavioural Ecology 22:993–1002.
Wilmé, L., S.M. Goodman, and J.U. Ganzhorn. 2006. Biogeographic evolution of Madagascar's microendemic biota. *Science* 312:1063–1065.
Wollenberg, K.C., D.R. Vieites, A. Van Der Meijden, F. Glaw, D.C. Cannatella, and M. Vences. 2008. Patterns of endemism and species richness in Malagasy cophyline frogs support a key role of mountainous areas for speciation. *Evolution* 62:1890–1907.
Wright, J.W., and D.G. Broadley. 1973. Chromosomes and the status of *Rhampholeon marshalli* Boulenger (Sauria: Chamaeleonidae). *Bulletin of the Southern California Academy of Science* 72:164–165.
Yoder, A.D., and M.D. Nowak. 2006. Has vicariance or dispersal been the predominant biogeographic force in Madagascar? Only time will tell. *Annual Review of Ecology and Systematics* 37:405–31.
Zachos, J.C., G.R. Dickens, and R.E. Zeebe. 2008. An early Cenozoic perspective on greenhouse warming and carbon-cycle dynamics. *Nature* 451:279–283.
Zachos, J.C., M. Pagani, L. Sloan, E. Thomas, and K. Billups. 2001. Trends, rhythms, and abberations in global climate 65 Ma to present. *Science* 292:686–693.
Zachos, J.C., M.W. Wara, S. Bohaty, M.L. Delaney, M.R. Petrizzo, A. Brill, T.J. Bralower, and I. Premoli-Silva. 2003. A transient rise in tropical sea surface temperature during the Paleocene-Eocene thermal maximum. *Science* 302:1551–1554.
Zani, P.A. 2000. The comparative evolution of lizard claw and toe morphology and clinging performance. *Journal of Evolutionary Biology* 13:316–325.
Zarcone, G., F.M. Petti, A. Cillari, P. Di Stefano, D. Guzzetta, and U. Nicosia. 2010. A possible bridge between Adria and Africa: New palaeobiogeographic and stratigraphic constraints on the Mesozoic palaeogeography of the Central Mediterranean area. *Earth-Science Reviews* 103:154–162.
Zari, T.A. 1993. Effects of body mass and temperature on standard metabolic rate of the desert chameleon. *Journal of Arid Environments* 24:75–80.
Zerova, G.A., and V.M. Chkhikvadze, 1984. Review of Cenozoic lizards and snakes of the USSR. *Izvestiya Akademii Nauk Gruzinskoi SSR, Seriya Biologicheskaya* 10:319–326. [in Russian].
Zhou, L., R.E. Dickinson, P. Dirmeyer, A. Dai, and S.-K. Min. 2009. Spatiotemporal patterns of changes in maximum and minimum temperatures in multi-model simulations. *Geophysical Research Letters* 36:L02702.
Zippel, K.C., R.E. Glor, and J.E.A. Bertram. 1999. On caudal prehensility and phylogenetic constraint in lizards: the influence of ancestral anatomy on function in *Corucia* and *Furcifer*. *Journal of Morphology* 239:143–155.
Zoond, A. 1933. The mechanism of projection of the chameleon's tongue. *Journal of Experimental Biology* 10:174–185.
Zoond, A., and J. Eyre. 1934. Studies in reptilian colour response. I. The bionomics and physiology of pigmentary activity of the chameleon. *Philosophical Transactions of the Royal Society of London, Series B* 223:27–55.

PHOTO CREDITS

Cover	Michele Menegon
1.1	Michele Menegon
1.2	Michele Menegon
1.3	Krystal Tolley
1.4	Michele Menegon
1.5	Michele Menegon
1.6	Krystal Tolley
1.7	Michele Menegon
5.1	Marius Burger, Tania Fouche, Krystal Tolley
6.1	Adnan Moussalli
6.2	Devi Stuart-Fox
6.3	Devi Stuart-Fox and Adnan Moussalli
6.4	Devi Stuart-Fox
8.1	Henrik Bringsøe
8.2	Krystal Tolley
8.3	Marius Burger
8.4	Marius Burger
8.5	Krystal Tolley
8.6	Marius Burger
8.7	Michele Menegon
8.8	William Branch
8.9	Krystal Tolley
8.10	Michele Menegon
8.11	Michele Menegon

INDEX

Figures cited without page numbers appear in the color insert.

abundance, 7, 91, 92, 102, 105, 110, 212
accessory palmar/plantar spines, 169
accommodation, 1, 44, 57–58, 116, 128
Acrodonta, acrodontan, acrodont iguanian, 175, 178–83, 187–88 (fig. 9.3), 189–92
Acrodont dichotomy, 189
Acrodonty, acrodont dentition, 13, 151, 179–81, 183
adrenocorticotropic hormone, 118
aestivation, 96, 103
Africa, 2, 4–5, 63, 85–86, 93, 95, 112, 131–35, 137–38, 143, 145–50, 152, 155–56, 161, 164, 175–76 (table 9.1), 185, 187, 188 (fig. 9.3), 190–92, 194, 196 (table 10.2), 197, 203 (table 10.4), 204 (table 10.5), 210–11 (table 10.8), 214–15 (fig. 10.4)
  Central, 98, 144, 149
  East, 4, 91, 93, 136–37, 139, 143, 145–46, 153, 155, 158–59, 166–67, 173, 176 (table 9.1), 190–92, Fig. 5.1
  North, 132, 134, 145
  South, 68, 71, 102, 105, 112, 133, 135, 144, 147, 153, 159–60, 166–67, 177 (table 9.2), 183, 186, 194–95, 197, 199 (table 10.3), 210, Fig. 1.6, Fig. 5.1
  Southern, 85, 93, 94, 98–99, 112, 134, 140–41 (table 7.1), 143–44, 147–48, Fig. 5.1
  sub-Saharan, 110, 113, 148, 195
  West, 42, 91, 93, 96, 110, 136, 144, 146, 149, 154–55, 172–73, 204 (table 10.5), 213
Afromontane, 135, 142–44, 146–47, 149
Agama, 30–31, 39, 66, 183
Agamidae, agamid, 3 (fig. 1.8), 13, 16, 25, 59–61, 126, 131, 151, 175–76 (table 9.1), 179, 180–83, 187, 188, (fig. 9.3), 189–92, Fig. 7.1
Agaminae, 179
aggressive display, 123–24
aggressive rejection, 123–24
Albertine Rift, 135–37, 140–41 (table 7.1), 143, 144, 147, 167
allopatric, 93–94, 135, 160
allopatry, 99
amnion, 62
Anguidae, anguid, 181
Anguimorpha, anguimorph, 178, 188 (fig. 9.3), 189
Anhuisaurus, 181
Anquingosaurus, 183
antipredator behaviour, 115, 126–27, 129–30
arboreal, 1–2, 4–5, 25, 30, 31, 49, 55, 63–64, 66, 68, 70, 72–73, 85–87, 89–90, 93, 96, 98, 101, 106, 109–12, 121, 126–28, 132–33, 135, 137–38, 149, 151–52, 157, 213–14
Archaius, 8, 11, 39, 47, 137, 153, 158, 188 (fig. 9.3), 191, 201–03, fig. 7.1, fig. 7.2
  tigris, 101, 106, 138, 145, 170, 191, 195, fig. 8.1. See also *Calumma tigris*
arrested development, 98
Asia, 63, 132, 134–35, 145, 149, 155, 164, 176 (table 9.1), 185, 187, 189–92, 206, 207 (fig. 10.3)
  Central, 189
auditory signal, 116

auditory system, 1, 57–58, 116
Australia, 130, 182–83, 189

barriers, 92, 212
Belgium, 183, 207
Bharatagama, 180, 187, 189, 192
bicuspid claws, 39–40, 169–70
biodiversity hotspot, 142, 147, 214
Bioko Island, 136, 146
bite force, 94, 104, 106–07, 125–26
body size, 87, 97, 100, 102–03, 106, 115, 126, 138
Bradypodion, 8, 10–12, 14–16, 25 (fig. 2.3), 30-31, 39–40, 45, 48, 51, 54, 65 (fig. 4.1), 68, 86, 93–94, 98–99, 112, 119, 121, 124, 125–27, 129, 133, 135, 140–41 (table 7.1), 144, 147, 152, 158–60, 164, 166–68, 171, 177 (table 9.3), 186, 188 (fig. 9.3), 194–95, 199 (table 10.3), 201, 203 (table 10.4), 210, 213–14, Fig. 5.1, Fig. 6.3, Fig. 7.1, Fig. 7.2
  damaranum, 37, 94
  pumilum, 8, 9 (fig. 2.1), 14, 49, 60, 71, 94, 99, 101–07, 112, 118, 121–23, 125–26, 141 (table 7.1), 153, 158–59, 160, 167, 214
  transvaalense, 94, 112, 124, 129, 141, 158, 213, Fig. 1.6, Fig 6.1
brain, 44, 49, 50, 59
Brevidensilacerta, 182
Brookesia, 8, 10–12, 14, 25 (fig.2.3), 26–27, 29, 37–41, 45, 47, 51, 53–54, 63, 72, 86–87, 90–92, 96–98, 100–02, 106, 110, 112, 117, 120, 126–29, 132–33, 136–40 (table 7.1), 146, 152, 155–57, 159–62, 170, 188 (9.3), 190–92, 194, 198–99 (table 10.3), 201, 203 (table 10.4), 209–10, 214, Fig. 7.1, Fig. 7.2, Fig. 8.3
  superciliaris, 9 (fig. 2.1), 14, 25 (fig. 2.3), 102, 110, 127, 160–61, Fig. 8.3
Brookesiinae, 152, 155–57, 165
burrows, 97
bushes, 93, 96, 129, 134–35

Calotes, 180, 183
Calumma, 8, 10–12, 36, 39, 51, 54, 63, 86–87, 89, 91–93, 98, 102, 106, 110–12, 121, 126, 133, 137, 138–40 (table 7.1), 146, 152, 156, 158, 162–63, 166–68, 185, 188 (fig. 9.3), 191, 194, 198–99 (table 10.3), 201, 203 (table 10.4), 104, 209–10, 213
  brevicorne, 11, 92, 102–03, 111, 121, 162–63
  globifer, 162, 185, 194
  oshaughnessyi, 87, 102, 121, 162
  tigris, 158, 191. See also *Archaius tigris*
camouflage, 3, 85, 94, 115, 119, 126–28, 130
Canary Islands, 146
cannibalism, 101, 108, 115
casque, 7, 11–12, 14–15, 38, 40, 95, 125–26, 159, 164, 166, 168, 171
Cenozoic, 188 (fig. 9.3)
Chamaeleo, 4 (fig. 1.9), 8, 10–12, 15, 25 (fig. 2.3), 27, 31, 36, 38–39, 44–46, 48,51, 54, 58, 66, 67, 86, 91, 93, 96, 98, 112, 117, 123, 133–35, 143–45, 147, 152, 159, 161, 163–65, 167–68, 177 (table 9.2), 178, 184–85, 187, 188 (fig. 9.3), 201–04 (table 10.5), 208
  andrusovi, 177 (table 9.2), 184–85
  bavaricus, 178 (table 9.2), 184
  bitaeniatus, 153, 185. *See also Trioceros bitaeniatus*
  calyptratus, 36, 53 (fig. 2.7), 60, 62, 67 (fig. 4.2), 68, 70, 82 (fig. 4.6), 100–01, 108, 117, 121–24, 130, 134, 145, 163, 185, 209
  caroliquarti, 177 (table 9.2), 184–85
  chamaeleon, 48–49, 96–98, 100–01, 105–06, 108, 111, 115, 120, 122–23, 126, 129, 134–35, 146, 163, 177 (table 9.2), 185, 187, 213
  dilepis, 49, 72, 95, 111, 118, 120–21, 134, 143–45, 163–65, 203–04 (table 10.5)
  intermedius, 177 (table 9.2), 185
  jacksonii, 187. See also *Triocerus jacksonii*
  namaquensis, 60, 96–97, 103, 105, 107–09, 111, 113, 120, 127, 134–35, 144, 163–65
  pfeili, 177 (table 9.2), 184
  simplex, 177 (table 9.2), 184
  sulcodentatus, 184
Chamaeleonidae, 3 (fig. 1.8), 7, 26, 105, 117, 119, 126, 130–31, 151–54, 155–57, 160, 166, 172, 174, 177 (table 9.2), 179, 183, 185, 188 (fig. 9.3), 190
Chamaeleoninae, 152, 155, 156
Chamaeleonoidea, 179
Changjiangosaurus, 181
China, 180–83, 190
Chlamydosaurus, 183
chromatophore, 61, 117
CITES appendix, 201, 209
cladistic, 153
climate change, 169, 211–13, 214, 216

clutch size, 100
color, 2–3, 37, 51–52, 61, 86, 93–94, 96, 100, 115–30, 132–33, 138, 148, 165, 201
Comoros Islands, 139, 148
conservation, 193–5, 197, 201, 210, 216
  status, 4, 193–194
contest, 119, 121, 125–26, 129
copulation, 53, 120, 122–23
courtship, 89, 117, 121–26, 129
  rejection, 122
Cretaceous, 4, 131–32, 162, 176 (table 9.2), 180–83, 188 (fig. 9.3), 189–192
critically endangered, 196–97
Crotaphytidae, 179
Czech Republic, 145, 177 (table 9.2), 183–85, 209

death-feigning, 129
desert, 2, 60–63, 81, 93, 96–97, 134, 143, 144
development, 8, 31–32, 45, 48, 54, 57, 62, 74, 88, 97–98, 100–01, 137, 155, 158, 169, 171, 196
dispersal, 134, 136, 138–39, 145–46, 161, 191–92, 212–13
distribution, 7, 87, 91–92, 95, 98, 113, 117, 134–35, 137, 139, 142, 144–50, 160, 169, 171, 185, 187, 190–91, 193, 195, 197, 202, 212–14
divergence dates, 190–91
diversity, 4, 64, 68, 70–71, 86, 91, 95, 125–26, 130, 138–39, 140 (table 7.1), 141–44, 146–48, 150, 163, 175, 187, 190, 192, 211, 214
dominant coloration, 119
Dorsetisaurus, dorsetisaur, 188
Draco, 180
draconine agamids, 180, 187
dry forest, 93, 138–39, 144
dwarf chameleons, 68, 72, 93–94, 119–20, 123, 129, 153, 159–60

ear, 42, 45–46, 58
East Usambara Mountains, 89, 92, 136
Eastern Arc Mountains, 135–37, 140–41 (table 7.1), 142, 147, 149, 214
Eastern Highlands, 144
ecomorph, 105–07, 125, 160, 174
ecotones, 90–91, 95, 107, 137
edge effect, 91
egg, 52, 62, 85, 88, 97–102, 109–11, 115
egg retention, 98
Egypt, 182
embryo, 62, 85, 88, 97, 98–99
embryonic diapause, 62, 97–98
endangered, 194–97, 200–01, 210
endemic, 134, 136, 138–39, 142–145, 147, 149, 152, 158, 194–95, 197, 203, 213–14
endemism, 4, 139, 142, 146, 148–49, 214, Fig. 7.3
England, 188
Eocene, 132, 137, 158, 176 (table 9.2), 181–83, 188–92, Fig. 7.1
epinephrine, 119
erythrophore, 61, 117–18
Ethiopian Highlands, 140–41, 143, 146, 148–49
Europe, 4, 63, 85, 96, 134–35, 145, 149, 155, 164, 175–76 (table 9.2), 181–85, 187–88 (fig. 9.3), 190–92, 206–07 (fig. 10.3), 210, Fig. 7.2
exports, 202–05 (fig. 10.2), 206, 209–10
eye, 1, 7, 13, 16, 40, 43–45, 47, 49–50, 57–59, 76, 85, 111, 116–17, 128, 132, 151, 180, Fig. 1.2

feeding, 1, 13, 63–64, 72–82, 89, 105–06, 195
fertilization, 122
fire, 87, 93
forest canopy, 90
fossil record, 4, 5, 131, 154, 175–76, 179, 181, 183, 187, 191
France, 138, 182–83, 187, 207
Furcifer, 8, 10–11, 27, 39, 42 (fig. 2.5), 45, 51, 53 (fig. 2.7), 54, 63, 85–86, 90–93, 95–96, 98–99, 103, 106–08, 110, 112, 122–26, 133, 137–40 (table 7.1), 145, 152, 156, 162, 165–68, 172, 188 (fig. 9.3), 194, 198, (table 10.3), 201–02 (fig. 10.1), 203 (table 10.4), 204 (table 10.5), 209–10, 214
  labordi, 85, 88, 96, 98, 103, 122–25, 138, 165–66
  lateralis, 10–11, 91, 96, 110, 122, 214, (fig. 8.6)
  pardalis, 11, 62, 89, 90–92, 95, 104, 107, 138, 146, 156, 165, 172, 185, 204 (table 10.5), 208–10
  verrucosus, 11, 87–88, 96, 99, 123–25, 138, 165, 210

gardens, 96, 102, 195, 214
Germany, 145, 177–78 (table 9.2), 183–85, 206, 207 (table 10.7)
global change, 193, 211, 216

Gondwana, 176 (table 9.1), 188 (fig. 9.3), 189–91
Gonocephalus, 180
grassland, 2, 86, 93–94, 96, 102, 112, 126–27, 133, 135, 137, 143
Greece, 183, 185
grip, 64, 66, 85, 89–90, 94, 112, Fig. 1.1
ground-dwelling, 85, 89–90, 100, 106
guanophores, 117
guilds, 85–87, 90
Guinean-Congolian forest, 144
gular, 21, 38, 41, 51, 111, 125–26, 129, 159, 164, 167–68, 170–71
  pouch, 51, 159, 164, 168

habitat alteration, 4, 214
hatchling size, 100
head bobs, 121
head shake, 121, 125
heathland scrub, 93
hemipenal, 50, 53, 152, 155–56, 159, 162, 164–65, 168–70
hemipenal apical ornamentation, 164
Holocene, 4, 177 (table 9.2), 183, 187
home range, 5, 120, 121
hotspot, 141–42, 147, 214
Huadiansaurus, 182
Huehuecuetzpalli, 189
Hungary, 183

Iguania, 25, 175, 178, 181, 187–88 (fig. 9.3), 189
Iguanidae, 25, 126, 131, 178–79
imports, 203–04, 206–07 (fig. 10.3), 208–09
incubation periods, 98
India, 63, 96, 134–35, 138, 145–46, 148, 175–76 (table 9.1), 180–83, 188 (fig. 9.3), 190, 192, Fig. 7.4
iridophores, 117–18
Isodontosaurus, 181, 188 (fig. 9.3), 192
Israel, 108, 111, 177 (table 9.2), 183, 187

Jacobson's organ, 48. *See also* vomeronasal organ
Jurassic, 131, 176 (table 9.1), 179–80, 187–88 (fig. 9.3), 190

Kazakhstan, 181, 182
Kenya, 72, 99, 107, 112, 137–39, 142–43, 146–47, 149, 155, 167, 173, 177 (table 9.2), 183, 185–86 (fig. 9.2), 197–98 (table 10.3), 202–04 (table 10.5), 205 (table 10.6, fig. 10.2), 206
Kenyan highlands, 99, 112, 135–36, 140 (table 7.1), 142, 167
Kinyongia, 8, 11, 39, 51, 54, 86, 92, 95, 100, 106, 133, 135, 137, 140 (table 7.1), 141 (table 7.1), 142, 147, 149, 152, 158, 163, 166–68, 188 (fig. 9.3), 198 (table 10.3), 200 (table 10.3), 201–02 (fig. 10.10), 203 (table 10.4), 204 (table 10.5), 210, 214
Kyrgyzstan, 179, 182

lateral compression, 7, 121, 125, 127, 129, 132
lateral display, 121, 125, 126
Laudakia, 183
Laurasia, 132, 176 (table 9.1), 189–90, 192
leaf chameleons, 88–90, 92, 96, 106, 131, 190, Fig. 5.1
least concern, 194–96 (table 10.1), 203–04 (table 10.5)
Lebanon, 177 (table9.2), 183, 187
Leiolepidinae, 179, 188 (fig. 9.3)
Leiolepis, 179, 182, 189–90
Lepidosauria, lepidosaurian, lepidosaur, 179, 180, 187
life-history, 85, 97–99, 102–03, 130, 212
limb, 2, 31, 34–36, 38–39, 59–60, 63–65 (fig. 4.1), 66–67 (fig. 4.2), 68–69 (fig. 4.3), 70–71, 112, 157, Fig. 2.4
locomotion, 2, 31–32, 34, 59, 63–64, 66, 68, 70–72
longevity, 103, 104
lung diverticulae, 51, 159, 168, 172–73
lung type, 156–57, 163–64, 166–67, 172–74

Madagascar, 2, 4, 5, 63, 85–99, 107, 109–12, 131–34, 137–40 (table 7.1), 143, 145–50, 152, 155, 163, 166, 176 (table 9.1), 177 (table 9.2), 183, 187–88 (fig. 9.3), 190–92, 194–98 (table 10.3), 200 (table 10.3), 203 (table 10.4), 204 (table 10.5), 205 (table 10.6, fig. 10.2), 211 (table 10.8), 213–14, 216, Fig. 5.1, Fig. 5.7, Fig. 7.2, Fig. 7.4
male harassment, 123
male-male competition, 124–25
Maputo-Pondo-Albany, 144
Mascarene islands, 155
mate choice, 122, 124

mate guarding, 88, 99, 100, 120, 121
mating system, 120, 121
Mediterranean, Mediterranean islands, 2, 61, 99, 134, 145–46, 148–49, 185, 192
melanophore, 61, 117–18, 128
melanophore-stimulating hormone (MSH), 118
melatonin, 119
Mergenagama, 182
Mesozoic, 132, 176 (table 9.1), 179, 188 (fig. 9.3), 190–91
metabolism, 60
Mexico, Mexican, 176 (table 9.1), 189, 206, 212
microcomplement fixation of albumin, 154
microendemism, 163
Middle East, 4, 155, 164, 175, 187, 190, 192
migration, 91, 102, 134, 185, 191
Mimeosaurus, 180, 183
Miocene, 93, 133–36, 145, 160, 175–76 (table 9.1), 177 (table 9.2), 182–86 (fig. 9.2), 188 (fig. 9.3), 189–91, Fig. 7.1
Miocene Climate Optimum, 191
mite pockets, 170. *See also* axillary and/or inguinal pits
Molecular Assumption, 154,
molecular phylogenetics, 154, 157, 174
molecular phylogeny, 5, 138, 172, 174
Mongolia, 176 (table 9.1), 180–82, 189
montane fynbos, 195
montane habitats, 98, 137
Morocco, 146, 182, 190
movement-based camouflage, 128
Mulanje, 136, 199 (table 10.3)
muscle, 2, 7, 13–14, 16–19 (table 2.1), 20 (fig. 2.2), 21–24, 26–37, 43–44, 51–52, 54, 59–60, 70–79, 81, 82, Fig. 2.4
muscle physiology, 59–60, 81

Nadzikambia, 39, 51, 136, 140–41 (table 7.1), 152, 166–68, 188 (fig. 9.3), 199 (table 10.3), 201, 202, 202 (table 10.4), Fig. 7.1, Fig. 7.2, Fig, 8.8
Namib desert, 143
Namibia, 111, 135, 144, 148, Fig. 5.1
natural selection, 126
near threatened, 4, 194–95, 196 (table 10.1), 200 (table 10.3)
Neogene, 136, 142, 176 (table 9.1), 180, 182, 184, 186–87, 188 (fig. 9.3), 192
neurophysiology, 57
nocturnal activity, 88–89, 112, 128
norepinephrine, 119
North America, 63, 181, 188, 190
numerical taxonomy, 153

oceanic dispersal, 138–39, 145–46
Oligocene, 133–35, 143, 145, 149, 158, 176 (table 9.1), 182, 188 (fig. 9.3), 189–91, Fig. 7.1
Oman, 134, 149, 182
open habitat, 71, 86, 93, 94–95, 105, 125
Opluridae, 179
origins, 34, 98, 187–91
ornament, ornamentation, ornamented, 3, 7, 37, 40–41, 53, 93, 125–26, 130, 155–56, 159, 164, 166–72, Fig. 1.6, Fig. 6.2
osteological, 152
oviparouos, 2, 98–100, 168, 172–73
oviposition, 62, 98

Palaeochamaeleo, 182–83
Paleobiogeography, 175
Paleocene, 132, 176 (table 9.1), 179, 181–83, 188 (fig. 9.3), 189–90, Fig. 7.1
Paleogene, 132, 176 (table 9.1), 179, 181, 188 (fig. 9.3), 190–92
parallax, 116, 128
parental care, 115
perch size, 101
phenetic assemblages, 157, 160
photoreceptor, 44, 118, 128
phylogeny, 5, 94, 98, 138, 151–53, 155–56, 158, 165, 172, 174, 178
Physignathus, 182, 183
pigment, pigmentation, 61, 117–18, 127, 159, 161, 166, 170, 173
Pleistocene, 144, 176 (table 9.1), 177 (table 9.2), 182–83, 187, 192
pleuroacrodont, 181
Pleurodonta, pleurodont iguanian, 178–79, 181, 188 (fig. 9.3), 189
Pleurodontagama, 180
Pliocene, 93, 145, 176 (table 9.1), 177 (table 9.2) 183, 186, 188 (fig. 9.3)
Polychrotidae, 179
polygamous, 120
Portugal, 183, 188

predation, 3, 87–89, 97, 107, 109–12, 115, 123–24, 130, 132
predator, 1, 3, 57, 89, 93–94, 101, 104, 109–13, 116–19, 126, 127–29, 212
prey abundance, 105
Priscagama, 180
Priscagamidae, 181, 188 (fig. 9.3)
Pseudotinosaurus, 182

Qianshanosaurus, 181
Quercygama, 182

range-restricted, 134, 137–41 (table 7.1), 143, 147, 149, 193
receptivity, 118, 120, 123
REM, 59
reproduction, 2, 99–100, 102, 115, 120, 123, 125, 130, 168, 170, 172–73
reproductive diapause, 96
reproductive status, 122–23
reproductive success, 120, 126
Réunion Island, 146, 150, 155
Rhampholeon, 8, 10–11, 38–40, 45, 49, 51, 53 (fig. 2.7), 54, 63, 75 (fig. 4.4), 86, 89–92, 98–99, 102, 106, 110–12, 117, 120, 126, 128, 132–33, 136–37, 140–41 (table 7.1), 142, 144, 146–47, 149, 152, 155–57, 161–62, 168–71, 177 (table 9.2), 186 (fig. 9.2), 188 (fig. 9.3), 195, 199–200 (table 10.3), 201–03 (table 10.4), Fig. 1.4, Fig. 5.1, Fig. 7.1, Fig 7.2, Fig. 8.9
gorongosae, 120
Rhynchocephalia, rhynchocephalian, 179–80, 182, 187–88 (fig. 9.3)
Rieppeleon, 8, 10–12, 38–39, 45, 51, 53 (fig. 2.7), 54, 63, 75 (fig. 4.4), 76–77 (fig. 4.5), 86, 93, 96, 107, 117, 132, 136–38, 145, 147, 152, 156–58, 161, 168–71, 186, 188 (fig. 9.3), 191, 201, 203 (table 10.4), Fig. 7.1, Fig. 7.2, Fig. 8.10
Rift Valley, 137, 142–43, 169, 186
riparian vegetation, 91
Romania, 183
roost, 88–90, 97, 101, 121, 128
roosting, 87–91, 95, 101–02, 110, 121, 128
roosting height, 89–90
roost-site fidelity, 121
rostral appendage, 125–26, 165, 169
rostral horn, 41, 126, 171, Fig. 6.4

salt gland, 54, 60
savannah, 118
scincomorph, 181
Scleroglossa, 175, 178, 188 (fig. 9.3)
seasons, 85–88, 102–03, 105, 144, 212
sensory physiology, 57
sexual differences, 95
sexual dimorphism, 120, 126, Fig. 6.2
sexual maturity, 123
sexual selection, 3, 37, 40, 61, 120, 125–26, 130
Seychelles, 63, 101, 106, 134, 137–38, 145, 152, 155, 158, 176 (table 9.1), 191, 194–95, 217
skin, 13, 25, 37, 41–42, 44–45, 54, 61, 64, 117–18, 128–29, 167
sleep, 59, 101
social behavior, 115, 130
Socotra, 134–35, 145, 194, 219
sound, 45–46, 58, 129, 262
Spain, 96, 100–01, 106, 108, 177 (table 9.2), 183, 185, 187, 192, 207 (table 10.7), 213
species assemblages, 87, 91
species diversity, 86, 143–44, 147
species richness, 4, 139, 142, 144, 146–49, 214, Fig. 7.3
sperm storage, 99, 121–22
sprint speed, 2, 59, 81
Sqamata, sqaumate, 3 (fig. 8.1), 7, 25, 52, 97–98, 101, 116, 131–32, 154, 157, 175–76 (table 9.1), 179, 181, 187–88 (fig. 9.3), 189–91
Sri Lanka, 63, 96, 134, 146, 175
starch gel electrophoresis, 154
stem-acrodontan, 180, 182, 188 (fig. 9.3), 189, 191–92
stem-chameleon, 183, 191–92
subcaudal lamellae, 156
submissive coloration, 119, 124–25
Sulcatidens, 182
supercontraction, 22, 74
Switzerland, 145, 178 (table 9.2), 183–85
sympatry, 86, 93, 137, 147
symplesiomorphy, symplesiomorphic, 161, 164, 166, 169, 157
synapomorphy, 151, 153, 155, 157, 159, 161, 164–66, 170–72, 174
synchronous hatching, 88, 97–98

Talosaurus, 182
Tanzania, 86, 136–37, 139, 142, 145–47, 149, 155, 177 (table 9.2), 183, 187, 195, 197, 200 (table 10.3), 204 (table 10.5), 205 (table 10.5, fig. 10.2), 211 (table 10.8), Fig. 1.4, Fig. 1.7
temperature, 2, 60–62, 81–82 (fig. 4.6), 88–89, 95–98, 100–01, 118, 132, 211–13
temperature-dependent colour change, 118
temporal gland, 54, 116
terrestrial, 2, 25, 31–32, 55, 66, 68, 70, 72, 90, 106, 110–12, 127, 130, 132–35, 137, 139, 144, 146, 152, 157, 191, Fig. 1.4
territorial, territoriality, territory, 120–21
thermoregulation, 2–3, 61, 82, 96, 119, 130
Tikiguania, 179–80, 187
Tinosaurus, 181–83, 188 (fig. 9.3), 189
tongue, 1, 2, 7, 16, 20 (fig. 2.2), 21–24, 47–50, 55, 57, 59–61, 63, 72–75 (fig. 4.4), 76–77 (fig. 4.5), 78–82 (fig. 4.6), 83, 85, 104, 109, 132, 151
trade, illegal, 210–11
trade, legal, 201–03 (table 10.4), 204, 210–11
tree falls, 90, 91
Triassic, 179, 187–88 (fig. 9.3)
trigger species, 197, 198–200 (table 10.4)
Trioceros, 8–9 (fig. 2.1), 10–12, 14, 20 (fig. 2.2), 25, 31, 39, 41–42, 45–47, 49, 51, 54, 58–59, 63, 66, 86, 89–91, 93, 95, 98–107, 116–18, 120, 122, 125–126, 133, 137, 140–41 (table 7.1), 142–43, 145–47, 149, 152, 164, 166–68, 171–73, 177 (table 9.2), 187–88 (fig. 9.3), 198–200 (table 10.3), 201–02 (fig. 10.1), 203 (table 10.4), 204 (table 10.5), 206, 209–10, Fig. 1.1, Fig. 1.2, Fig, 1.5, Fig. 7.1, Fig. 7.2, Fig. 8.11
  ellioti, 118, 171–73
  hoehnelii, 31, 41, 46 (fig. 2.6), 49, 58, 95, 99–100, 102–04, 120, 171–73, 203–04 (table 10.5)
  jacksonii, 47, 58, 61, 72, 95, 99–100, 102–06, 118, 120, 125–26, 171–173, 177 (table 9.2), 187, 203–04 (table 10.5), 209, Fig. 6.2, Fig. 6.4. See also *Chamaeleo jacksonii*
trogonophidae, trogonophid amphisbaenian, 179
Turkey, 185

Udzungwa Mountains, 147, 200 (table 10.3)
Ukraine, 183, 209
ultraviolet, 58, 61–62, 124
undisturbed forest, 91–92
Uromastyx, 179, 182–83, 189–90

Vastanagama, 182
vibration, 89, 117, 121–22, 129
vicariance, 133–34, 136, 145, 169
vision, 58, 115–16, 127
visual system, 1, 57–58, 116, 119, 124, Fig. 1.3
vitamin D, 61
viviparity, 85, 95, 98, 159, 166, 172,
vulnerability, vulnerable, 4, 81, 93, 100, 109–12, 128–29, 195–96 (table 10.1), 200 (table 10.3), 212–13

water, 37, 52, 60, 86, 93, 104, 108, 136, 144–46, 148, 191
weighted endemism, 148, Fig. 7.3

xanthophore, 61, 117–18
Xianglong, 180

Zephyrosaurus, 182
Zimbabwe, 89, 103, 112, 137, 144, 148